NATURAL DISASTER RESEARCH, PREDICTION AND MITIGATION SERIES

TSUNAMIS: CAUSES, CHARACTERISTICS, WARNINGS AND PROTECTION

Natural Disaster Research, Prediction and Mitigation Series

The Phoenix of Natural Disasters: Community Resilience
Kathryn Gow and Douglas Paton (Editors)
2008. ISBN: 978-1-60456-161-6

Natural Disasters: Public Policy Options for Changing the Federal Role in Natural Catastrophe Insurance
U.S. Government Accountability Office
2008. ISBN: 978-1-60456-717-5

Cyclones: Background, History and Impact
Terrance G. LaBeau (Editor)
2009. ISBN: 978-1-60692-064-0

Cyclones: Background, History and Impact
Terrance G. LaBeau (Editor)
2009. ISBN: 978-1-60876-711-3
(Online Book)

Hurricane Katrina: Impact, Recovery and Lessons Learned
Nessa P. Godfrey (Editor)
2009. ISBN: 978-1-60692-478-5

National Emergency Responses
Paul B. Merganthal (Editor)
2009. ISBN: 978-1-60692-355-9

Earthquakes: Risk, Monitoring and Research
Earl V. Leary (Editor)
2009. ISBN: 978-1-60692-648-2

Solar Activity and Forest Fires
Milan Rodovanovic and Joao Fernando Pereira Gomes
2009. ISBN: 978-1-60741-002-7

Indigenous Knowledge and Disaster Risk Reduction: From Practice to Policy
Rajib Shaw, Anshu Sharma and Yukiko Takeuchi (Editors)
2009. ISBN: 978-1-60741-574-9

Indigenous Knowledge and Disaster Risk Reduction: From Practice to Policy
Rajib Shaw, Anshu Sharma and Yukiko Takeuchi (Editors)
2009. ISBN: 978-1-60876-674-1
(Online Book)

Forest Fires: Detection, Suppression and Prevention
Eduards Gomez and Kristina Alvarez (Editors)
2009. ISBN: 978-1-60741-716-3

Forest Fires: Detection, Suppression and Prevention
Eduards Gomez and Kristina Alvarez (Editors)
2009. ISBN: 978-1-61668-409-9 (Online Book)

Meltdown: Climate Change, Natural Disasters and other Catastrophes – Fears and Concerns of the Future
Kathryn Gow
2009. ISBN: 978-1-60876-153-1

Landslides: Causes, Types and Effects
Ernest D. Werner and Hugh P. Friedman (Editors)
2010. ISBN: 978-1-60741-258-8

Dust Storms: Elemental Composition, Causes and Environmental Impacts
Siniša Brstilo and Quentin Madunic (Editors)
2010. ISBN: 978-1-60876-456-3

Dust Storm Identification via Satellite Remote Sensing
D.G. Kaskaoutis, H.D. Kambezidis; K.V.S. Badarinath and Shailesh Kumar Kharol
2010. ISBN: 978-1-60876-906-3

Haiti: Earthquake and Response
Rachael A. Donlon (Editor)
2010. ISBN: 978-1-61668-998-8

Tsunamis: Causes, Characteristics, Warnings and Protection
Neil Veitch and Gordon Jaffray (Editors)
2010. ISBN: 978-1-60876-360-3

NATURAL DISASTER RESEARCH, PREDICTION AND MITIGATION SERIES

TSUNAMIS: CAUSES, CHARACTERISTICS, WARNINGS AND PROTECTION

NEIL VEITCH
AND
GORDON JAFFRAY
EDITORS

Nova Science Publishers, Inc.
New York

LIBRARY OF CONGRESS CATALOGING-IN-PUBLICATION DATA

Tsunamis: causes, characteristics, warnings and protection / Neil Veitch and Gordon Jaffray, editors.
p. cm.
Includes index.
ISBN 978-1-60876-360-3 (hardcover)
1. Tsunamis--Environmental aspects. I. Veitch, Neil. II. Jaffray, Gordon.
GC221.2.T785 2009
363.34'94--dc22

2009037469

Published by Nova Science Publishers, Inc. ✢ New York

CONTENTS

PREFACE

A tsunami is a series of water waves that is caused when a large volume of a body of water, such as an ocean, is rapidly displaced. Earthquakes, volcanic eruptions and other underwater explosions, landslides, bolide impacts, and other disturbances above or below water all have the potential to generate a tsunami. Furthermore, Tsunamis and storm surges have killed more than one million people, and some three billion people live with a high risk of these disasters that are becoming more frequent and devastating worldwide. This book presents field survey results on tsunami arrival times, wave runup heights, inundation distances and damage to properties on beaches due to tsunamis. The main injuries of survivors (i.e., aspiration and trauma) are also analyzed. Effects of coastal forests on tsunami run-up heights are discussed as well. Other chapters in this book highlight topics such as tsunamis and poisonous gases generated by asteroid impact in the Black Sea, tsunami simulation research, coastal protection measures for tsunami disaster reduction, case studies from the Sri Lanka tsunami, tsunami monitoring and detection, and early warning systems.

A simple model is proposed to evaluate the effects on coastal regions of an asteroid impacting the Black Sea. The initial kinetic energy of the asteroid is mainly transferred to the seawater and to the seafloor. Therefore both the water and the sea-bottom are deformed and also their internal energy increases. Chapter 1 focuses on the dynamics of the initial cavity created in the seawater. This initial cavity constitutes 'the source' for the all subsequent phenomena. The two different phenomena that may affect the coastal regions start simultaneously, at the place of the asteroid impact. First, the initial seawater cavity constitutes the source for a tsunami wave. Also, the water ejected by the asteroid impact is broken up into mist-like droplets during both the ascending and descending parts of the trajectory. The gases dissolved in the water are transferred to the atmosphere. We will limit our discussion to the hydrogen sulfide dissolved in the Black Sea waters, although in some places the release of much larger volumes of methane, which would explosively burn might pose an even more serious threat. Thus, the H_2S expelled into the atmosphere (H_2S is denser than air) ''falls'' with a lower speed than the falling water droplets and finally creates a gaseous ''cloud'' or ''blanket'' on the sea surface. This H_2S cloud moves and disperses in the mean atmospheric wind field. The two phenomena have their own dynamics. Their effects on the coastal regions are different and depend on many factors among which the most important is impacting asteroid size.

Chapter 2 presents field survey results on tsunami arrival times, wave runup heights, inundation distances and damage to properties on beaches in Penang and Kedah due to the 26

December 2004 Andaman tsunami. This onsite survey results are used to calibrate and validate a tsunami simulation model TUNA developed by the authors. Simulation results indicate that TUNA performs satisfactorily and can indeed reproduce salient features of tsunami characteristics observed in beaches in Penang and Kedah post 2004 Andaman tsunami. To understand quantitatively the role of mangrove in reducing the impact of tsunami, a numerical simulation model is developed and successfully applied to a stretch of mangrove forests in Penang. A main theme of this chapter is the desire to develop coastal communities that are tsunami resilient. With this in mind the authors outline briefly a series of workshops, conferences and other research development activities undertaken by key research institutions and government agencies in the past five years towards helping to achieve tsunami resilient communities in the regions, with particular reference to the South China Sea. It is hoped that this presentation will further enhance active collaboration with other research and operational institutions worldwide towards tsunami resilient communities.

Around 10:00 AM, 26 December 2004, Tsunami, the name that was unknown to most Thai people, had moved to the Andaman Sea having started from Phuket, Phang-Nga, Ranong, Krabi, Trang and Satun provinces. The main injuries of the survivors were aspiration and trauma. Trauma to parts of the body and extremities were sustained when the wave hit and swept back. Patterns of injury can be varied depending on the tidal velocity and size of particles that hit the body.

By the rolling and sweeping motion of the huge wave, victims were surged or heaved along an undetermined path and direction similar to a bullet being fired through surrounding targets with destructive potential. High impact velocity acting on a hard heavy object could directly cause permanent damage such as crushed, cavitation and tissue loss to any part of the body. Those lucky enough to survive were left with several injuries. The observed injury patterns and wounds are depended on the magnitude of the direct impact, indirect or secondary injury and condition of the landing surface. The variety of wounds seen in the Tsunami incident was enormous. Damage of skin and underlying soft tissues can range from self treated to life threatening condition which requires immediate resuscitation. The wounds had characteristics of multiple site involvement; any part of the body can be injured. Wound types categorized by the size and depth can be varied from area to area even in the same extremity. Extensive skin loss with severe contamination and multi-organ involvement are nearly similar to those of war wound injuries. All wounds can be graded by severity and contamination into *first degree* that involves only the covering skin and underlying soft tissues. *Second degree* has more involvement of skin and underlying soft tissue with retained contamination and skeletal involvement. In *severe degree,* there are extensive damages to the skin and underlying vital structures such as bone, joint, nerve and vessels. The small penetrating wound is the most dangerous with high incidence of sepsis and death. Fractures, dislocation and tendon injury were also common. Another characteristic finding was the early development of wound infection. Spreading of infection and sepsis were strongly related to the emergency aspect of the situation, which overwhelmed the available resource.

Care begins at the scene with immediate first aid and life saving measures by trained people. Patients should be taken to the hospital with resuscitative capacities and rescue pain control. Contaminated wound should be temporarily cleaned, covered and immobilized immediately. Hospital care begins with triage on arrival at the front gate. Primary wound dressing can be temporarily made by using tap water and then cover the wound with a sterile green towel while waiting for definite surgical intervention. For major life threatening wounds, the damage control surgery should be applied to minimize high rate of amputations and mortality. A wound with a small penetrating entrance should be considered dangerous. Aggressive debridement and thorough irrigation should be performed as soon as possible and the wound should be left open. Fasciotomies should be carried out to release pressure within the compartment. Broad spectrum antibiotics should be administered once cultures have been processed. Wounds should then be further re-explored and redebrided 24–48 hours after the initial procedures. Immediate stabilization of bone fractures with the least amount of stable splints to the most stable fixator is a basic principle of orthopedic care. Patients with more complex wounds and significant injuries should be stabilized and then immediately transported through a more advancing levels where definitive procedures can be effectively performed. Close surveillance and additional therapy for possible wound infection and early sepsis are supremely important. Skin or tissue reconstruction of high energy open wounds should be attended to as soon as possible once the wound bed is clean.

With 2004 tsunami, it is quite clear that the immediate Multiple Casualty Incident (MCI) must be handled at the time and place of impact with proper equipment, supplies and variety of experts in emergency management. Chapter 3 demonstrated that wounds had affected people and health care staffs. A better medical response can be definitively provided with a proper or well prepared planning and exercising. The mainstay of primary closure of wounds and increase use of antibiotics in civilian practice is not appropriate for any disaster incident. Initial wound assessment, appropriate multiple debridement, soft tissue and bone stabilization, further assessment and duration of the exposed wound are all of crucial factors that affect incidence of infection and healing.

Tsunamis and storm surges have killed more than one million people, and some three billion people live with a high risk of these disasters that are becoming more frequent and devastating worldwide. Effective mitigation of such disaster is possible via healthy coastal forests, which can reduce the energy of the wave. Many independent studies are conducted to evaluate the control functions of coastal forest against tsunami and several recommendations are proposed to improve the mitigative effects of such natural shield. However, in many instances the ideas diverse in their view points. In order to synthesise the research findings and to propose a best practicable and optimal solution, a comparative study was conducted among available results and recommendations. Especial focus was given to the Tsunami 2004 in Indian Ocean. Chapter 4 reviews previously produced numerical and experimental results and compares them with field observation. The relationship between the degree of damage reduction and associated parameters especially width, size and density of coastal vegetation are discussed.

Tsunami may be generated by earthquake triggered movement of the sea bottom, landslides and collapses. It has caused great impacts on human life and coastal environments, including massive loss of human life, devastation of coastal ecosystems and settlements, and damage to infrastructure and facilities. In Chapter 5, tsunami protection works will be investigated as hard structures likes tsunami breakwaters, sea walls etc. and natural barriers including coastal forest etc. except soft approaches (education, awareness, evacuation, etc.). Although hard structures may provide highly effective protection, they may have high cost and may also cause large amount of negative environmental impact on the coastal areas. That is why, natural coastal barriers which have lower environmental impact and higher additional natural value can be considered as a protection measure against tsunami effects.

Tsunami damage occurs mostly in the nearshore zone and in the coastal area behind the coastline because of the tsunami hydrodynamics during the run-up period. Therefore, tsunami run-up height was also investigated. Tsunami run-up heights for impermeable and permeable (not armored and armored) beaches were examined and empirical equations suggested. Following, coastal protective measures, to reduce tsunami damage on coastal areas were analyzed as hard structures and natural barriers. Furthermore, the effects of coastal forests on tsunami run-up heights were discussed and empirical equations suggested.

The Indian Ocean tsunami of December 2004 transmitted multiple lessons: whereas certain strips were destroyed in totality, others remained unaffected or practically intact. Detailed post-tsunami field surveys conducted along the coast of Tamil Nadu, south India, in April 2005 and January 2006, confirmed that coastal landforms and vegetation played a significant role in neutralizing the force of virulent waves. Impact on casuarina forests was restricted to a maximum of 25 meters from the dune line. Only frontal casuarina strips were attacked, bent and stripped of their leaves by wave up-rush. Whereas dune creepers and herbs were uprooted, coconut and palm trees remained in position. This phenomenon was verified along the entire Tamil Nadu and Pondicherry coasts. Evidence of minimal damage to casuarina plantations and coconut groves supports the view that biological buffers can serve as efficient energy dissipaters during powerful oceanographic events. Villages located behind dense plantations remained safe. In 2006, natural restoration was identified in the form of rejuvenated and healthy vegetal species. Dune vegetation had bounced back and bent casuarinas had sprouted.

In Chapter 6, the need for a protective coastal buffer zone is proposed. Its levels of effectiveness will depend on a progression of species landward from the shore. Casuarinas should not be located on dunes, but planted further backshore. Herbs – shrubs – bushes – trees form a gradation of species and a natural slope that is inferred to offer protection as natural shelter belt against any eventual extreme event. Future plantation strategies will have to consider natural bio-zonation rather than haphazard patterns that are observed at present. Green belts are beneficial for several reasons: control of erosion, stabilization of shores, alleviation of wind energy, effective buffer against the force of waves, preservation of biodiversity and advantage in terms of food, shelter and income.

It is established that physical and geological processes are intense along the open ocean. Manmade structures thus experience extreme processes as such sites become vulnerable to natural hazards. In comparison, the forested hinterland is sufficiently protected from physical forces as vegetation attenuates energy from onrushing waters. Therefore, elevated coastal stretches with protective vegetation are the only environments where risks due to extreme oceanographic events are modest.

As explained in Chapter 7, the tsunamigenic Sumatra-Andaman earthquake (moment magnitude $M_W = 9.3$) of December 26/2004 caused destruction and human casualties in many coastal Indian Ocean countries with more than 35,000 deaths in the island of Sri Lanka. Tsunami waters started inundating many coastal regions of the island at different times after the event. The initial water movement was characterized by a rapid drawdown or lowering of the sea surface at the coast as waters moved into the area of seabed displacement. In different parts of Sri Lanka, the drawdown resulted in recession of the sea about 500 m from the present coastline. According to eyewitness accounts, this phenomenon had lasted for about 20 minutes before the return of massive turbulent tsunami waves inland with speeds of about 30 to 40 km per hour bringing behind them long trains of water from deep sea environments. The waters carrying deep sea materials such as dark suspensions of fine sediments encroached inland via embayments such as harbours, bays, estuaries, lagoons and upstream along rivers before depositing them. Run up heights of about 3 meters or more were reported from different coastal locations. The tsunami event has left significant geological signatures with changes in coastal geomorphology and deposition of sediments along and across the coast.

The sediments consisting of discontinuous sheets of differing thickness are found at coastal depressions and similar low lying areas. Such sediments are poorly sorted and contain heterogeneous mixtures of debris of buildings and vehicles, tree trunks, shells of organisms etc., derived from coastal and deep marine environments. However, at other locations such as depressions where water had been stagnant for a few hours before retreating back to the sea , thin films of brownish clayey silt materials from suspensions had been deposited. Sediments with comparable textures and compositions were located in previously empty bottles, stacked on a bench of about 50 cm from the floor of a house close to the beach, which had been subsequently filled with tsunami waters. The recent tsunami sediments when studied under the Scanning Electron Microscope (SEM) showed the occurrence of foraminiferal facies and other deep sea fauna. Drilling carried out deeper into coastal depressions in different locations revealed sediments with textures, structures and microfossils characteristic of 2004 deposits in at least two lower horizons signaling the occurrence of paleo-tsunami events. Although historical evidence also indicates at least two tsunami events, radiocarbon dating yielded anomalous results.

On December 26, 2004, a magnitude 9.0 undersea earthquake off the west coast of northern Sumatra, Indonesia, unleashed a tsunami that affected more than 12 countries throughout south and southeast Asia and stretched as far as the northeastern African coast. Current official estimates indicate that more than 160,000 people are dead and millions of others are affected, including those injured, missing, or displaced, making this the deadliest tsunami on record. News reports suggest that the death toll may be well above 200,000. Sections of Indonesia, Sri Lanka, India, and Thailand have suffered the worst devastation. Eighteen Americans are confirmed dead, with another sixteen presumed dead, and 153 remain unaccounted for.

In response, the United Nations, the United States, and other donor nations have organized what some have called the world's largest relief and recovery operation to date. President Bush pledged $350 million in aid and mobilized the U.S. military to provide logistical and other assistance.

Funding the Indian Ocean tsunami relief and reconstruction effort is likely to be a challenge faced by the 109th Congress. Even before the disaster struck, Congress was expected to struggle to find the resources to sustain U.S. aid pledges amid efforts to tackle rising budget deficits by, among other measures, slowing or reducing discretionary spending. Congress also may wish to consider debt relief as a means of helping those nations hit by the tsunami to recover economically. Additionally, there have been calls to institute a tsunami detection and warning system in the Atlantic and/or Indian Oceans, both of which would require allocations of funds. The large-scale U.S. response to the tsunami is unlikely to reverse the decline in the U.S. image abroad since the September 11 attacks, because this decline primarily is due to American policies in the Middle East. However, the scale and scope of U.S. assistance could provide a positive example of U.S. leadership and military capabilities. Additionally, the disaster relief cooperation between the U.S. and Indonesian militaries is likely to be mentioned during the annual congressional deliberations over renewing restrictions on U.S.-Indonesian military-to-military relations, which the Bush Administration has sought to restore since the September 11, 2001 attacks.

Chapter 8 summarizes the extent of the disaster and relief effort and includes descriptions of the U.S. and international assistance efforts. It also examines protection mechanisms for children and separated orphans. A section is devoted to the situation in each of the affected countries followed by an analysis of selected issues for Congress. A *Most Recent Developments* section at the beginning of the chapter is updated through February 10, 2005. The rest of the chapter is updated through January 21, 2005.

As discussed in Chapter 9, recently, some in Congress have become concerned about the possible vulnerability of U.S. coastal areas to tsunamis, and about the adequacy of early warning for coastal areas of the western Atlantic Ocean. Those concerns stem from the December 26, 2004, tsunami that devastated many coastal areas around the northern Indian Ocean, where few tsunami early warning systems currently operate. Caused by a strong underwater earthquake off the coast of Sumatra, Indonesia, the tsunami disaster is estimated to have claimed at least 150,000 lives. Affected nations, assisted by others, are pursuing a multilateral effort to develop a detection and warning network for the Indian Ocean. Also, some Members of Congress and the Bush Administration have proposed a tsunami warning network for the U.S. Atlantic seaboard. Although instrumentation costs could run into the millions of dollars, existing weather buoys and state and local coastal and ocean observation networks might serve as possible platforms for the instrumentation. The European Union, Canada, and the United States may consider multilateral efforts to establish coverage for the North Atlantic. This report will be updated as warranted.

The word "tsunami" comes from the Japanese language. It means harbor (" tsu") and wave ("namis"). A tsunami is a very long wave generated by different methods.

In Chapter 10, the authors are looking for the causes and the generation of tsunamis, as well as the mathematical modeling that simulates the tsunamis [*and the water waves impacts*]. The most famous origin of generation of tsunamis comes from a sudden displacement of the ocean bottom. This happens due to some kind of earthquake in the ocean. Two different models exist for the generation of tsunamis by the underwater earthquake. Another origin for the generation of tsunamis is falling cosmic bodies.

The second part of this work deals with the mathematical modeling. The authors start with a very simple model, then the authors continue with other different more complicated models, like the Boussinesq model and other nonlinear gravitational wave equations. Another model represented by the well-known KdV equation is also studied. This represents the famous nonlinear equation that describes the nonlinear behavior of this killer physical phenomenon, i.e., tsunamis.

In: Tsunamis: Causes, Characteristics, Warnings and Protection ISBN: 978-1-60876-360-3
Editors: N. Veitch and G. Jaffray, pp. 1-28

Chapter 1

TSUNAMIS AND POISONOUS GASES GENERATED BY ASTEROID IMPACT IN THE BLACK SEA

***V. Badescu*[1]*, D. Isvoranu*[2]*, R.B. Cathcart*[3] *and R.D. Schuiling*[4]**

[1] Candida Oancea Institute, Polytechnic University of Bucharest, Bucharest 060042, Romania

[2] Faculty of Aeronautics, Polytechnic University of Bucharest, Bucharest 060042, Romania

[3] Geographos, Burbank, California 91506, USA,

[4] Faculty of Earth Sciences, Utrecht University, 80.021, 3508 TA, Utrecht, The Netherlands

Abstract

A simple model is proposed to evaluate the effects on coastal regions of an asteroid impacting the Black Sea. The initial kinetic energy of the asteroid is mainly transferred to the seawater and to the seafloor. Therefore both the water and the sea-bottom are deformed and also their internal energy increases. We shall focus on the dynamics of the initial cavity created in the seawater. This initial cavity constitutes 'the source' for the all subsequent phenomena. The two different phenomena that may affect the coastal regions start simultaneously, at the place of the asteroid impact. First, the initial seawater cavity constitutes the source for a tsunami wave. Also, the water ejected by the asteroid impact is broken up into mist-like droplets during both the ascending and descending parts of the trajectory. The gases dissolved in the water are transferred to the atmosphere. We will limit our discussion to the hydrogen sulfide dissolved in the Black Sea waters, although in some places the release of much larger volumes of methane, which would explosively burn might pose an even more serious threat. Thus, the H_2S expelled into the atmosphere (H_2S is denser than air) ''falls'' with a lower speed than the falling water droplets and finally creates a gaseous ''cloud'' or ''blanket'' on the sea surface. This H_2S cloud moves and disperses in the mean atmospheric wind field. The two phenomena

[1] E-mail address: badescu@theta.termo.pub.ro. Spl. Independentei 313, Bucharest 060042, Romania.

[2] E-mail address: ddisvoranu@gmail.com.

[3] E-mail address: rbcathcart@gmail.com. 1300 West Olive Avenue, Suite M, Burbank, California 91506, USA

[4] E-mail address: schuiling@geo.uu.nl.

have their own dynamics. Their effects on the coastal regions are different and depend on many factors among which the most important is impacting asteroid size.

1. Introduction

The USA's National Geophysical Data Center (NGDC) "World-wide Tsunamis, 2000 B.C – 2007" list records 1,700 known tsunami coastal events by location and date. About 3% of these events happened in the Black Sea Basin, which has a coastline length of about ~4546 kilometers, during that 4000+ year time period. The NGDC alleges that <1% of the historically documented 9600 tsunami shore run-up places are situated in the Black Sea Basin (NGDC, 2009). Both percentages are remarkable because the Black Sea is only ~0.13% of the Earth's ocean area.

Normally the consequences of an impact of an asteroid in a water body are discussed in terms of the beach run up of impact-generated waves. However, an additional poisonous effect may happen when the impact place is the Black Sea, the largest anoxic water body on Earth. Two different phenomena that may affect the coastal regions start simultaneously at the place of the asteroid impact. First, the initial water cavity constitutes the source for a tsunami wave. Second, the H_2S released into the atmosphere (which is denser than the air) finally creates a gaseous "cloud" or "blanket" on the sea surface. This H_2S cloud moves and disperses in the mean atmospheric wind field. The two phenomena have their own dynamics and their effects on the coastal regions are different.

Asteroid impacts in the Black Sea are apparently unrecorded by historians. In this chapter we present a scenario concerning the consequences of an asteroid impacting the Black Sea. Results previously published in Badescu (2008) are summarized. Also, we develop the tsunami propagation model shortly described in Schuiling et al. (2007) and a number of quantitative estimates are presented.

2. Contemporary Black Sea Conditions

The Black Sea is an oval-shaped body of seawater situated between 40.55 and 46.32 deg North latitude by 27.27 and 41.42 deg East longitude; it has a coastline of 4090 km (~1.4% of the world's coastline) and a maximum water depth of 2,200 m.

Whilst about 9400 years ago the Black Sea was ~30 m below its present-day level (Giosan et al., 2009), this current low area-coverage of ~0.13% is unlikely to increase markedly even owing to possible future 21st Century global sea level rise, nowadays estimated at ~1 m maximum. In alphabetical order, the length of relevant national coastline facing the Black Sea are: Bulgaria, 354 km; Georgia, 310 km; Romania, 225 km; Russia, 475 km (without the Azov Sea coastline); Turkey, 1400 km and Ukraine, 2782 km (Figure 1). Seawater circulation in the Black Sea's toxic layer is dominated by regional wind circulation. The very low connectivity—the fraction, ~0.06%, of the coastline that is an opening to the world-ocean—restricts seawater flushing of the Black Sea by the world-ocean.

The distinguishing physical peculiarity of the Black Sea is the presence of a ''permanent'' halocline situated at a depth of 100–200 m, a stratification generated mostly by freshwater inputs from rivers flowing into the Black Sea. Of its total volume of water—

approximately 547000 km^3—some 475000 km^3, or 87%, is anoxic, without dissolved oxygen and therefore lifeless except for anaerobic bacteria. The Black Sea is Earth's biggest single reservoir of hydrogen sulfide (H_2S), an oceanographic fact only discovered circa 1891. H_2S is generated by bacterial reduction of sulfate both in the water column and in the Holocene seafloor sediments deposited since the connection of the Black Sea with the Mediterranean Sea.

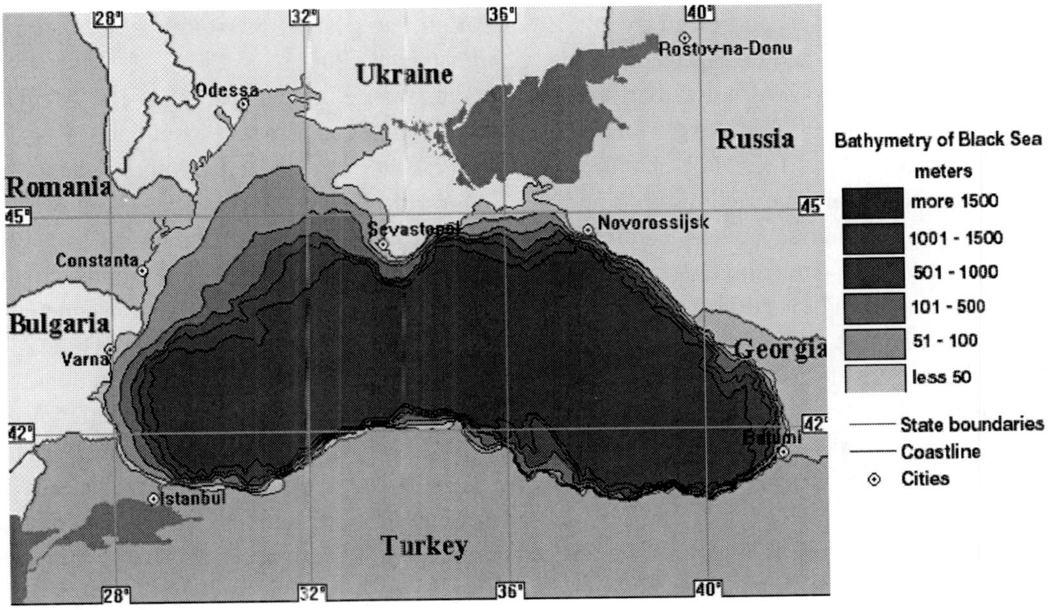

Figure 1. Geographical place and bathymetry of Black Sea.

The future of the human populations residing and working on lands that are immediately adjacent to the Black Sea depends on the future stable "balance" of the oxic and anoxic volumes of seawater in the Black Sea.

3. Recent Black Sea Tsunami Run-Up Events

Tectonically-induced tsunamis—generated by the impulsive vertical displacement of seawater that is dispersive in radiation pattern from the place of origination—occurred in the Black Sea during the 20th Century (Nikonov, 1997). Most of these observed tsunamis were initiated close to the coastline. Earthquakes generated major tsunamis in the Black Sea during 1927, 1939 and 1966 (Dotsenko and Ingerov, 2007). Many tsunami reports center on the coasts of the Crimean Peninsula, the Caucasus and the northern beaches of Turkey. The Black Sea's seafloor topography, particularly the extent of shallow sea-bottom, is a major factor for this noteworthy geographical distribution (Ignatov, 2008); the topographic capture of tsunami energy by the Black Sea shelf accounts for this phenomenon. The maximum heights of all properly recorded Black Sea tsunamis are not remarkably impressive—probably ~4 meters—and the most recent known significant major earthquake-induced tsunami occurred during 1970 (Yalciner et al., 2004).

4. Asteroid Impacts on Earth

Near-Earth objects (NEOs), which include near Earth asteroids (NEAS) are usually defined as objects with perihelion distance 1.3 astronomic units (AU) and aphelion distance 0.983 AU (Ipatov, 1998, 1999). Most analysis so far has emphasized telescopic searches for large (i.e. more than 1 km diameter) near-Earth asteroids. NASA's stated goal was to find 90% of the large NEAs by 2008 (Pilcher, 1998). Several categories of potentially threatening NEOs remain mostly unaddressed by concentration on these larger asteroids: the remaining 10%, smaller NEAs, and long-period comets (Chapman et al., 2001).

Rough estimates give more than 30 NEOs larger than 5 km in diameter, 1500 NEOs larger than 1 km and 135000 larger than 100 m (Rabinovitz et al., 2004). About half of NEOs are Earth-crossers and there is a chance to collide the Earth in near or far future. The main sources of NEOs are the asteroid belt and the Edgeworth–Kuiper belt (EKB). Other Earth-crossers are long-period comets coming from the Oort cloud. Icy bodies can also migrate inside the Solar System from the regions located between the EKB and the Oort cloud.

Almost all of what is known about the potential environmental and societal consequences of asteroids impact on Earth has been obtained from numerical simulations (Chapman et al., 2001). Some conclusions were also derived (for smaller impacts) from extrapolations of nuclear weapons tests (Glasstone and Dolan, 1977) and (for larger impacts) from inferences from the geological record for the Cretaceous /Tertiary (K/T) impact about 65 Myr ago. The environmental consequences from asteroid impacts are usually classified in three size ranges (Toon et al., 1997): (i) regional disasters due to impacts of multi-hundred meter objects; (ii) civilization-ending impacts by multi-km objects and (iii) K/T-like cataclysms that yield mass extinctions. Recently, a number of researchers are arguing a forth size range should be added, namely (iv) multi-ten meters impactors like Tunguska (see Jewitt (2000), Foschini (1999) and reference therein).

There is considerable uncertainty about the environmental consequences of larger impacts (categories (i) and (ii)). It is expected that they have diverse physical, chemical, and biological consequences, which dominate the Earth ecosphere in ways that are difficult to imagine and model. Atmospheric perturbations due to dust and aerosols lofted by impacts are some of these effects that have been studied by using global circulation and climate models. The “asteroidal winter” may be a consequence, deriving from a strong injection of dust in the atmosphere (Cockell and Stokes, 1999). Impacts may also induce chemical changes in the atmosphere, mainly by injection of sulphur into the stratosphere. These are related to the vaporization of both the impactor and a part of the target. Large impact events may inject enough sulphur to produce a reduction in temperature of several degrees and a major climatic shift (Foschini, 1999). Additional effects on atmospheric chemistry are the potential for the destruction of the ozone layer from shock heating atmospheric nitrogen and the injection of fluorides from the vaporized impacting body (Foschini, 1999). The greatest danger from smaller (category (iii)) impacts are tsunamis, which transfer the effects of a localized ocean impact into dangerous, breaking "tidal waves" on distant shores (Ward and Asphaug, 2000). The current philosophy of impact hazard considers the danger from small asteroids is negligible. However, several facts claim for a revision of this philosophy. The impactors in category (iv) may have major local consequences near ground zero. Also, they could generate social effects, political ramifications and fallout from the public. In fact, there are several scientists suggesting that small asteroids might be even more dangerous than larger bodies (Foschini, 1999).

5. Frequency of Asteroid Impacts in Black Sea

The impact is a random process in geographical space but also in time. Estimates of such impact rates based on the number of asteroids and dynamical considerations are rather uncertain, so it may be more robust to determine them from the historical impact records (Ivezic et al., 2001). On the other hand, the latter method suffers greatly from the small number statistics and unknown sample completeness. In practice, evaluation of impact frequency is made by using empirical or semi-empirical formulas. The scarce existing data yield often contradictory results.

A rough estimate for the impact frequency as function of impactor size is given in (Chapman et al., 2001): (i) multi-hundred meter objects impact Earth every 10^4 years (ii) multi-km objects impacts occur on a million-year timescale; (iii) K/T-like cataclysms occur on a 100 Myr timescale. Also, tens of meters impactors collide with the Earth on timescales comparable to or shorter than a human lifetime.

Some conclusions concerning the impact frequency may be obtained from impact crater studies. To-date, over 160 impact craters have been identified on Earth (Earth Impact Database, 2006). From estimates of the terrestrial cratering rate, the frequency of K-T-sized events on Earth is of the order of one every 50-100 million years. The formation of impact craters 20 km in size occur on Earth with a frequency of two or three every million years. Events such as Tunguska occur on a time-scale of 100's of years. The frequency of Earth impact by NEOs is estimated to be larger by a factor 2, 14, 24 and 30 as compared to Venus, Mars, Moon and Mercury, respectively (Bottke et al., 1994). Shoemaker et al. (1990) showed that asteroidal impacts probably dominated the production of craters on the Earth with diameter smaller than 30 km, whereas cometary impacts were responsible for the craters with diameter larger than 50 km. The impact probability of a long-period comet with the Earth per revolution is estimated as $2\text{-}3 \cdot 10^{-9}$ (Marsden and Steel, 1994). Note that more recent results show a decrease by a factor of 4 in the estimated number of asteroids with diameter larger than 1 km (Ivezic et al., 2001). This implies that the time until the next collision of an asteroid of this size with the Earth may be at least four times longer than previously thought (estimated to be $1\text{-}2 \cdot 10^5$ years, see e.g. Rabinowitz et al. (2000)).

The estimated frequency of Earth collisions by space debris objects of tens-of-meters diameters differs from author to author. The number of all 100-m NEOs is evaluated in (Foschini, 1999) to be about 70,000-160,000. One expects such objects collide the Earth on average ones in 600-1400 yr. For size less than 70 m the frequency of collisions is greater by a factor of 2 than that for size larger than 100 m. Under this assumptions, Tunguska-size objects collide the Earth once in several hundreds (500) years. By using data from lunar craters, Chapman and Morrison (1994) consider a time interval of 250 yr for the Tunguska–like events. Farinella and Menichella (1998) found that the impact frequency is 1 per 100 yr by studying the interplanetary dynamics of Tunguska–sized bodies. The impact frequency can be once in 17 or 40 for a 1 Mton airburst explosion (ReVelle, 1997; Nemtchinov et al., 1997). This implies a Tunguska event (12.5 Mton) once in 100 or 366 yr. Hunt et al. (1960) evaluated a smaller explosion for Tunguska (10 Mton). Then the impact frequency decreases to 88 or 302 yr. The other Tunguska–like events in South America in 1930 and 1935 reported in Steel (1995) may lead to the impact frequency value of one per 100 yr (or less).

Recent studies focused on the atmospheric penetration of small NEAs. The asteroid composition proved to be very important. The new models allow for energy loss by stony impactors below 200 m diameter, with complete disintegration (an airburst) below a diameter of about 100 m (Bland and Artemieva, 2003). The results indicate that asteroids with a diameter greater than 200 m will hit the surface approximately once every 160,000 years - way down on previous estimates of impacts every 2,500 years. However, the number of impacts for iron asteroids of similar size was comparable with that predicted by older models.

A simple way to evaluate the frequency of asteroid impacts in the Black Sea is to multiply the estimations above (in years between successive collisions with similar size objects) by the ratio between the Earth and Black Sea surface (which is 1208, for an Earth surface of 510,065,600 km^2) (Paine, 1999). However, the possible impactors are grouped by families according to their origin, as shown before. One may speculate rather similar dynamical and trajectory properties for objects of the same family and this may decrease the randomness degree of impact spatial distribution. As an example, the impact crater distribution on Europe surface shows a larger number of impacts around Baltic Sea and (interestingly) in the north of the Black Sea. But this interpretation must be taken with caution because most of the terrestrial impact craters have been obliterated by other terrestrial geological processes (Grieve, 1990).

6. Effects of an Asteroid Impacting the Black Sea

Very accurate numerical simulations provided valuable information about the interaction between larger size asteroids and the atmosphere, the seawater and the underwater medium (Gisler et al., 2003). A brief review of these findings follows. Usually, less than 0.01 of the impactor's kinetic energy is dissipated during the atmospheric passage. The remaining part of the kinetic energy is absorbed by the ocean and seafloor within less than one second. The water immediately surrounding the impactor is vaporized, and the rapid expansion of the vapor excavates a cavity in the water. This cavity is asymmetric in case of oblique incidence angles, and the splash, or crown, is higher on the side opposite the incoming trajectory. The collapse of the crown creates a precursor tsunami that propagates outward. The higher part of the crown breaks up into droplets that fall back into the water. The hot vapor from the cavity expands into the atmosphere. When the vapor pressure diminishes enough, water begins to fill almost symmetrically the cavity from the bottom. The filling water converges on the center of the cavity and generates a jet that rises vertically in the atmosphere to a height comparable with the initial cavity diameter. It is the collapse of this central vertical jet that produces the principal tsunami.

Modeling the initial water displacement by asteroid impact in a water body is a daunting task and various computational set-up scenarios are described in Shuvalov et al. (2002), Crawford and Mader (1998), Mader and Gittings (2003), Gisler et al. (2003). All of them predict water disturbances of a characteristic length scale comparable with water depth at impact point.

Due to complex water movement at impact source, the usual approach consists in designing an equivalent water cavity as in modeling waves generated by underwater explosions (Mehaute and Wang, 1996) or explosions of underwater volcanoes (Mirchina and Pelinovsky, 1988). Ward and Aspaugh (2002) suggested a relation between the radius R_c and depth D_c of the cavity of the form

$$D_c = qR_c^{\alpha} \tag{1}$$

where q and α are parameters depending on asteroid properties. It is assumed that only a fraction ε of asteroid kinetic energy is consumed in the cavity formation process, hence the depth of the cavity is given by

$$D_c = \begin{cases} \sqrt{\dfrac{2\rho_i \varepsilon R_i^3 V_i^2}{\rho_w g R_c^2}}, & D_c \le h \\ h & \end{cases}, \tag{2}$$

with ρ_i, V_i, R_i the density, velocity and radius of the impactor, ρ_w the seawater density, h still water depth and g the gravitational acceleration. From Eqs. (1) and (2) one gets the diameter d_c of the cavity

$$d_c = 2R_i \left(\frac{2\varepsilon V_i^2}{gR_i}\right)^{\delta} \left(\frac{\rho_i}{\rho_w}\right)^{\delta} \left(\frac{1}{qR_i^{\alpha-1}}\right)^{2\delta}. \tag{3}$$

with $\delta = 0.5/(1+\alpha)$. From laboratory investigations, $\alpha = 1.27$.

Assuming central symmetry for the the equivalent water cavity at impact point and initial moment $t = 0$, the parabolic shape of the water displacement is given by (Kharif and Pelinovsky, 2005):

$$\eta(r, t=0) = \begin{cases} D_c\left(1 - \dfrac{r^2}{R_c}\right), & r \le R_D \\ 0, & r > R_D \end{cases} \tag{4}$$

with $R_D = \sqrt{2}R_c$ and η is the water displacement relative to the underformed sea level. This result implies that all water deposits on the border lip of the cavity.

A model was developed to evaluate the effects of an asteroid impacting Black Sea (Badescu, 2007). The initial kinetic energy of the asteroid is mainly transferred to the seawater and to the seafloor ground. Therefore both the water and the bottom ground are deformed and also their internal energy increases to some extent. We shall focus on the dynamics of the initial cavity created in the water. This initial cavity constitutes "the source" for the subsequent phenomena. The water ejected by the asteroid impact is broken up into drops during both the ascending and descending parts of the trajectory. The gases dissolved in the water are transferred to the atmosphere. We will limit our discussion to the hydrogen sulfide dissolved in the Black Sea waters, although in some places the release of much larger volumes of methane, which would explosively burn might pose an even more serious threat. Two different phenomena that may affect the coastal regions start simultaneously at the place of the asteroid impact. First, the initial water cavity constitutes the source for a tsunami wave. Second, the H_2S expelled into the atmosphere (H_2S is denser than air) ''falls'' with a lower speed than the falling water and finally creates a gaseous ''cloud'' or ''blanket'' on the sea surface. This H_2S cloud moves and disperses in the mean atmospheric wind field. The two phenomena have their own dynamics. Their effects on the coastal regions are different and depend on many factors among which the most important is of course the asteroid size.

7. Tsunami Dynamics

In order to describe the tsunami wave propagation, modified Navier-Stokes equations, including bottom friction effects, are used. Neglecting the Coriolis effect due to relatively small size of Black Sea and dispersion term, the wave equations in spherical coordinates describing the temporal and spatial evolution of sea level are (Dao and Tkalich, 2007; Kowalik et al., 2005):

$$\frac{\partial \eta}{\partial t}+\frac{1}{R\cos\phi}\left[\frac{\partial M}{\partial \lambda}+\frac{\partial}{\partial \phi}(N\cos\phi)\right]=0$$

$$\frac{\partial M}{\partial t}+\frac{1}{R\cos\phi}\frac{\partial}{\partial \lambda}\left[\frac{M^2}{D}\right]+\frac{1}{R}\frac{\partial}{\partial \phi}\left[\frac{MN}{D}\right]+\frac{gD}{R\cos\phi}\frac{\partial \eta}{\partial \lambda}+\frac{\tau_\lambda^b}{\rho}=0 \tag{4}$$

$$\frac{\partial N}{\partial t}+\frac{1}{R\cos\phi}\frac{\partial}{\partial \lambda}\left[\frac{MN}{D}\right]+\frac{1}{R}\frac{\partial}{\partial \phi}\left[\frac{N^2}{D}\right]+\frac{gD}{R\cos\phi}\frac{\partial \eta}{\partial \phi}+\frac{\tau_\phi^b}{\rho}=0$$

where $D=h+\eta$ is total water depth, R the average Earth radius, λ longitude and ϕ is latitude. M and N are the depth averaged water discharges in the longitude and latitude directions.

$$M=\int_{-h}^{\eta} v_\lambda dr=\overline{v}_\lambda D$$

$$N=\int_{-h}^{\eta} v_\phi dr=\overline{v}_\phi D \tag{5}$$

where v_λ and v_ϕ comes from the spherical Navier-Stokes equations. The last terms in Eqs. (4) are given by:

$$\tau_{\lambda}^{b} = f\frac{M}{D^2}\sqrt{M^2+N^2}$$
$$\tau_{\phi}^{b} = f\frac{N}{D^2}\sqrt{M^2+N^2} \qquad (6)$$

They represent bottom friction terms which become important in shallow waters. The friction coefficient f can be computed from Manning's roughness n

$$n = \sqrt{\frac{fD^{1/3}}{2g}}$$

such that Eqs. (6) become (Dao and Tkalich, 2007):

$$\tau_{\lambda}^{b} = \frac{2gM}{D^{7/3}}\sqrt{M^2+N^2}$$
$$\tau_{\phi}^{b} = \frac{2gN}{D^{7/3}}\sqrt{M^2+N^2} \qquad (7)$$

7.1. Numerical Approach

The system of partial differential nonlinear hyperbolic nonconservative equations is solved using TsunamiClaw code embodying many features and subroutines from Clawpack package available at (Clawpack, 2009). An extensive explanation of its features and numerical approach can be found in (LeVeque, 2002).The code is based on a finite difference technique using second-order Godunov flux-splitting scheme (Stegger and Warming, 1981) and Roe's approximate Riemann solver with entropy fix (Roe, 1981) for convective terms and 2-stage Runge Kutta method for evaluating source terms. The boundary conditions of the computational domain are free boundary conditions (zero order extrapolation) such that to capture flooding phenomena around shore lines. The computational grid containing both water and land domain and is equally spaced in E-W and N-S directions in steps of $h_{\lambda} = R\cos\phi\Delta\lambda$ in longitude and $h_{\phi} = R\Delta\phi$ in latitude.

7.2. General Features of Tsunamiclaw

The code allows modeling tsunamis and inundation on either latitude-longitude grid or on Cartesian grid with a diverse range of temporal and spatial scales. This is accomplished by using up to two coarse levels grids for entire domain and evolving rectangular Cartesian sub-grids of higher refinement level that track moving waves and flooding around shoreline (Berger and LeVeque, 1998; Berger and Rigoutsos, 1991). At any given time in the computation, a particular level of refinement may have numerous disjoint grids associated with it. User may specify the refinement ratios such that, starting with a coarse 10^2 km grid cell that tracks the long wavelength of the deep water tsunami he is able to resolve the shore line and innundation potential with a level of refinement up to 10^2 m or even lower. Imposing shallow water depth (typical value of 100 m) one can indicate which areas are to be refined close to coastal lines. Friction may be important for realistic run-up heights and typical value for Manning coeficient is 0.025. A combined bathymetry and topography file in standard GIS format for the Black Sea (Smith and Sandwell, 1997) is used throughout.

7.3. Results

Three zones of asteroid impact locations were considered in Badescu (2008) (Figure 2). Zone A is situated in the Southwest region of Black Sea, at longitudes between those of the large cities Istanbul and Izmit. Only results for the impact position A3 in that paper (latitude $42.55°N$, longitude $29.05°E$, sea depth -2137 m) are reported in this section. The shortest distance between location A3 and the coast is about 150 km.

An asteroid of diameter $D_i = 250$ m and impact velocity 20000 m/s is considered as an example. The asteroid material is assumed to be either dunite (mass density $\rho_i = 3000$ kg/m^3) as a mockup for typical stony asteroids. Taking into account Eqs. (2) and (3) we obtained cavity depth $D_c = 2137$ m and cavity diameter $d_c = 4790$ m for $q = 0.1$ and $\varepsilon = 0.15$.

The salient features of the overall Black Sea tsunami simulation are presented in Figure 3. Complex pattern of waves stands out mostly due to the almost confined character of Black Sea that generates wave interference. Early warning signs (receding water) along beaches of Bulgaria and North Turkey are apparent after 600 s from impact. The global map of tsunami (Figure 3) propagation provides clues regarding major flooding risk beach areas. Among them, we point out, Varna and Burgas surrounding areas on Bulgaria coastline, Zonguldak and Eregli on north beach of Turkey and Mangalia area on Romanian shore.

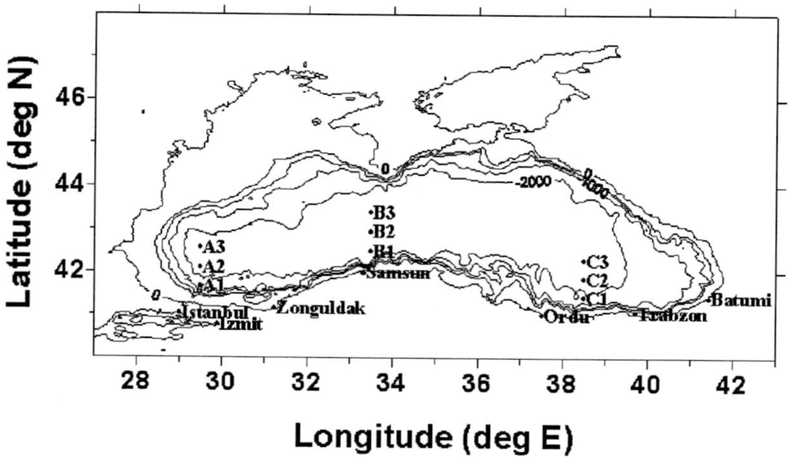

Figure 2. Asteroid impact positions (A1-A3,B1-B3, C1-C3) and major cities on and near the Southern coast of Black Sea.

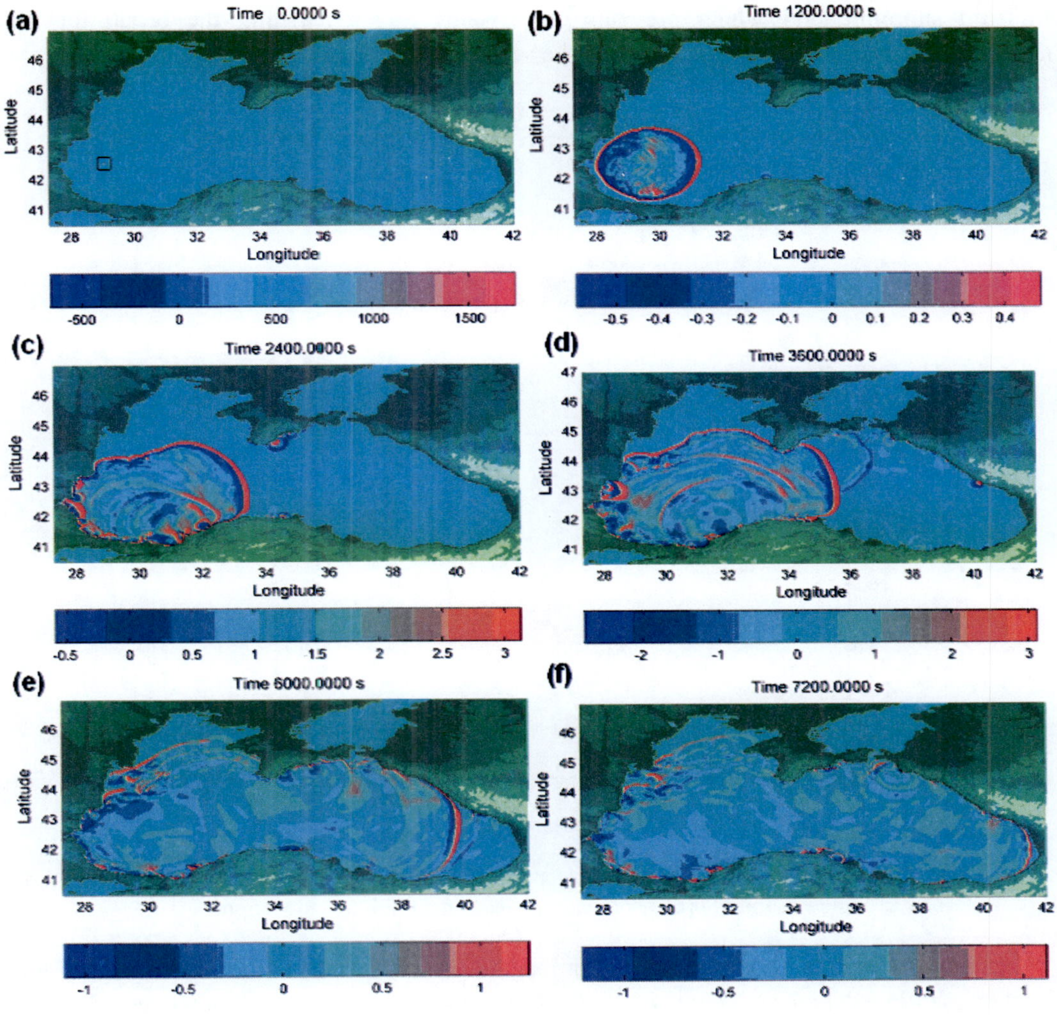

Figure 3. Time snapshots of tsunami propagation. The unit on the colour map is meter.

The initial sea level perturbation is presented in Figure 4 showing a peak of only 1750 m after interpolation procedures on the impact area. This observation is relevant concerning the potentially higher damages of such an impact may have on beach areas besides those reported.

In order to accommodate the most important features of tsunami impact with coastal regions we have performed a separate simulation only for the west basin of Black Sea with emphasis on flooding risk beaches.

In this case the coarsest grid had a number of 210x150 cells while further grids were refined in ratios of 2, 4 and 8 for the finest grid covering the above mentioned flooding risk areas. One particular shore in the Eregli-Zonguldak area (41.15° N, 31.24° E) has attracted our attention. The bathymetry profile on the azymuth line connecting the impact point and a location of geographical coordinates 41.51° N and 31.51° E is illustrated in Figure 5. The particular feature is the underwater spur that rises up to 500 m bellow sea level at around 70 km from shoreline and which prevents deep water waves reaching the beach though enhancing the short waves amplitude as can be seen from Fig 7 and Fig 10.

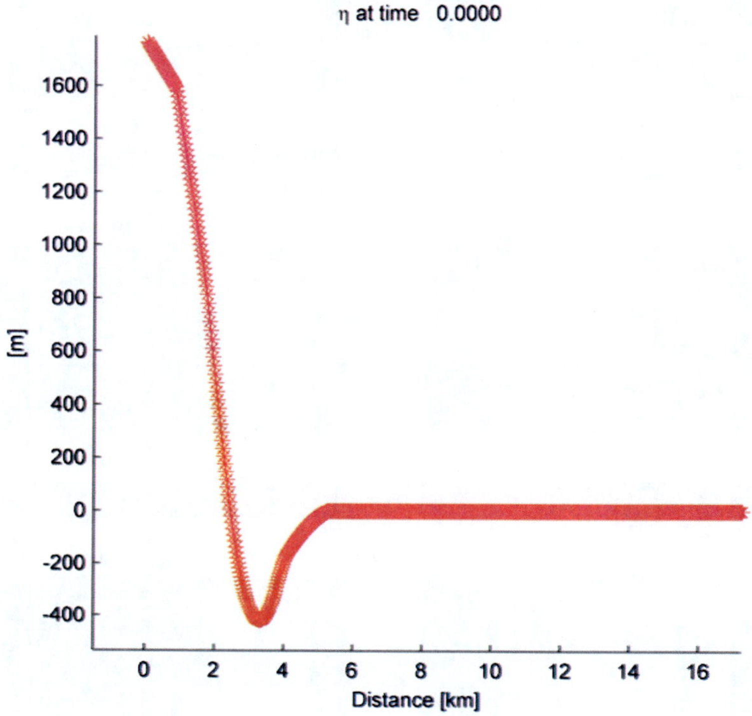

Figure 4. Initial profile of water displacement at impact location.

Figure 6 presents a slice through the propagating wave at 600 s after impact along the azymuth mentioned before. There are several striking features that characterize the water motion. First, the main wave of up to 18 m high and 20 km in length followed by a wave packet of 5 km wave length and maximum amplitude of 0.5 m. The second feature is the generation of a beach wave of up to 7 m high that arrives much earlier than the main wave. Around 1020 s from impact (Figure 8) this secondary wave produces a maximum run-up height of 6.76 m at shoreline and penetrates the beach over almost 0.5 km in straight line. Three minutes later, water recedes for more than 600 m from initial sea level (Fig 9). The major wave impact with shoreline is produced at 2340 s from asteroid collision (Fig 10). Now, we can see that behind it there is no wave structure well organized like in Figure 7 due to interference and dispersion with beach reflected waves. The wave structure is presented in Figure 11. It has trough at -6.7 m and the crest rises to 8.3 m. The water displacement extends over 0.5 km on the beach. After 3 minutes the water recedes another 0.5 km from the beach line (Figure 12).

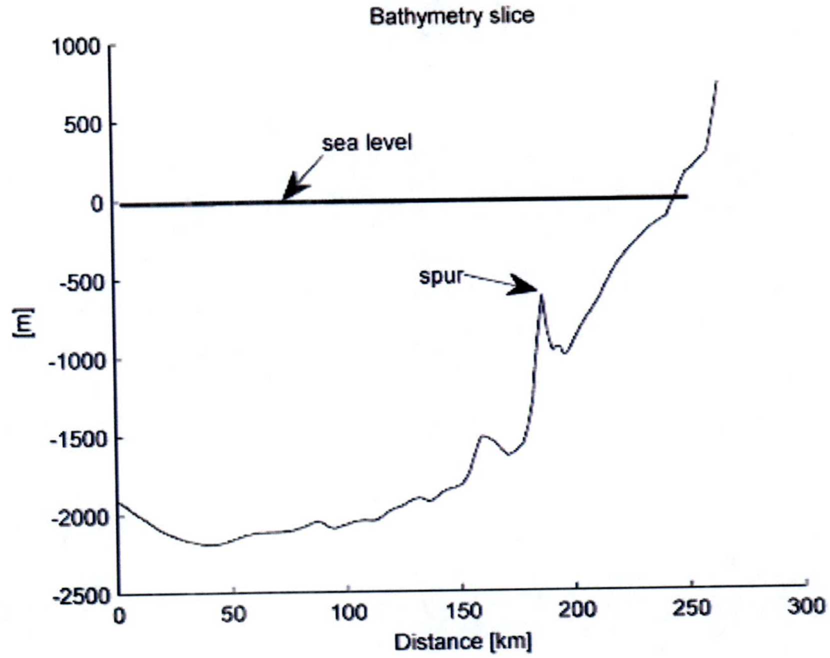

Figure 5. Bathymetry and topography profile on the azymuth between impact location and Eregli-Zonguldak area.

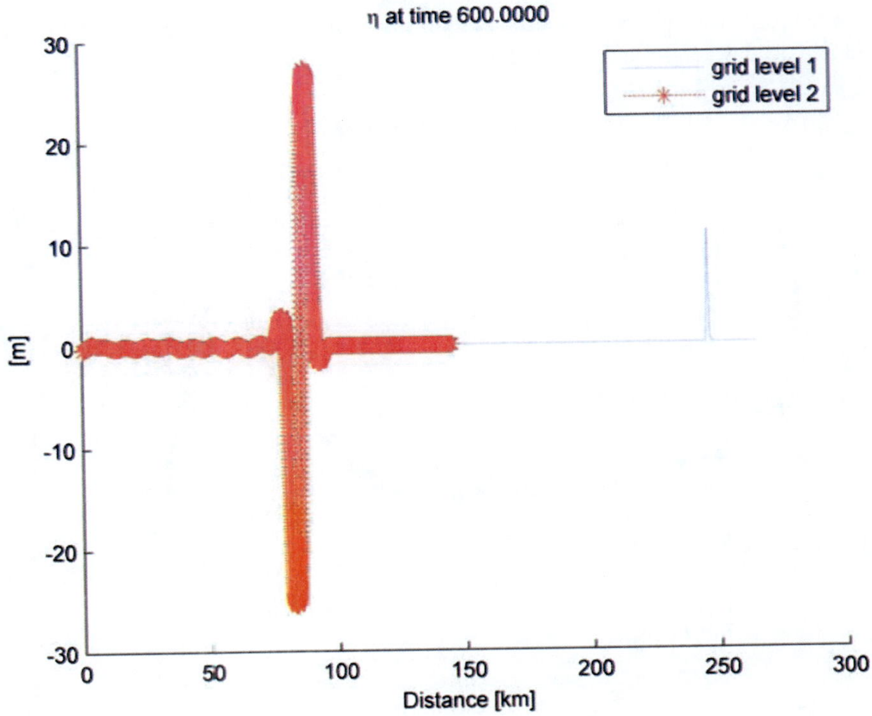

Figure 6. Water displacement at t=600 s from impact.

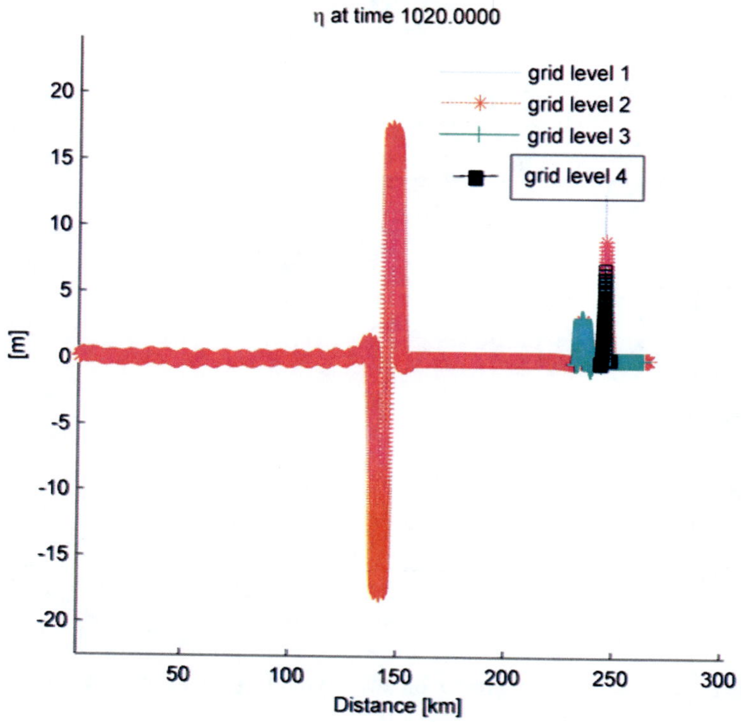

Figure 7. Wave profile function of distance at t=1020 s from impact for different grid refinement levels. Level 1 and 2 in deep waters; 3 and 4 for beach area.

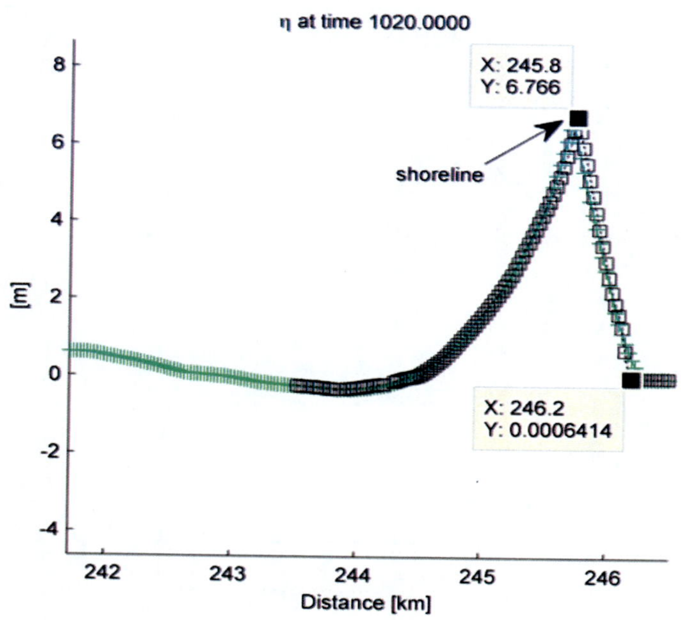

Figure 8. Run-up height at t=1020 s. The inland extent of water flooding is comprised between the shoreline and x-coordinate given in the right textbox.

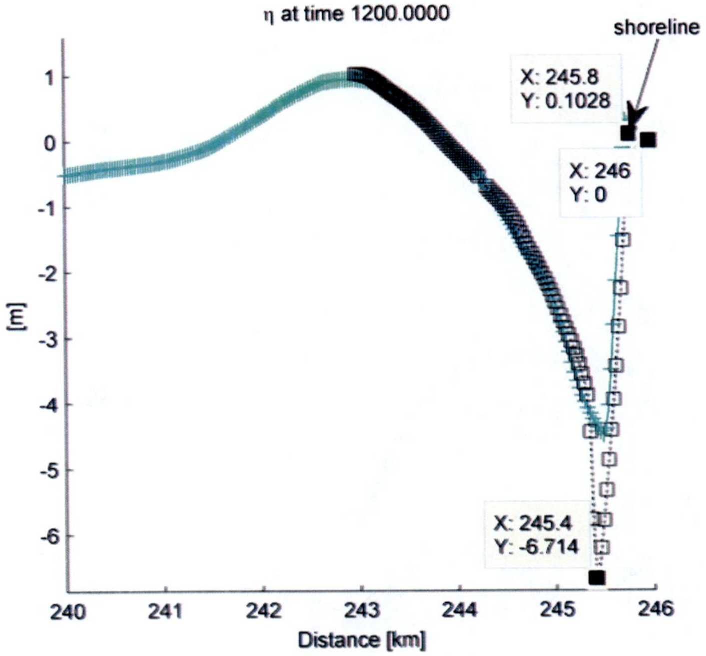

Figure 9. Run-off height at t=1200 s. Water recedes from shoreline to x-coordinate given in the bottom textbox.

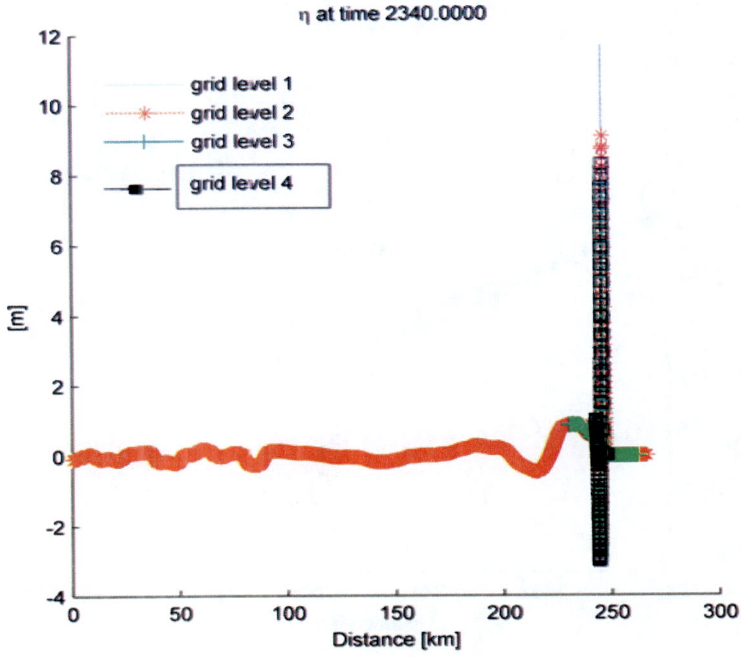

Figure 10. Wave profile function of distance at t=2340 s from impact for different grid refinement levels. Level 1 and 2 in deep waters; 3 and 4 for beach area.

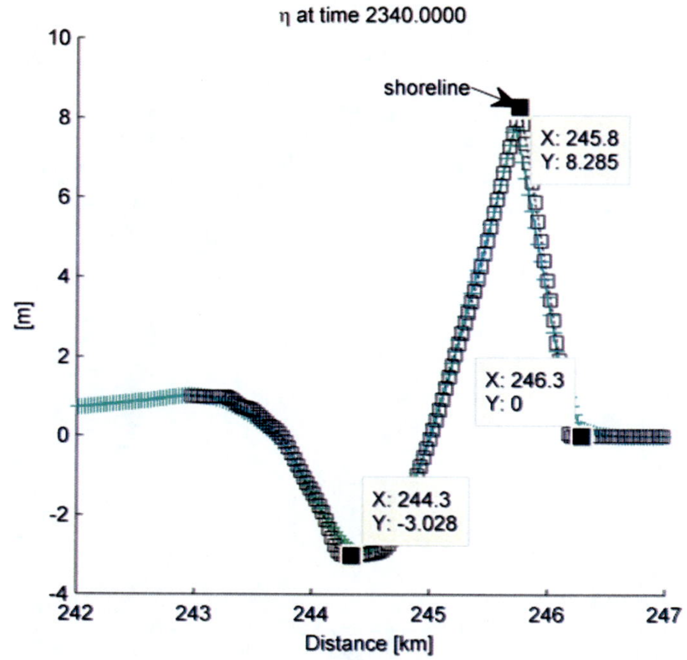

Figure 11. Run-up height at t=2340 s. The inland extent of water flooding is comprised between the shoreline and x-coordinate given in the right textbox.

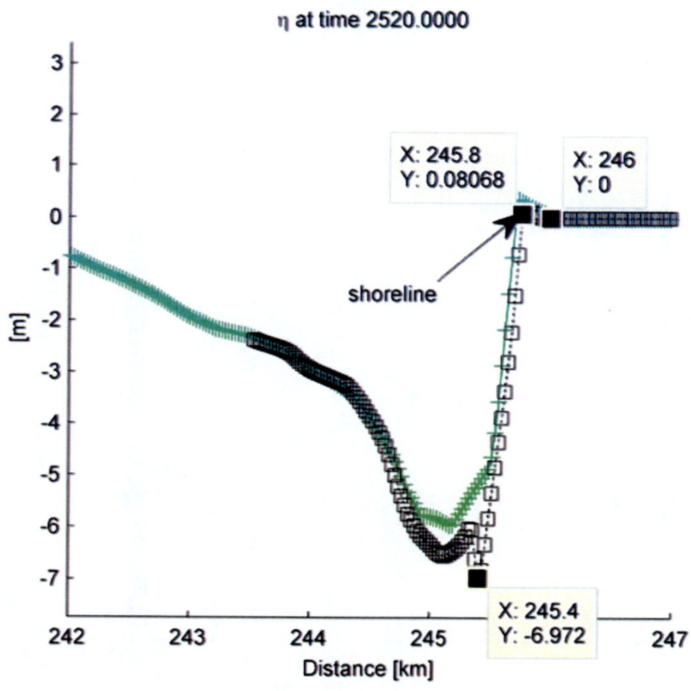

Figure 12. Run-off height at t=2520 s. Water recedes from shoreline to x-coordinate given in the bottom textbox.

The beach sweeping ceases only one hour after the impact moment. In between, beside the snaphots already presented, there are numerous series of waves hitting the coastline of less important amplitudes.

8. Hydrogen Sulfide Cloud Dynamics

8.1. Information about Hydrogen Sulfide

The relative molecular mass of hydrogen sulfide is 34.08. At room temperature H_2S is a colorless, flammable gas with a characteristic odor of rotten eggs. Its vapor pressure is 1929 Pa (at 21.9 °C). Hydrogen sulfide is soluble in water, alcohol, ether, glycerol, gasoline, kerosene, crude oil, and carbon disulfide. Its water solubility (at 20 °C) is 1 g in 242 ml (Chou, 2003). The taste threshold for hydrogen sulfide in water is between 0.05 and 0.1 mg/liter (WHO, 1993). The Henry's law constant at 20 °C is 468 atm/mole fraction (ATSDR, 1999). Hydrogen sulfide may evaporate easily from water. In general, low pH and high temperature tend to favor evaporation (HSDB, 1998). Other physical and chemical properties of hydrogen sulfide can be found in the International Chemical Safety Card (IPCS, 2000).

Releases to the environment are primarily in emissions to the atmosphere, where it usually remains for less than 1 day, but in winter may persist for as long as 42 days. Hydrogen sulfide in the air is oxidized by molecular oxygen and hydroxyl radicals, forming the sulfhydryl radical and ultimately sulfur dioxide or sulfate compounds (Hill, 1973; NSF, 1976). Sulfur dioxide and sulfates are eventually removed from the atmosphere through precipitation or through absorption by plants and soils. The concentration of hydrogen sulfide in air in unpolluted areas is very low, between 0.03 and 0.1 μg/m^3. Ambient air concentrations of hydrogen sulfide as a result of natural sources range in USA from 0.14 to 0.4 μg/m^3 (US EPA, 1993). Ground level samples taken around a sulfurous New Zealand lake had hydrogen sulfide levels of 175–5500 μg/m^3 (Siegel et al., 1986). [Note the conversion factor for hydrogen sulfide in air (20 °C, 101.3 kPa) is 1 mg/m^3 = 0.71 ppm].

A number of experimental studies exist on the effect of hydrogen sulfide on animals. Usually, single inhalation exposures to hydrogen sulfide result in death and respiratory, immunological/lymphoreticular, cardiovascular, and neurological effects. Short-term exposures effects include ocular, cardiovascular, neurological, metabolic, hepatic, and developmental effects. Medium-duration inhalation studies have reported respiratory, neurological, and olfactory effects. It seems no long-term inhalation studies in animals exist. The most sensitive target organ for medium-term exposure in animals is the nasal olfactory mucosa. The no-observed-adverse-effect level (NOAEL) for Sprague-Dawley CD rats was 14 mg/m^3. This NOAEL is used as a basis for the development of a medium-term tolerable concentration.

Most data about the effects of hydrogen sulfide on humans are derived from acute poisoning case reports, occupational exposures, and limited community studies. Reported health effects in humans following exposure to hydrogen sulfide include death and respiratory, ocular, neurological, cardiovascular, metabolic, and reproductive effects. The geometric mean odor threshold for a large population is 11 μg/m^3. At concentrations greater than 140 mg/m^3, olfactory paralysis occurs. A few breaths at 700 mg/m^3 can be fatal. There are many case reports of human deaths after single exposures to high concentrations (larger than 700 mg/m^3) of hydrogen sulfide gas (Beauchamp et al., 1984). Most of them have occurred in relatively confined spaces; the victims lost consciousness quickly after inhalation of hydrogen sulfide, sometimes after only one or two breaths (the so-called "slaughterhouse sledgehammer" effect). Quantitative results about human health effects at various hydrogen sulfide concentrations may be found in Table 2 of Chou (2003).

The lowest-observed-adverse-effect level (LOAEL) is 2.8 mg/m^3 in asthmatic individuals for respiratory and neurological effects. This LOAEL is used as a basis for the development of a short-term tolerable concentration. For short-term (for exposure durations of 1–14 days) and medium-term (for exposure durations of up to 90 days) inhalation exposures, the tolerable H_2S concentrations in air is 100 μg/m^3 and 20 μg/m^3, respectively.

The respiratory tract is the major target organ of hydrogen sulfide toxicity. Therefore, humans with asthma, the elderly and young children with compromised respiratory function represent sensitive human subpopulations

8.2. Hydrogen Sulfide Cloud Generation

The model developed in Badescu (2007) is briefly described now. One denotes by $k_{H_2S\uparrow}$ a coefficient (valued between 0 and 1) showing how much H_2S is released in the atmosphere from the ejected sea water. The mass participation of H_2S in the cloud (which is a mixture of air, hydrogen sulfide and water vapor) is denoted by g_{H_2S} . g_{H_2S} is valued between 1 (for a pure H_2S blanket) to near 0 (for very small quantities of H_2S in the air). The dynamics of the hydrogen sulfide release in the atmosphere and the formation of the H_2S cloud are very complex processes. Therefore, the evaluation of $k_{H_2S\uparrow}$ and g_{H_2S} is not a simple problem. Two models based on different mechanisms of hydrogen sulfide release were proposed in Badescu (2007) to provide a rough solution. A combination of these (and other) mechanisms will probably arise in reality. In the following only model B is described. One assumes most of the hydrogen sulfide is released from the falling liquid seawater when it is splashed by collision with sea surface. In this case the hydrogen sulfide release is a more violent, surface process. One expects the release degree to be higher and the H_2S cloud thickness to be much smaller than in the first scenario. In computations we shall use $k_{H_2S\uparrow} = 0.7$ and one shall assume the extreme case when the cloud consists of pure hydrogen sulfide ($g_{H_2S} = 1$). This scenario was already used in Schuiling et al. (2006).

8.3. Model of Hydrogen Sulfide Down-Wind Dispersion

For low-momentum dense gas releases at ground level on uniform, level terrain with unobstructed atmospheric flow, the buoyancy-dominated, stably-stratified, and passive dispersion regimes can be modeled with DEGADIS (Dense Gas DISpersion) model (Havens and Spicer, 1985; Havens, 1988). A few details about this model follow (Spicer and Havens, 1992). If the primary source (gas) release rate exceeds the maximum atmospheric take-up rate, a denser-than-air gas blanket is formed over the primary source. This near-field, buoyancy-dominated regime is modeled using a lumped parameter model of a denser-than-air gas "secondary source" cloud which incorporates air entrainment at the gravity-spreading front using a frontal entrainment velocity. If the primary source release rate does not exceed the maximum atmospheric take-up rate, the released gas is taken up directly by the atmosphere and dispersed downwind. For either source condition, the downwind dispersion phase of the calculation assumes a power law concentration distribution in the vertical direction and a modified Gaussian profile in the horizontal direction with a power law specification for the wind profile. The source model represents a spatially averaged concentration of gas present over the primary source, while the downwind dispersion phase of the calculation models an ensemble average of the concentration downwind of the source.

DEGADIS is a lumped parameter model of the formation of the denser-than-air gas source cloud or blanket, which may be formed from a primary source such as an evaporating liquid pool or otherwise specified ground-level emission source, or by an initially specified gas volume of prescribed dimensions for an instantaneous release. The gas blanket is represented as a cylindrical gas volume which spreads laterally as a density-driven flow with entrainment from the top of the source blanket by wind shear and air entrainment into the advancing front edge. The source blanket will continue to grow over the primary source until the atmospheric take-up rate from the top is matched by the air entrainment rate from the side and, if applicable, by the rate of gas addition from under the blanket. The model treats dispersion of gas entrained into the wind field from an idealized, rectangularly shaped source. The circular source cloud is represented as an equivalent area rectangle with equivalent fetch.

The general application of the model involves formation of a "secondary" gas source, the subsequent entrainment of gas from that secondary source by the wind field, and down dispersion of the gas plume or cloud. DEGADIS incorporates heat transfer and water transfer when applicable from the underlying surface to the cloud. As a standalone model, DEGADIS application is limited to the description of atmospheric dispersion of denser-than-air gas releases at ground level onto flat, unobstructed terrain o water. The desired released gas properties include the molecular weight, the release temperature, and two constants, which describe the heat capacity A constant heat capacity can also be specified. The ambient wind field is characterized by a known velocity u_0 at a given height z_0 and the atmospheric Pasquill stability class.

8.4. Model Implementation

Various scenarios concerning asteroid impact were considered in Badescu (2008). The Southern Black Sea coasts were envisaged in that paper as possible regions to be affected by the asteroid impact. These coasts have a rather dense population, which increases in some particular areas during summer, as a result of tourist affluence.

8.4.1. Population Distribution

Figure 13 shows the spatial distribution of population on or near the Southern coast of Black Sea.

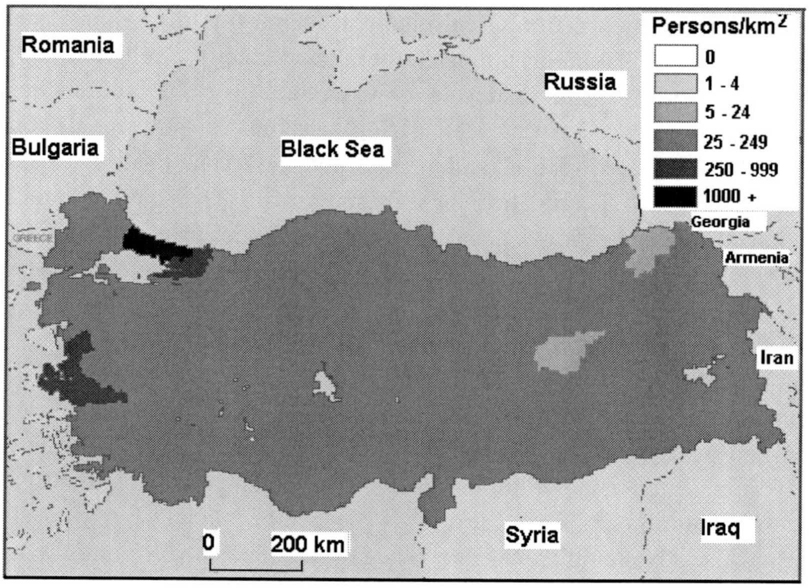

Figure 13. Population distribution on the South coast of Black Sea (Turkey). Source: CIESIN (2000).

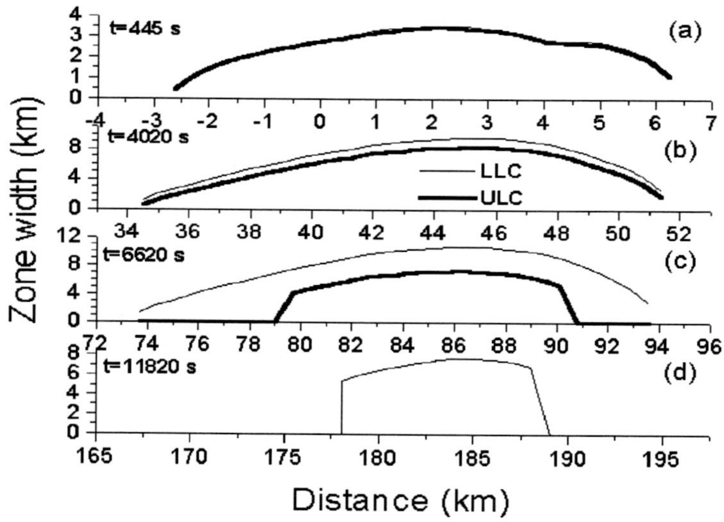

Figure 14. Time and space dependence of the ULC and LLC zones width from the cloud centerline, respectively. (a) 445 s after asteroid impact; (b) 4020 s; (c) 6620 s; (d) 11820 s. All data refer to the elevation of 2 m above sea level. ULC = H2S upper limit concentration; LLC – H2S lower limit concentration. The distance is measured downwind from the impact location. A 250 m diameter impactor was considered. The wind speed is 10 m/s.

The most populated cities are Istanbul (latitude $41.02°N$, longitude $29.00°E$, with 9760000 people in 2002), Trabzon ($41.05°N, 39.72°E$, 979081 people in 2000), Samsun ($41.28°N, 33.33°E$, 396900 people in 2004), Kocaeli (Izmit) ($40.76°N, 29.91°E$, 195699 people in 2000), Zonguldak ($41.15°N, 31.24°E$, 116725 people in 1990), Ordu

($40.59°N, 37.53°E$, 112525 people in 2000) in Turkey and Batumi ($41.39°N, 41.44°E$, 121806 people in 2002) in Georgia. Except for the densely populated large cities and for a thinly populated area in the East, the population density on the Southern regions of the Black Sea ranges in average between 25 and 249 persons/km^2.

8.4.2. Wind Data

The interaction of the poisoning gas cloud with the population on the coast depends strongly on the direction and speed of the wind. Measured wind speed and wind direction data for the Black Sea may be found in the SeaWinds database (RSS, 2005). The SeaWinds database includes daily and time averaged geophysical data. The data are organized according to observation date. Daily orbital data is mapped to 0.25 deg longitude by 0.25 deg latitude Earth grid. Weekly averages and monthly averages of all data within the calendar month are also reported. For time averaged scatterometer data, the wind speeds are scalar averaged and the wind directions are vector averaged. Missing data generally affects the daily data, but can also reduce the number of observations in weekly and monthly averages. Data gaps are generally due to missing data upstream from the processing facility, such as the instrument being turned off. The SeaWinds dataset spans about six months from 10 April 2003 through 24 October 2003.

8.4.3. Results and Discussions

Computations were made for downwind dispersion of the cloud consisting of hydrogen sulphide, air and water vapor. The initial time ($t = 0$ s) diameter of the hydrogen sulphide cloud depends on asteroid size. It is about 2580 m, 4916 m and 12288 m, for $D_i = 70$ m, 250 m and 1000 m, respectively, whatever the impact location is. Larger clouds are generated by larger size asteroids, as expected. Also, both the cloud thickness and the H_2S mass it contains are larger at impact positions with deeper seawater. However, for small size asteroids ($D_i = 70$ m) the cloud thickness does not depend on impact position. This is explained by the fact that the depth of the water cavity ($D_c = 892\ m$) created by the impact is smaller than the sea depth for all the nine impact locations.

In the analysis reported below two values are considered when referring to hydrogen sulphide concentrations: the lower concentration limit (LLC) of 19.88 ppm and the upper concentration limit (ULC) of 497 ppm. These values are associated to the following effects on humans. The lower concentration limit (LLC) corresponds to fatigue, loss of appetite, headache, irritability, poor memory, dizziness (Ahlborg, 1951). The upper concentration limit (ULC) corresponds to death after single exposures (Beauchamp et al., 1984). The lowest concentration of interest (i.e. LLC = 19.88 ppm) is the concentration at which the calculations by DEGADIS model are stopped.

The dynamics of the hydrogen sulphide cloud generated at a given impact location was simulated by using DEGADIS model. Figure 14 shows sample results for the status at various moments of the H_2S cloud generated by an asteroid of diameter $D_i = 250$ m at impact position B1($42.48°N, 33.50°E$, -2186 m). The wind speed is $u_0 = 10\ m/s$ at $z_0 = 10\ m$

height. It may be considered as a rather common value and should not be seen as an extreme. The adopted atmosphere stability Pasquill class needed by the DEGADIS model is F.

All results refer to the elevation of 2 m above sea level. The width from the cloud centerline to the indicated concentration levels at the indicated height is shown in case of the lower limit concentration (LLC) and upper limit concentration (ULC), respectively. At short time after the impact the asymmetry induced by wind direction is already obvious and the two cloud dimensions (i.e. parallel and perpendicular on wind direction) are comparable in size (Figure 14a). Also, the ULC cloud core is comparable in size with the LLC cloud core. The nearest large locality to impact position B1 is Samsun. Thus, the most dangerous wind direction is towards this town. This case is considered next. Figure 14b refers to a time moment when H_2S concentrations higher than ULC entered 1 km the coast where Samsun is placed. This happens about 1 h and 7 minutes after the impact. The cloud is more diluted as reported in case of Figure 14a. This is shown by the rather important difference between the widths of the ULC zone and LLC zone, respectively. The LLC zone width from the cloud centerline is about 1.5 km larger than the corresponding ULC zone width. The cloud "length" on wind direction is about 16 km while its maximum "length" perpendicular on wind direction is again of the order of 16 km. Figure 14c shows the last moment (for a time step of 290 s) when the DEGADIS model shows the existence of the ULC core in the cloud. This core, of size about 6 km x 10 km entered the land about 40 km in about 40 minutes. By using the terminology used in tsunami's research area, one may call "run-in" the travel of the hydrogen sulphide cloud over the land. Note that DEGADIS model account for the complex down-wind dispersion process and the "traveled" distance by the cloud margins is of course larger than the distance evaluated by the usual formula (wind speed x time interval). During its run-in, the ULC cloud core covered about 400 km^2 of land. Figures 2 and 13 show the population density around Samsun is about 25 to 249 people per km^2. Thus, the population affected by the ULC cloud core may range between 10000 and 99600 people. Figure 14d shows the last moment when the DEGADIS model shows the existence of the LLC core in the cloud. This core entered the land about 127 km in about 2 h. Its last size is around 10 km x 14 km. The LLC cloud core covered during its run-in is about 1778 km^2 and the number of people affected may range between 44450 and about 442000. All these population estimates do not include Samsun's habitants.

The hazard plays an important role in all phenomena where atmospheric factors are involved. Simulations for different Pasquill degrees of atmospheric stability were performed, for the same asteroid sizes and impact locations. The results obtained are different, as expected, but they keep the main already outlined features. Note that the dry land topography was not taken into account when modeling the dispersion process during the run-in. It may have important effects on the results.

9. Risks for Nuclear Explosions

The impact by asteroids of the size analyzed in this paper may be considered as a remote risk. There are, however, other risks with greater proximity for a poisonous gas cloud to be released. We are referring here to the situation when very large explosions occur in the deep seawater of the Black Sea. This may happen for example in case nuclear weapons are deliberately used or in case of accidental nuclear explosions on submarines.

The energy released by nuclear weapons is usually measured in ton of Tri -Nitro-Toluene (TNT). One ton of TNT is the amount of energy released by 1 short ton (or 0.907 metric ton) of TNT (in SI units, $4.184 \cdot 10^9$ J). The bomb dropped at Hiroshima was equivalent of 15 kton of TNT (63 TJ) while that dropped at Nagasaki was equivalent of 20 kton (84 TJ) (Encyclopedia Americana, 1995). 'Castle/Bravo' was the largest nuclear weapon ever detonated by the United States (Bikini, February 28, 1954). It was expected to explode with an energy equivalent of about 8 Mton of TNT but it produced almost twice that explosive power - equivalent to 15 Mton of TNT (US ACDA, 1975). The world's most powerful hydrogen bomb was detonated by USSR (Novaya Zemlya, October 30, 1961). The bomb released the energy of 50 Mton of TNT (210000 TJ) (Khariton and Smirnov, 1993). Nuclear explosive devices with quite different yields have been built during the years. Some older bombs had yields of about 20 Mton TNT (84000 TJ) but today most nuclear devices have yields of less than 1 Mton TNT (<4000 TJ) (World Book Encyclopedia, 1999).

We may compare the energy released by a nuclear explosion with the kinetic energy carried by an asteroid, which is proportional to the third power of its diameter. As a rough approximation we associate to the Tunguska event a 70 m diameter asteroid and a 10 Mton explosion. If a nuclear (deliberate or accidental) explosion would occur these days, it is likely the energy released would be less than 1 Mton TNT, i.e. more than ten times smaller than the Tunguska explosion. This explosion would correspond to the impact by an asteroid of 33 m diameter. Thus, the results we presented for the 70 m impactor may be seen as the most pessimistic scenario for a "common" nuclear explosion. The very large 50 Mton TNT nuclear explosion is the equivalent of a 120 m asteroid impact. The effects of this explosion would range between those associated to the 70 m and 250 m asteroids, respectively, considered here.

The effects of the nuclear accident which occurred on April 26, 1986 at Chernobyl may be compared with those predicted by the present model. This may give a perspective to our findings. The explosion at Chernobyl was chemical, driven by gases and steam generated by the nuclear reactor core runaway, not by nuclear reactions (Rhodes, 1993). Based on the official reports, near 8,400,000 people in Belarus, Ukraine and Russia were exposed to the radiation. About 155,000 km^2 of territories in the three countries were contaminated. Agricultural areas covering nearly 52,000 km^2 were contaminated with cesium-137 and strontium-90 (UN, 2004). One can see that the surface area affected by the Chernobyl accident is much larger than that affected by the poisonous gas cloud released into the atmosphere by a 1 km impact asteroid. However, the effects on human population (in lives losses) may be much more important in case of the asteroid impact. Also, note that here we did not consider the other destructive effects that an asteroid impact in the sea may have, such as a tsunami generation.

10. Possible Social Impacts and Prevention

Though an asteroid impact in the Black Sea has a very low probability on a short time period, an integrated approach to the science, technology, and public policy aspects of the impact hazard is necessary, taking account of its destructive potential.

Very few steps have been done on this line. Several schemes have been proposed which might diminish the H_2S content of the Black Sea in the long run (see Schuiling et al. (2006)

and references therein). Even if implemented, their impact on the H_2S inventory of the Black Sea is likely to be small and slow. There is some technical cooperation in the Black Sea region, fostered by the Convention on the Protection of the Black Sea Against Pollution (1992), the Odessa Ministerial Declaration (1993) and the Black Sea Strategic Action Plan (1996). However, there is no single planning authority in the region surrounding the Black Sea, which could develop a contingency plan in case an asteroid impact in the Black Sea would take place.

Some activities to be done have been suggested in a more general context by Chapman et al. (2001). Formal procedures should be prepared for linkage between the astronomic entities who would announce discovery of a potential impacting body and the national social institutions that might undertake preventive measures. Formal procedures should be stated for evaluating information about potential impacts, to outline what should be done in various cases, and to assign responsibilities to various institutional bodies. Plans should be developed for activities associated to the impact and different scenarios should be implemented into such rescue and relief plans. Procedures should be designed in advance to calm public fear and panic arising from the prediction itself.

The most likely international disaster that would result from an impact in the Black Sea is a large tsunami. However, there is no entity in the region responsible for warning about tsunamis. Once established, such an entity should be responsible for warning about dangerous dispersive hydrogen sulfide clouds, too. If the impact time and location are rather accurately predicted, might merit evacuation of people along the nearby land areas, even in case of small size impacts. It is likely that a small impact would happen without advanced warning. Therefore, a protocol should be established within the tsunami/hydrogen sulfide cloud detection and warning systems for recognizing an impact-produced tsunami and understanding how its associated effects (mainly the occurrence of a poisonous hydrogen sulfide cloud) might differ from the effects of the tsunamis produced by more usual causes (e.g., earthquakes).

11. Conclusion

The hydrogen sulfide-rich waters of the Black Sea pose a potential danger for the surrounding land regions. The impact of an asteroid exceeding tens of meters in size may cause both a tsunami wave and a catastrophic poisonous gas release in the atmosphere. The effects of these phenomena on the coastal regions were evaluated.

In the case of the hypothetical asteroid impact location chosen we conclude that no major consequences are to be expected for the population located in the flooding risk areas previously mentioned. Although high run-up wave heights have been found from the numerical simulation, the inland extent of the flooding waters is quite limited due to steep coastal topographical profile. Early warning signs of tsunami hazard are predicted in our analysis allowing population to migrate towards higher locations. Of course, for a different asteroid impact location choice (somewhere in front of Romanian coast) the damages could have been of greater proportions taking into account the extent of smooth, low-level beaches dominating the Romanian border.

In the Black Sea belt-like region with continental shelf seawater less than 200 m in depth, no hydrogen sulfide cloud is generated at the impact, due to the small or even null content of dissolved hydrogen sulfide. At the other impact positions, the initial diameter of the hydrogen sulfide cloud depends on asteroid size. It is about 2580 m, 4916 m and 12288 m, for diameters $D_i = 70$ m, 250 m and 1000 m, respectively, whatever the impact location is. The initial thickness of the poisonous cloud depends, in addition, on sea depth at impact location and on the model adopted for cloud formation.

Two hydrogen sulfide concentrations in the cloud are of particular interest: the lower concentration limit (LLC) of 19.88 ppm (associated to fatigue, loss of appetite, headache, irritability, poor memory, dizziness) and the upper concentration limit (ULC) of 497 ppm (associated to death after single exposures).

The maximum distance traveled by the ULC and LLC cloud cores increases by increasing the asteroid size and wind speed. The influence of the impact position on the distance traveled by hydrogen sulfide clouds is rather weak, as far as the seawater depth does not change significantly.

Results obtained for the particular impact position adopted in this chapter are summarized now. A 250 m size asteroid is considered. At a wind speed value of 10 m/s the ULC cloud core entered the land about 40 km in about 40 minutes. This core has a size of about 6 km x 10 km in its final stages. The population affected by the ULC cloud core may be as large in number as 99600 people. The LLC cloud entered the land 127 km in about 2 h. Its last stages size may be about 10 km x 14 km. The land surface area covered by the LLC cloud core during the run-in is about 1778 km^2. This may affect between about 44450 and 442000 people. These evaluations do not include the population of the towns on the sea shore and may be a few times underestimated for some particular wind directions.

No single planning authority exists in the Black Sea region which could develop a contingency plan in case an asteroid impact would take place. Formal procedures should be stated to assign responsibilities to various institutional bodies and plans should be developed for activities associated to the impact. Also, different scenarios should be implemented into such plans. The existing international technical cooperation (for example, the Black Sea Strategic Action Plan) might be extended to cover these issues.

References

Ahlborg G. *Arch Ind Hygiene Occup Med.* 1951. vol. 3. 247–266.

ATSDR (Agency for Toxic Substances and Disease Registry), 1999. *Toxicological profile for hydrogen sulfide.* US Dept of Health and Human Services, Atlanta, GA.

Badescu V. *Earth Interactions* 2007. vol.11. 1-27.

Badescu V. *Stoch Environ Res Risk Assess* 2008. vol. 22. 461-476.

Beauchamp RO Jr., Bus JS., Popp JA, Boreiko CJ, *Andjelkovich Critical Rev. Toxicology* 1984. vol. 13. 25–97.

Berger M.J. LeVeque R.J. (1998) *SIAM J Numer Anal* 1998. vol. 35. 2298-2316.

Berger M.J. Rigoutsos I IEEE *Trans. Sys. Man. and Cyber. 1991 vol.* 21. 1278-1286.

Bland P.A. Artemieva N.A. *Nature* 2003. vol. 424. 288-291.

Bottke W.F. Jr., Nolan M.C., Greenberg R. and Kolvoord R.A., 1994, *Collisional Lifetimes and Impact Statistics of Near-Earth Asteroids.* In: T. Gehrels (Ed), Hazards Due to Comets and Asteroids, (Tucson and London: The University of Arizona Press), pp. 337-357

Chapman C.R. and Morrison D. *Nature* 1994. vol. 367. 33-40.

Chapman C.R., Durda D.D. Gold R.E. 2001 *The Comet/Asteroid Impact Hazard;* www.boulder.swri.edu/clark/beowp.html

Chou C-H Selene J. 2003. Hydrogen sulfide: human health aspects, *Concise International Chemical Assessment Document* 53, World Health Organization, Geneva. Switzerland

CIESIN (2000) Gridded Population of the World: Future Estimates (GPWFE), Palisades, NY, Socioeconomic Data and Applications Center (SEDAC), Columbia University, Center for International Earth Science Information Network (CIESIN), Columbia University; United Nations Food and Agriculture Programme (FAO); and Centro Internacional de Agricultura Tropical (CIAT); http://sedac.CIESIN.columbia.edu/gpw

Cockell C.S. Stokes M.D. *Clim Chng* 1999.vol. 41. 151-173.

Crawford D.A. Malder C.L. *Sci. Tsunami Haz* 1998. vol. 16. 21-30.

Clawpack (2009) http://www.amath.washington.edu/~claw.

Dao M.H. Tkalich P. *Nat. Haz Earth Syst Sci* 2007. vol. 7. 741-754.

Dotsenko, S.F. Ingerov A.V. Phys. *Ocean* 2007. vol. 17. 269-277.

Earth Impact Database 2006 Planetary and Space Science Centre. Department of Geology. University of New Brunswick CA. http://www.unb.ca/passc/ImpactDatabase/europe.html

Encyclopedia Americana, 1995, Danbury, CT: Grolier, p. 532.

Farinella P. Menichella M. *Planetary and Space Science* 1998. vol. 46. 303-309.

Foschini L. Astron Astrophys 1999. vol. 342. L1-L4.

Giosan, L. Filip F. Constantinescu S. *Quart Sci Rev* 2009. vol. 28. 1-6.

Gisler G. Weaver R. Mader C.L. Gittings M. *Sci Tsunamis Hz* 2003. vol. 21. 119-134.

Glasstone S. and Dolan P.J., 1977, *The Effects of Nuclear Weapons* (3rd edition) (Washington DC: U.S. Government Printing Office).

Grieve R.A.F. Sci Amer 1990. vol. 262. 66-73.

Havens JA, Spicer TO. *Development of an atmospheric dispersion model for heavier-than-air gas mixtures.* Final Report to US Coast Guard, CG-D-23-80, USCG HQ, Washington DC, May 1985.

Havens JA.*A dispersion model for elevated dense gas jet chemical releases.* Volumes I and II, US Environmental Protection Agency, Office of Air and Radiation, Office of Air Quality Planning and Standards, Research Triangle Park, North Carolina 27711, April 1988.

Hill FB. *Brookhaven Symposia in Biology* 1973. vol. 30. 159–181

HSDB (Hazardous Substances Databank) US National Library of Medicine, National Toxicology Program, *Bethesda, MD*, 25 February 1998.

Hunt J.N. Palmer R. Penney W. *Phil. Trans. Roy. Soc. London* 1960. vol. 252A. 275–315.

Ignatov, E.I. *The Black Sea.* Springer: Berlin. 2008. pp. 47-62.

Ipatov S.I., 1998, Migration of Kuiper–Belt Objects Inside the Solar System. In: L.M. Celnikier and Tran Thanh Van (Eds) Planetary Systems — the Long View. *Proc. 9th Rencontres de Blois* (June 22–28, 1997), (Gif sur Yvette: Editions Frontieres), pp. 157-160

Ipatov S.I. *Celest Mech Dyn Astron* 1999. vol 73. 107-116.

IPCS (International Programme on Chemical Safety), 2000. *International Chemical Safety Card — Hydrogen sulfide* (ICSC 0165). World Health Organization, Geneva, Switzerland.

Ivezic Z. Juric M. Lupton R.H. Tabachnik S. Quinn T. *Astron J. 2001.* vol. 122. 2749-2784.

Jewitt D. *Nature* 2000. vol. 403. 145-146.

Kharif C., Pelinovsky E. C.R. *Physique* 2005. vol. 6. 361-366.
Khariton Y. Smirnov Y. *Bulletin of the Atomic Scientists*. May 1993, p. 25.
Kowalik Z., Knight W., Logan T., Whitmore P. *Sci. Tsunamis Hz* 2005. vol. 23. 40-56.
Le Mehaute B. Wang S. *Water waves Generated by underwater Explosion*, World Scientific, Singapore. 1996.
LeVeque R.J. *Finite Volume Methods for Hyperbolic Problems*. Cambridge University Press. 2002.
Malder C.L., Gittings M. *Sci. Tsunami Hz* 2003. vol. 21. 91-102.
Marsden B.G. and Steel D.I., 1994, *Warning times and impact probabilities for long-period comets*. In: T. Gehrels (Ed), Hazards Due to Comets and Asteroids, (Tucson and London: The University of Arizona Press), pp.221-239
Mirchina, N.R. Pelinovsky E.N. *Nat Haz* 1988. vol. 1. 277-283.
Nemtchinov I.V. Svetsov V.V. Kosarev I.B. Golub A.P. Popova O.P. Shuvalov V.V. Spalding R.E. *Jacobs C. Tagliaferri E. Icarus* 1997. vol. 130. 259-274.
NGDC 2009 http://www.ngdc.noaa.gov/.
Nikonov A.A. *Izvest. Phys .Sol .Earth*. 1997. vol. 33. 77-87.
NSF (National Science Foundation) 1976. Behavior of hydrogen sulfide in the atmosphere and its effects on vegetation. Report No. NSF/RA760398; *NTIS Publication* No. PB-262733, Research Applied to National Needs, Washington, DC, by Thompson RC, University of California, Statewide Air Pollution Research Center, Riverside, CA
Paine MP. 1999. *Sci Tsunami Hazards* 17(3): 155-166
Pilcher C., 1998, Testimony before House Subcommittee on Space and Aeronautics, [http://www.house.gov/science/Pilcher_05-21.htm]
Rabinowitz D. Helin E. Lawrence K. Pravdo S. *Nature* 2000. vol. 40. 165-166.
ReVelle D.O. *Annals of New York Academy of Sciences* 1997. vol. 822. 284–302.
Rhodes R. Nuclear Renewal, chapt 5, Toronto: Penguin Books, Toronto CA. 1993.
Roe P.L. J Comput Physics 1981. vol. 43. 357-372.
RSS 2005, SeaWinds data. A product of Remote Sensing Systems sponsored by the NASA Ocean Vector Winds Science Team [See: http://www.ssmi.com/qscat/qscat_description.html, viewable at May 17, 2005]
Shoemaker E.M., Wolfe R.F. and Shoemaker C.S., 1990, Asteroid and Comet Flux in the Neighborhood of Earth. In V.L. Sharpton and P.D. Ward (Eds), Global Catastrophes in Earth History. *Geological Soc. of America, Special Paper* 247. Boulder, pp. 155-170
Schuiling R.D. Cathcart R.B. Badescu V. Isvoranu D. Pelinovsky E. *Nat Hazards* 2007. vol. 40. 327–338.
Shuvalov V. Dypvik H. Tsikalas F. *J. Geophys. Res.* 2002. vol. 107. E7.
Siegel SM Penny P Siegel BZ Penny D. *Air and Soil Poll* 1986. vol. 28. 385–391.
Smith W.H.F. Sandwell D.T. *Science* 1997. vol. 277. 1956-1962.
Spicer TO, Havens J. 1992. User's Guide for the DEGADIS 2.1 Dense Gas Dispersion Model. EPA-450/4-89-019, November 1989. National Technical Information Service, Document no. PB90-213893. DEGADIS v2.1, 22 August 1992, 5285 Port Royal Rd, Springfield, VA 22161 [see also http://www.ess.co.at/HITERM/MODELS/degadis.html].
Steel D.I. WGN: *The J. of the Intl .Meteor. Org* .1995. vol. 23. 207-209.
Stegger J.L. Warming R.F. *J. Comput. Physics* 1981. vol. 40. 263-293.
Toon O.B. Zahnle K. Morrison D. Turco R.P. and Covey C. *Rev. Geophys.* 1997. vol. 35. 41-78.

UN 2004 United Nation Office for the Coordination of Humanitarian Affairs, 2004, *The United Nations Chernobyl Forum*, Vienna 10–11 March 2004.

US ACDA (US Arms Control and Disarmament Agency), 1975, *Worldwide effects of nuclear war,* p. 3.

US EPA (US Environmental Protection Agency), 1993. Report EPA/453/R93045. *NTIS Publication* No. PB941312240 to Congress on hydrogen sulfide air emissions associated with the extraction of oil and natural gas, Office of Air Quality Planning and Standards, Research Triangle Park, NC

Ward S.N. and Asphaug E. *Icarus* 2000. vol. 145. 64-78.

Ward S., Aspaugh E. *Deep-Sea Res.* Pt. II 2002. vol. 49. 1073-1079.

WHO (World Health Organization), 1993. *Guidelines for drinking-water quality* (2nd ed.) vol. 2, p. 48. Health criteria and other supporting information, Geneva, Switzerland.

World Book Encyclopedia. 1999, Chicago: World Book p. 597

Yalciner A. Pelinovsky E. Talipova T. Kurkin A. Kozelkov A. Zaitsev A. *J. Geophys. Res.* 2004. vol. 109. C12023.

In: Tsunamis: Causes, Characteristics, Warnings and Protection ISBN: 978-1-60876-360-3
Editors: N. Veitch and G. Jaffray, pp. 29-56

Chapter 2

TSUNAMI SIMULATION RESEARCH AND MITIGATION PROGRAMS IN MALAYSIA POST 2004 ANDAMAN TSUNAMI

Koh Hock Lye[1,*], Teh Su Yean[1], Philip L.-F. Liu[2,3] and Mohd Rosaidi Che Abas[4]

[1]School of Mathematical Sciences, Universiti Sains Malaysia, 11800 Penang, Malaysia.
[2] School of Civil and Environmental Engineering, Cornell University, USA
[3]Institute of Hydrological and Oceanic Sciences, National Central University, Taiwan
[4]Geophysics and Tsunami Division, Malaysian Meteorological Department, Malaysia

Abstract

This chapter presents field survey results on tsunami arrival times, wave runup heights, inundation distances and damage to properties on beaches in Penang and Kedah due to the 26 December 2004 Andaman tsunami. This onsite survey results are used to calibrate and validate a tsunami simulation model TUNA developed by the authors. Simulation results indicate that TUNA performs satisfactorily and can indeed reproduce salient features of tsunami characteristics observed in beaches in Penang and Kedah post 2004 Andaman tsunami. To understand quantitatively the role of mangrove in reducing the impact of tsunami, a numerical simulation model is developed and successfully applied to a stretch of mangrove forests in Penang. A main theme of this chapter is the desire to develop coastal communities that are tsunami resilient. With this in mind we outline briefly a series of workshops, conferences and other research development activities undertaken by key research institutions and government agencies in the past five years towards helping to achieve tsunami resilient communities in the regions, with particular reference to the South China Sea. It is hoped that this presentation will further enhance active collaboration with other research and operational institutions worldwide towards tsunami resilient communities.

---

* E-mail address: hlkoh@cs.usm.my. Fax: 6-04-6570910.

Introduction: 26 December 2004 Tsunami

An earthquake with magnitude of M_w= 9.3 occurred off the western coast of Banda Aceh, North Sumatera at 00:58:53 UTC (08:58:53 Malaysian time) on 26 December 2004. It was the largest earthquake ever recorded in the Indian Ocean and the fourth largest in the world. This undersea earthquake lifted upwards a large volume of water, which is estimated as in the order of magnitude of two hundred trillion tons. This column of water subsequently propagated away from the source, generating a mega tsunami that killed more than 200, 000 people in the affected coastal regions (Figure 1), particularly in Banda Aceh of Indonesia and Phuket of Thailand. Malaysia was not spared the agony inflicted by this mega tsunami, with a casualty of 68 deaths of which 52 happened in Penang Island. The earthquake occurred on the tectonic boundaries of the subduction zones, between the Indian plate and the Sunda plate. It is estimated that this earthquake ruptured a fault slip of over 1200 km long, extending from the tectonic plate boundary near North Sumatera to the Andaman Islands. However, several different initial estimates were also reported in the literature, with the smallest having a length of only 600 km. The instantaneous uplift of the sea floor caused the sea level to be lifted upward by as much as 12 meters at the source (Stein and Okal, 2007), triggering a mega tsunami.

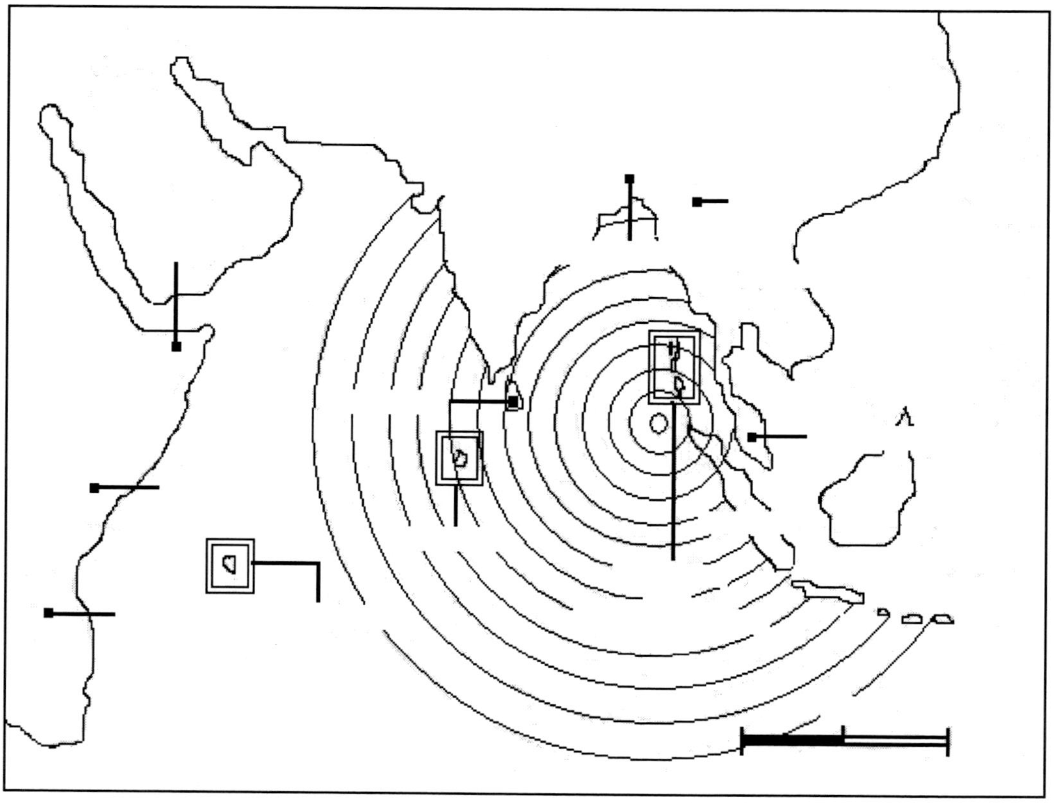

Figure 1. Countries affected by the 2004 Andaman Tsunami.

Post Tsunami Research Activities in Malaysia

Mindful of the deadly potential of future tsunamis inflicting its devastating toll on Malaysian coastal communities, Universiti Sains Malaysia (USM) immediately initiated a tsunami research program after the 2004 Andaman Tsunami. The goals of the program are to understand the mechanism of tsunami generation, propagation and runup, to develop the capability of predicting impacts of potential tsunamis on Malaysian coastal communities and to establish mitigation measures for reducing these impacts. Two field surveys were therefore conducted along the affected beaches in Penang and Kedah to assess the runup heights and inundation distances. These surveys have two objectives. First, we desire to have a comprehensive knowledge regarding areas that were badly affected by taking direct measurement of the extent of runup and inundation and their destructive consequences. Second, we wish to collate scientific data needed to calibrate and validate a tsunami simulation model (TUNA) developed in-house in USM. Survey results are tabulated in Table 1, where the survey locations, indicated by numbers, are shown in Figure 2.

Table 1. Survey runup heights and inundation distances for the 2004 Andaman tsunami

	Location Name	Date '05	Lat. (N)		Long. (E)		Run-up (m)	Inun. Dist. (m)
1a	B. Ferringhi (Teluk Bayu)	20/4	5	28.26	100	14.63	3.4	19.2
1b	B. Ferringhi (Miami Beach)	20/4	5	28.67	100	16.07	4.0	25.6
2a	Tanjung Tokong	20/4	5	27.62	100	18.48	3.6	35.8
2b	Tanjung Tokong	20/4	5	27.57	100	18.41	N/A	190.0
2c	Tanjung Tokong	20/4	5	27.70	100	18.50	2.6	18.3
3a	Tanjung Bungah	21/4	5	28.21	100	16.66	2.3	18.4
3b	Tanjung Bungah	21/4	5	28.20	100	16.65	2.9	36.2
4	Kuala Kedah	22/8	6	6.00	100	26.00	0.9	N/A
5	Yan (Kg. K.S. Limau)	22/8	5	53.00	100	21.00	1.2	12.9
6	Sg Udang	22/8	5	48.00	100	22.00	1.5	N/A
7	Tanjung Dawai	22/8	5	40.00	100	21.00	0.3	75.3
8	Kota K. Muda	22/8	5	34.00	100	20.00	3.8	100.5
9	Kuala Kurau	23/8	5	0.00	100	25.00	1.9	N/A
10	Pantai Acheh	23/8	5	24.00	100	11.00	2.5	13.4
11	Pantai Tengah (Lanai Hotel)	24/8	6	15.00	99	43.00	3.6	44.5
12	Pantai Chenang (Pelangi Hotel)	24/8	6	17.00	99	43.00	3.7	54.7
13	Kuala Teriang	24/8	6	21.00	99	42.00	3.0	27.0
14a	Pantai Kok (Mutiara Beach Resort)	24/8	6	21.00	99	40.00	2.2	50.8
14b	Pantai Kok (Berjaya Hotel)	24/8	6	21.00	99	40.00	2.9	34.9

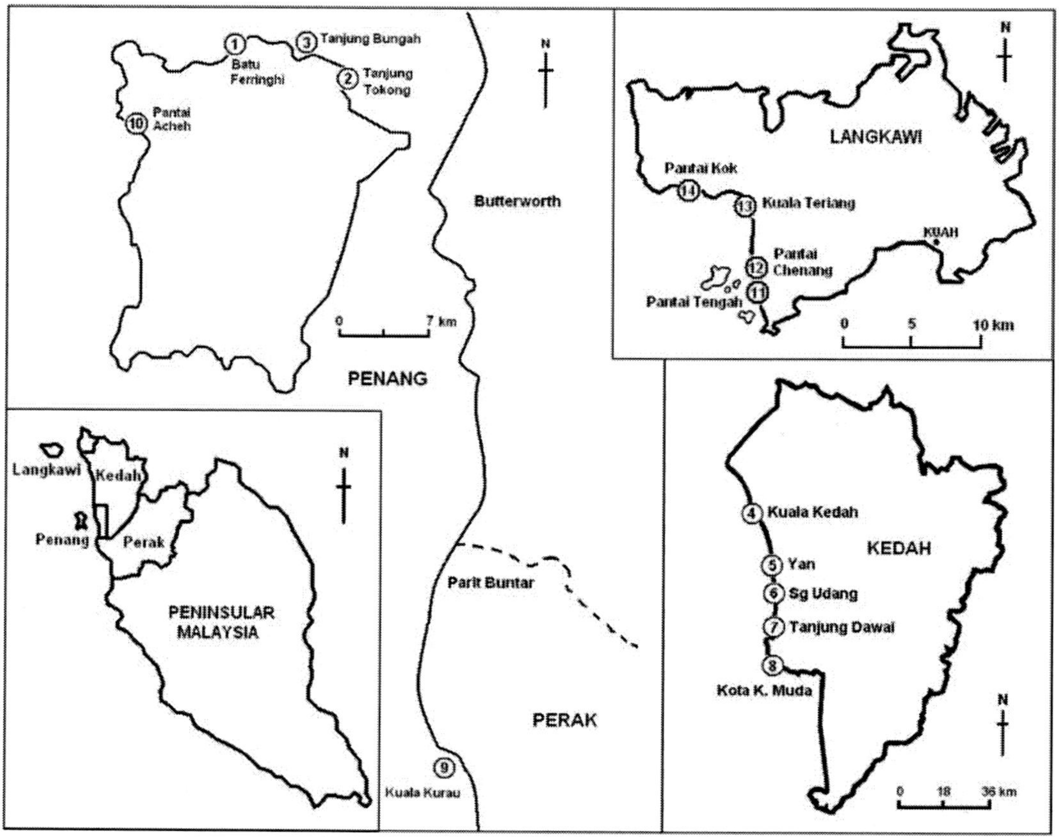

Figure 2. Survey locations in Malaysia corresponding to Table 1.

Concurrently, another group of researchers from Universiti Kebangsaan Malaysia (UKM) formed a team to conduct research on the socio-economic-cultural impact of this tsunami (UKM, 2006). Further, the Malaysian Meteorological Department (MMD) contributed significantly towards the establishment of early warning systems to serve as the front-line mitigation measures for future tsunamis in the region. To promote the science of tsunami, the Academy of Sciences Malaysia (ASM) has coordinated several research projects on earthquake and tsunamis. A series of international workshops, conferences and roundtables were held in Malaysia and other affected countries in Asia including Thailand, Indonesia, Singapore, Philippines, Taiwan, China, Australia, Sri Lanka and India to report research findings, to coordinate collaboration and to establish effective early warning systems in the affected regions. On an ongoing basis, researchers and community advocates devote their attention to issues concerning not only the Indian Ocean and the Andaman Sea, but also the South China Sea. Other international organizations such as UNESCO-IOC and USGS also actively participate in these research and advocacy activities. This series of activities and achievements has been reported briefly in an article published by the United Nations Asian and Pacific Centre for Transfer of Technology (UN-APCTT) to provide guidelines and resource for policy makers (Koh et al., 2007).

Arrival Time, Runup and Inundation

Immediately after the occurrence of the 2004 Andaman tsunami, another survey team was assembled by UKM (2006) to record the extent of damage inflicted by this tsunami. It was commonly observed during this survey that the sea level abruptly receded into the sea by distances exceeding 100 meters before the first tsunami waves arrived. This implies that the tsunami waves were leading depression N-waves. The tsunami waves arrived at Langkawi at 12:40 pm, Penang at 1:15 pm and Kota Kuala Muda, Kedah at 1:40 pm. This translates to an arrival time of about 3 hours and 40 minutes for Langkawi. In Penang and Langkawi, it was reported by reliable eyewitness that there were three tsunami waves, the first two occurring at approximately 1:15 pm and 1:30 pm, respectively while the timing of the third wave could not be deduced accurately from anecdotes. Kota Kuala Muda in Kedah was also reportedly struck by three tsunami waves, at approximately 1:40 pm, 1:50 pm and 2:00 pm respectively. In all these three locations, it was observed that the second tsunami waves were the biggest and the most destructive, claiming lives, causing grave injury and devastation to property. Along the shores, wave heights reaching 2 to 4 meters were commonly observed in all three locations, which were sufficiently high to cause significant damage and human fatality. The human casualty was often due to the injury arising from debris flows or drowning, as the current speeds on shore exceeded 12 m/s. The inundation distance in some places reached 350 m. Table 2 provides the averaged arrival times, tsunami wave heights and inundation distances surveyed in these three locations. A total of 68 deaths were reported in this tsunami, with 52 deaths in Penang, 12 occurring in Kota Kuala Muda, one in Langkawi and three in Perak.

Table 2. Averaged arrival times, wave heights and inundation distances

Location	Wave Arrival Times		Runup (m)	Inun. Dist. (m)
	First	Second		
Kota Kuala Muda	13:45	14:00	2.0-3.4	200-350
Penang	13:15	13:20	2.5-4.0	150-300
Langkawi	12:35	12:50	2.5-3.0	100-250

Damage along Malaysian Coasts

Surveys were conducted to assess damage to properties caused by the tsunami. These surveys also provided indirect assessment of the hydraulic forces induced by the tsunami. Damage and destruction of property occurred to rows of houses located tens of meters inland from the shoreline (Figure 3), providing concrete and direct evidence of inundation distances and runup heights along the respective beaches in some cases. In some locations where dense mangroves and other coastal vegetations such as wild nipah palm trees might have been effective in providing protection from the destructive waves, houses remained intact despite the high waves observed nearby. Destructions to boats and fishing equipments were commonly observed in all three surveyed areas, where some boats were overturned or lifted tens of meters inland and some were detached and displaced 30 m to 50 m from their mooring sites (Figure 4). These observations provide valuable information in estimating the hydraulic forces, wave heights and current velocities in the tsunami waves at the respective locations.

Figure 3. Houses in Kota Kuala Muda destroyed by the tsunami waves.

Figure 4. Boats and cars in Penang dislocated tens of meters by the tsunami waves.

Paddy fields and agricultural farms were contaminated with salt water, rendering the farms unsuitable for agricultural activities for a long duration (Figure 5). Salinity intrusion of coastal agricultural land as observed after the 2004 Andaman tsunami is also commonly seen in other places as well, following large storm surges or inundation by hurricanes. Hence the study team initiated a research program on simulating salinity intrusion and its impacts on coastal vegetation. For this purpose a simulation model known as MANHAM has been developed, with collaboration with scientists from South Florida (Teh et al., 2008a). The results will be presented in a later section.

In some locations layers of mud with thickness reaching 10 cm were deposited far inland, providing yet another indication of the force of the waves. It was also observed that bananas, papayas and newly planted young coconut trees were badly damaged in Kota Kuala Muda (UKM, 2006). Anecdotal accounts appear to suggest and arouse interests that some fruit trees such as mango produced abundant fruits one season after the tsunami, the reasons of which are presently not yet clear.

Figure 5. Paddy field killed by saltwater due to tsunami inundation.

Based upon this comprehensive compilation of onsite survey results, a computer simulation model TUNA is developed and verified. This model can be used to assess the impact of this 2004 Andaman tsunami and more importantly the potential impacts of future tsunamis on Malaysian coastal regions. The mathematical and numerical description of TUNA, based upon the shallow water equations, is briefly provided in the following sections, with details available elsewhere (Koh et al., 2009; Teh et al., 2009).

Shallow Water Equations

The propagation of tsunami in open ocean may be simulated by the depth-averaged two-dimensional shallow water equations (SWE) as recommended by the Intergovernmental Oceanography Commission (IOC, 1997). The SWE is applicable when the water depth is much smaller than the wavelength, which is typically satisfied by tsunami propagation and runup. Hence, the conservation of mass and momentum can be depth averaged (Hérbert et al., 2005) and may be written in the following forms.

$$\frac{\partial \eta}{\partial t}+\frac{\partial M}{\partial x}+\frac{\partial N}{\partial y}=0, \tag{1}$$

$$\frac{\partial M}{\partial t}+\frac{\partial}{\partial x}\left(\frac{M^2}{D}\right)+\frac{\partial}{\partial y}\left(\frac{MN}{D}\right)+gD\frac{\partial \eta}{\partial x}+\frac{gn^2}{D^{7/3}}M\sqrt{M^2+N^2}=0, \tag{2}$$

$$\frac{\partial N}{\partial t}+\frac{\partial}{\partial x}\left(\frac{MN}{D}\right)+\frac{\partial}{\partial y}\left(\frac{N^2}{D}\right)+gD\frac{\partial \eta}{\partial y}+\frac{gn^2}{D^{7/3}}N\sqrt{M^2+N^2}=0\,. \quad (3)$$

Here, the volume fluxes per unit length (M, N) in the x- and y- directions are related to depth averaged velocities u and v by the expressions $M = u(h+\eta) = uD$, and $N = v(h+\eta) = vD$, where h is the mean sea depth and η is the water elevation associated with tsunami. Further details are available elsewhere (IOC, 1997). The evolution of earthquake-generated tsunami waves has three distinct stages: generation of initial vertical water displacements at source, propagation of generated tsunami waves in open ocean and wave runup in coastal areas. There are several numerical models developed based upon Equations (1) to (3) to simulate tsunamis propagations, for example, the model TUNAMI-N2, developed by Imamura of Tohoku University (Imamura et al., 1988) and the MOST model (Titov, 1997). Both models are copyrighted by the developers. Therefore it becomes an urgent necessity to develop an in-house tsunami simulation model referred to as TUNA.

Numerical Model Tuna

The SWE form a hyperbolic quasi-linear system for which the explicit finite difference method has been known to be most efficient. The general shallow water model consisting of Equations (1) to (3) will therefore be solved by the explicit finite difference method with staggered grids (Koh et al., 2009). The locations of η, u and v are shown in Figure 6. This staggered scheme as illustrated by Figure 6 has been used in several tsunami simulation models (Yoon, 2002; IOC, 1997; Yoon and Liu, 1992) and is therefore employed to develop the tsunami simulation model named TUNA. The algorithm performs well provided that the time step Δt fulfils the Courant criterion, namely

$$\Delta t \leq \frac{\Delta x}{\sqrt{2gh}}, \quad (4)$$

where Δx denotes the finite difference grid size. Further details regarding the staggered scheme are available from Koh (2004); hence are omitted here.

Tuna vs. Comcot

Before TUNA can be confidently used to simulate real tsunami events it must be subject to validation so as to ensure that it performs as well as other more established models. For this purpose, the simulation results of TUNA have been tested by comparing them with those obtained from COMCOT (Cornell Multi-grid Coupled Tsunami Model), the results of which indicate satisfactory performance of TUNA.

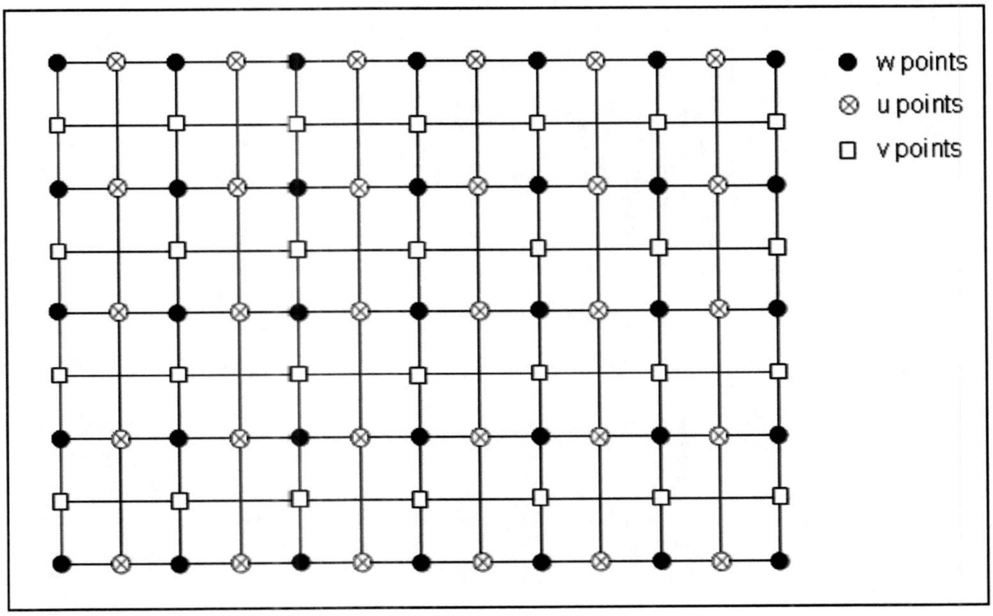

Figure 6. Staggered scheme in TUNA.

Details regarding the COMCOT model can be referred to Liu et al. (1998). A computational domain in the form of a rectangular channel of dimension 10 km by 40 km with a depth of 10 m is chosen for this comparative study. The solid boundary condition is imposed at the north, south and west ends of the channel to simulate complete reflection of waves. Open radiation boundary condition is imposed at the east end of the channel to allow waves to pass through without reflection. The initial source of maximum height of 1 m is chosen to be located at the middle of the channel, with a vertical distribution represented by a Gaussian hump given by $\eta = ae^{-(x/\sigma_x)^2} e^{-(y/\sigma_y)^2}$ with standard deviations σ_x of 1500 m and σ_y of 2500 m. A grid size of 50 m and a time step of 1.25 s are used. Several observation points are placed in the study domain to record the simulated time series for the comparisons between TUNA and COMCOT. Figure 7 shows the comparison between TUNA and COMCOT simulated time series for wave heights at four selected locations (the comparison at other locations not shown). The agreement between results simulated by TUNA and those simulated by COMCOT at all observation points is very good, indicating proper performance of TUNA. Hence we will now use TUNA with confidence regarding its performance and credibility.

Simulation Results

To efficiently simulate the impact of the 2004 Andaman tsunami on Malaysian coastal waters an appropriate computational domain needs to be chosen. The domain should be large enough to contain all areas of interest yet not too large as to impose high computational cost.

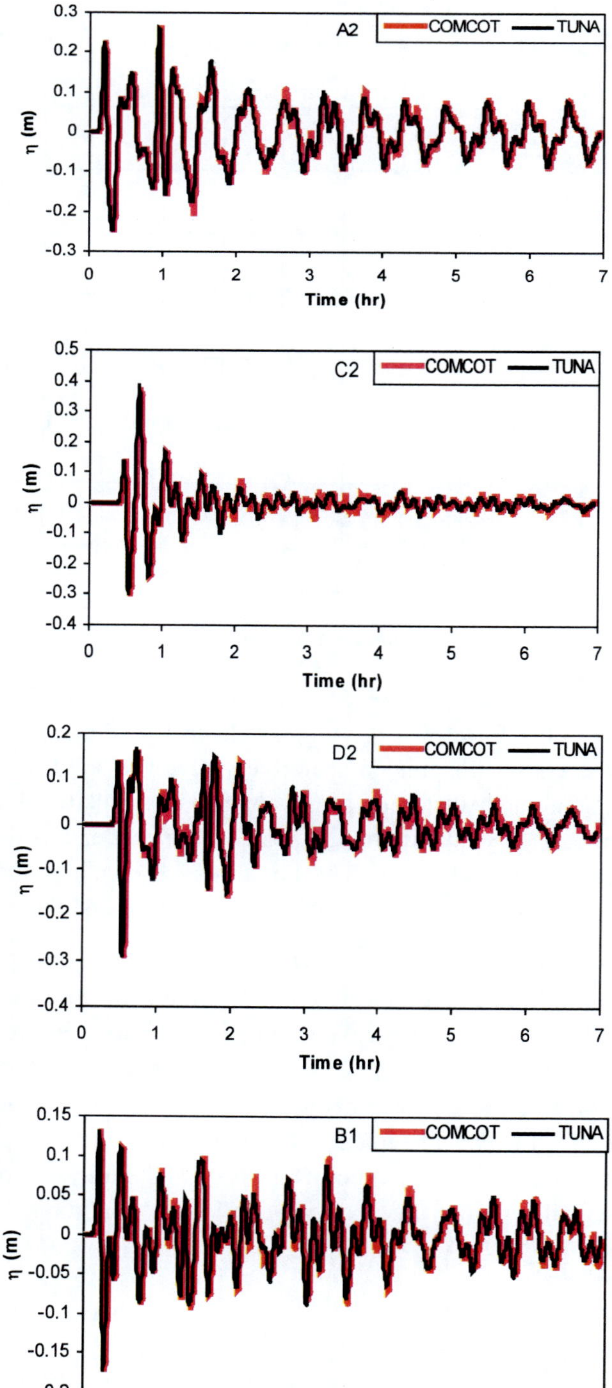

Figure 7. Wave height time series at several locations to compare TUNA and COMCOT.

The computational domain chosen is the sea contained within the rectangle of dimension 1200 km × 1500 km (The small rectangle within left panel of Figure 8).

This computational domain is suitable for the purpose of this study since the focus is the tsunami waves propagating towards the northwest of Peninsular Malaysia. The initial source term of 1200 km length (right panel in Figure 8) is generated by means of the Okada model (Okada, 1985). With a grid size of $\Delta x = 1000$ m, this computational scheme results in a total of 1.8 million nodes. To ensure stability, Δt is taken as 3 s. The radiation boundary condition for long waves is imposed on the open boundary to allow waves to pass through freely.

Snapshots of contour plots of the simulated tsunami propagation wave heights at eight time steps are shown in Figure 9, indicating wave propagations starting at the source region to wave arrival at Penang and Langkawi and into the Straits of Malacca. The waves appear to pass through the open boundary without reflection as desired. The first frame shows that the initial free surface displacement splits into two waves, one propagating towards the East, while the second westwards. After 1.8 hours, the waves begin to arrive offshore of Phuket of southern Thailand, following a leading depression. The simulated waves arrive offshore of Langkawi after 3 hours and at Penang after 3.6 hours, following a leading depression for both places too. These simulated results agree with eyewitness accounts, both in terms of arrival times as well as the shape of the waves in the form of leading depression N waves (UKM, 2006). The simulated maximum wave heights offshore at depths of 50 m are 3.6 m, 1.5 m and 1.2 m at Phuket, Langkawi and Penang respectively. These maximum simulated wave heights are about 3 times smaller that maximum runup heights measured along the beaches. For example in Penang the maximum runup height at some location may reach 3.5 to 4.0 m, while the maximum wave heights simulated by TUNA-M2 at offshore location in Penang is only 1.0 to 1.2 m, which is about one third of maximum runup heights observed.

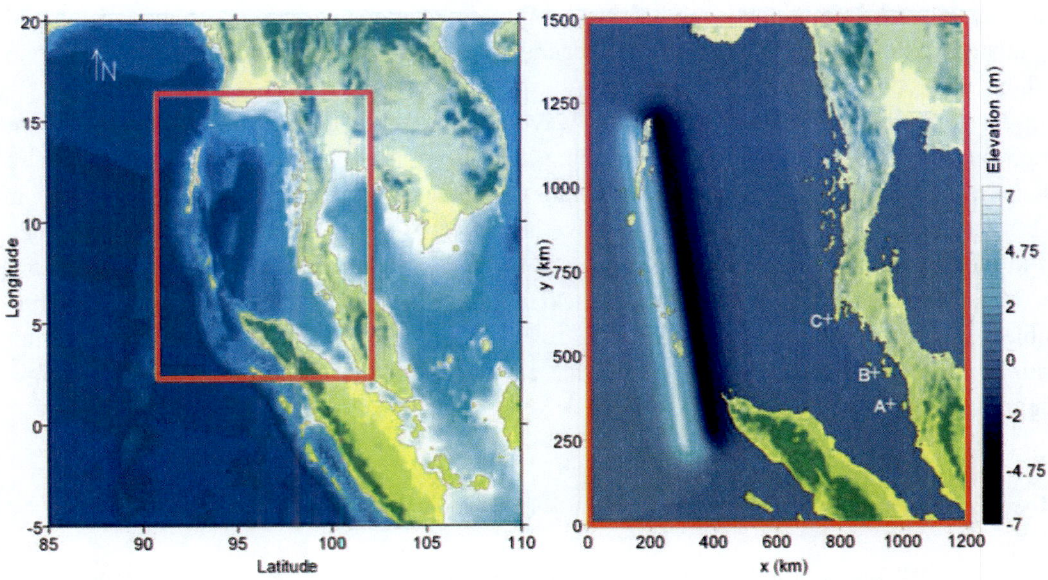

Figure 8. Computational domain and initial source for the 26 December tsunami.

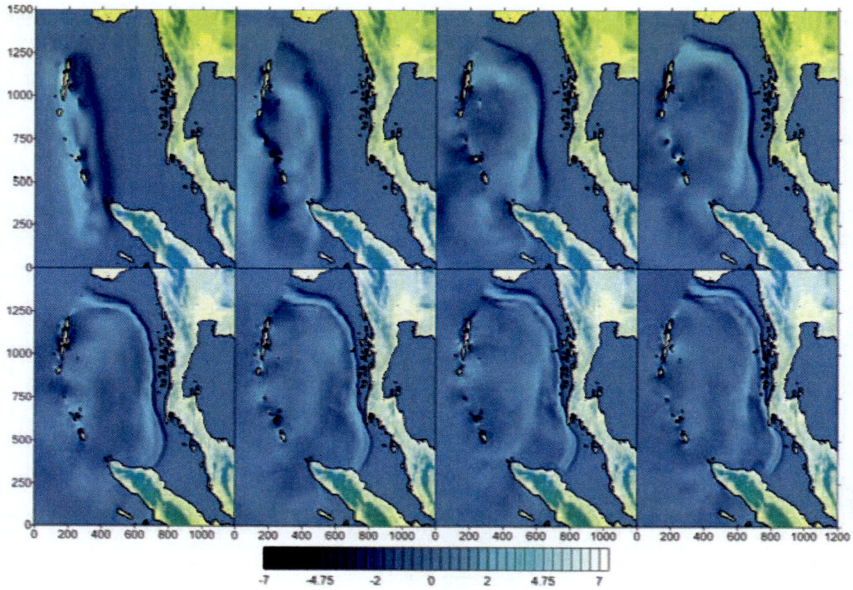

Figure 9. Propagation of the 26 December 2004 tsunami.

This is not totally surprising as TUNA-M2 simulates tsunami propagation from the open ocean to offshore locations with depth of about 50 m. The runup simulation from these offshore locations to the shore is performed by TUNA-RP, the runup component of TUNA. A brief account of this runup simulation will be provided later. The waves at 50 m depth may be amplified by a factor of up to 3.5 as they finally runup the shallow beaches (Teh et al., 2008b). It may be concluded that TUNA is capable of simulating the 2004 Andaman tsunami satisfactorily. The simulated wave heights at offshore locations with depth of about 50 m and simulated runup heights along selected beaches in Penang agree well with surveyed runup heights along corresponding beaches.

These simulated offshore and runup wave heights serve to provide initial assessment regarding the relative vulnerability of Penang, Langkawi and Phuket to the hazards of future tsunami that may be generated by earthquakes originating in northern tip of Sumatra. It is apparent that Penang, with an offshore wave height of 1.0 m to 1.2 m, is relatively shielded from the 2004 Andaman tsunami as compared to Phuket, which has an offshore wave of 5.5 m. This difference is because the location, extent and orientation of the tsunami source terms subject Phuket directly to the tsunami hazards by being directly in the tsunami propagation path, while Penang receives only waves that are refracted as the waves move over shallow region.

Beach Runup

TUNA-RP is a simplified one-dimensional model to calculate runup and inundation of the tsunami waves from offshore locations with depth of about 50 m onto the dry beaches. TUNA-RP applies the open radiation boundary on the sea boundary to allow free flow of currents without reflection. To simulate wave runup along the beaches, the on-offshore components of Non-Linear Shallow Water Equations (NLWE) are used.

$$\frac{\partial \eta}{\partial t} + \frac{\partial M}{\partial x} = 0, \tag{5}$$

$$\frac{\partial M}{\partial t} + \frac{\partial}{\partial x}\left(\frac{M^2}{D}\right) + gD\frac{\partial \eta}{\partial x} + \frac{gn^2}{D^{7/3}}M^2 = 0, \tag{6}$$

These NLWE equations are discretized by the explicit finite difference method. To avoid costly computational time, a study domain of about 4 km with grid size Δx of 2 m is chosen for these simulations. A soliton of 1 m amplitude and 1 km wavelength is initiated at x = 1500 m at the deep-sea region with depth h = 50 m as shown in Figure 10 (visually distorted due to scaling factor).

Figure 11 shows snapshots of the simulation results on the dry land domain using moving boundary condition (MBC) (Teh et al., 2008b) on the right hand side (land boundary) and the open boundary condition on the left hand side (ocean boundary). Figure 11 shows the wave propagation from the deep sea region (x = 0 to 3200 m) at depth of 50 m into the region (x = 3200 to 4000 m), decreasing from 50 m to 0 m in depth over this stretch as the wave runup onto the dry land, reaching a maximum runup height of about 3.1 m for this setup. After achieving the maximum runup height, the wave slides down from the dry land slope as formulated in the MBC. As it slides down from the dry land into the sea, the wave breaks into smaller waves that interact with each other to create more waves, not shown. During this wave interaction, some waves meet to amplify in momentum and heights. The waves then continue to propagate back into the sea. Many simulations are performed for several scenarios and locations in Penang, with different sea bottom slope and beach topography. The results of these simulations will be reported elsewhere.

This series of runup simulations produce a wide range of runup amplifications of wave heights, which serve to remind us regarding the uncertainty and inaccuracy of tsunami runup simulations with inadequate data regarding beach topography and bathymetry.

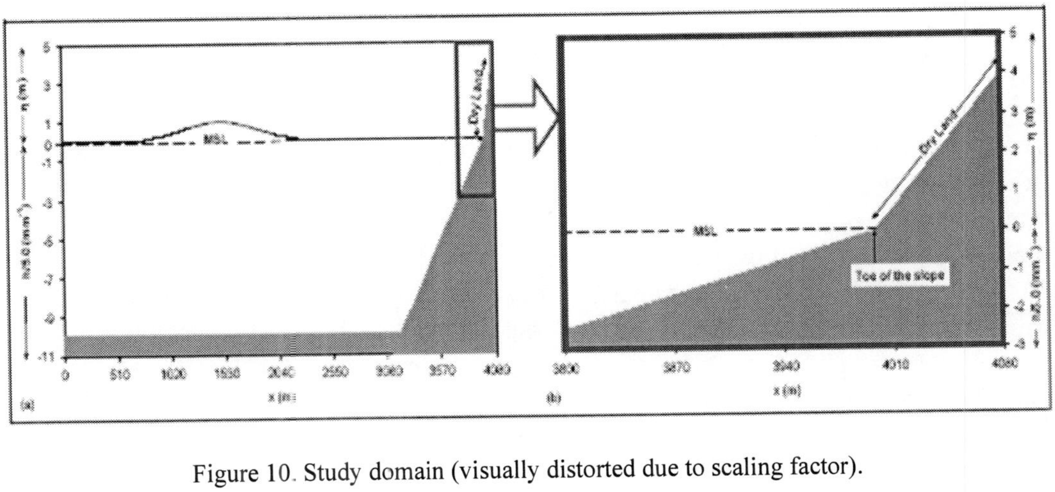

Figure 10. Study domain (visually distorted due to scaling factor).

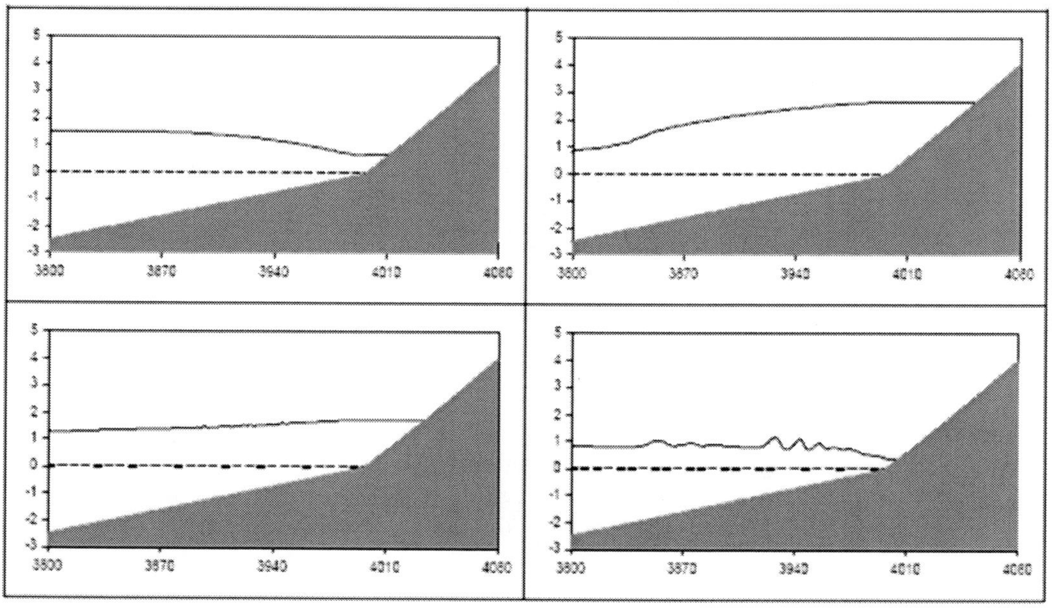

Figure 11. Snapshots of the wave runup onto a dry land slope of 0.05.

A maximum amplification factor of 3.3 to 3.5 and a maximum inundation distance of about 60 m to 80 m are observed in the simulations, depending on the bathymetry and other parameters used in these simulations. In general, simulated wave heights off the coast of Penang at depths of around 50 m may reach a maximum of 1.0 to 1.2 m, while runup wave heights may achieve maximum of 3.3 to 4.0 m. These simulated runup wave heights are within the range of the runup wave heights of 2.3 to 4.0 m (Table 1) surveyed along beaches in Penang after the 26 December 2004 tsunami. The simulated inundation distance of about 60 to 80 m is within the range of measured inundation distances of 20 to 100 m. Further refinement of TUNA RP simulations will be performed in the near future when more detailed topography and bathymetry for Penang and Kedah are available from the federal authorities.

Manham

It was observed during the surveys that significant stretches of coastal land in Kedah were exposed extensively to saline seawater due to the Andaman tsunami. This seawater inundation has resulted in sharp increase in salinity in the coastal soil, leading to the land being unsuitable for cultivation of crops such as rice. This salinity inundation is also obvious in Aceh of Indonesia. Some members of the study team are therefore motivated to pursue research to understand salinity intrusion in coastal land and its impact on coastal vegetations. In collaboration with a group of research scientists based in South Florida, we develop a simulation model known as MANHAM to simulate salinity intrusion mechanism following a massive inundation of saline seawater and the subsequent impact on vegetation succession induced by such salinity changes (Teh et al., 2008a; Sternberg et al., 2007). The initial success of this research provides the incentives to extend the current framework of MANHAM in order to link it with TUNA. This model linking will enable the assessment of the impact of tsunami and other severe storm surges due to typhoons and monsoons on the salinity regimes and coastal vegetations in coastal zones of Southeast Asia and South China Sea regions.

Role of Mangrove

Coastal forests have been recognized and reported in the literature as capable of providing several useful functions towards tsunami mitigation and control. They help to minimize debris drift and they reduce tsunami wave heights and energy. The effectiveness of coastal forests as tsunami mitigation measure depends on the forest density, diameters of tree trunks, and extensiveness of the root systems and the width of forest cover. It has been observed that an increase in forest width can minimize not only runup heights and inundation depths but can also reduce current speeds and hydraulic forces as well. However, it has also been recognized that coastal forests could be destroyed by a large tsunami with wave heights exceeding 4 m, in which case the floating debris may cause serious secondary damage (Harada and Imamura, 2005). In order to evaluate the effectiveness of coastal forests in reducing tsunami impact quantitatively, numerical simulation of tsunami runup and inundation as the waves travel through coastal forests has been conducted to assess the role of mangrove in tsunami mitigation in Penang. This forms the subject of discussion in the following section.

Numerical Algorithm

To simulate tsunami flow through coastal vegetation, one-dimensional continuity and momentum equations in flux forms in the flow direction are used in this research. The following equation describes the momentum equations in which the two last terms are the resistance force used to model the effects of coastal vegetation such as mangroves (Harada and Imamura, 2005; Massel et al., 1999; Mazda et al., 1997).

$$\frac{\partial M}{\partial t}+\frac{\partial}{\partial x}\left(\frac{M^2}{D}\right)+gD\frac{\partial \eta}{\partial x}+\frac{gn^2 M|M|}{D^{7/3}}+\frac{C_d}{2}A_0\frac{M|M|}{D^2}+C_m\frac{V_0}{V}\frac{\partial M}{\partial t}=0 \quad (7)$$

In the above,

M = flow flux, m^2/s;
$D = (\eta + h)$ = total water depth, m;
h = still water depth, m;
n = Manning coefficient;
g = gravitational acceleration, m/s^2;
C_d = drag coefficient;
C_m = inertia coefficient (dimensionless)
A_0 = projected area of trees under water surface, per 100 m^2;
V = a chosen control volume (m^3);
V_0 = volume of trees under water surface within V (m^3).

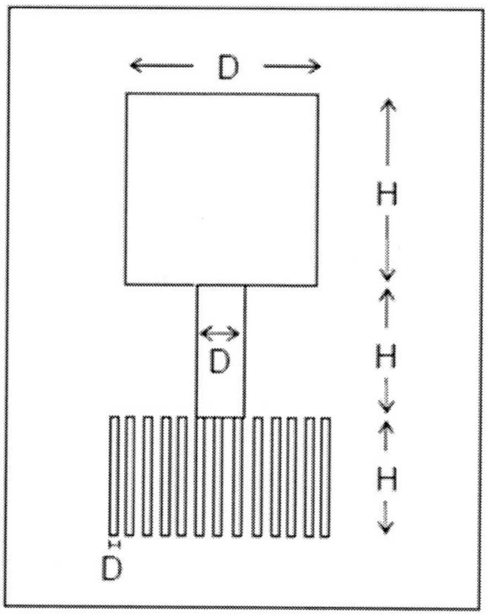

P_L = leaf porosity;
N_T = number of trees per 100 m^2;
N_R = number of prop roots per tree;
D_T = diameter of stem, m;
D_R = diameter of each prop root, m;
D_L = diameter of leaf part, m;
H_R = height of root part, m;
H_T = height of stem part, m;
H_L = height of leaf part, m.

Figure 12. Image of modeled unit mangrove tree.

The characteristics of the mangrove forest are represented by three parameters, namely A_0 (m^2 per 100 m^2), which is the projected area of trees under water surface (per 100 m^2) and V_0 (m^3) representing the volume of trees under water surface within a chosen control volume V (m^3). Equation (7) is discretized by the explicit finite difference method, the details of which are available in Teh (2008). The modeled mangrove tree is envisaged as a structure consisting of three parts: root system, stem (trunk) and leaf (canopy), formed by cylinders of different diameters and heights. A modeled image representing mangrove of the species *Rhizophora stylosa* is shown in Figure 12, in which the parameters used to describe the mangrove forest are also given on the right.

The projection area A_0 and volume V_0 are derived from three components of the root, stem, and leaf part. The effective projection area A_0 and volume V_0 occupied by trees under water surface depend on the water height. Note that A_0 and V_0 are measured as per 100 m^2 of forests. The meaning of terms used is given in Figure 12 right.

An Illustrative Example

As a start, we present a simplified simulation of the impact of mangroves by lumping the friction effects into one single term consisting of Manning friction by setting $C_D = 0.0$, and choosing n = 0.3 to reflect strong friction due to good mangrove covers. The incident solitary wave is a positive half sine curve with a wavelength L = 12000 m, period T = 0.15 hour and amplitude a = 1 m. This wave enters the computational domain at time 0 hour and distance 0 m from the left, with a depth h = 50 m, and travels a distance of 10000 m along this flat seabed before it begins its climb at the toe of the slope at 10000 m. The concave slope, with an average slope of 1:40, has a horizontal length of 2000 m, climbing 49 m along this stretch. The velocity of this incident wave is given by the shallow water theory, namely $\eta\sqrt{g/h}$ =

$\eta\sqrt{10/50} = 0.45\ \eta$ m/s. The mangroves are located between 12000 m and 13000 m, represented by Manning n = 0.3 $sm^{-1/3}$.

Figure 13 shows snapshots of wave heights η at intervals of 0.05 hours. The presence of a patch of mangrove would slightly amplify wave heights in front of the mangrove due to compression of waves as they move up the shallow shore with reduced velocity. However, as the waves pass through the forest the energy is reduced by the presence of the forests giving rise to much-reduced waves behind the forests.

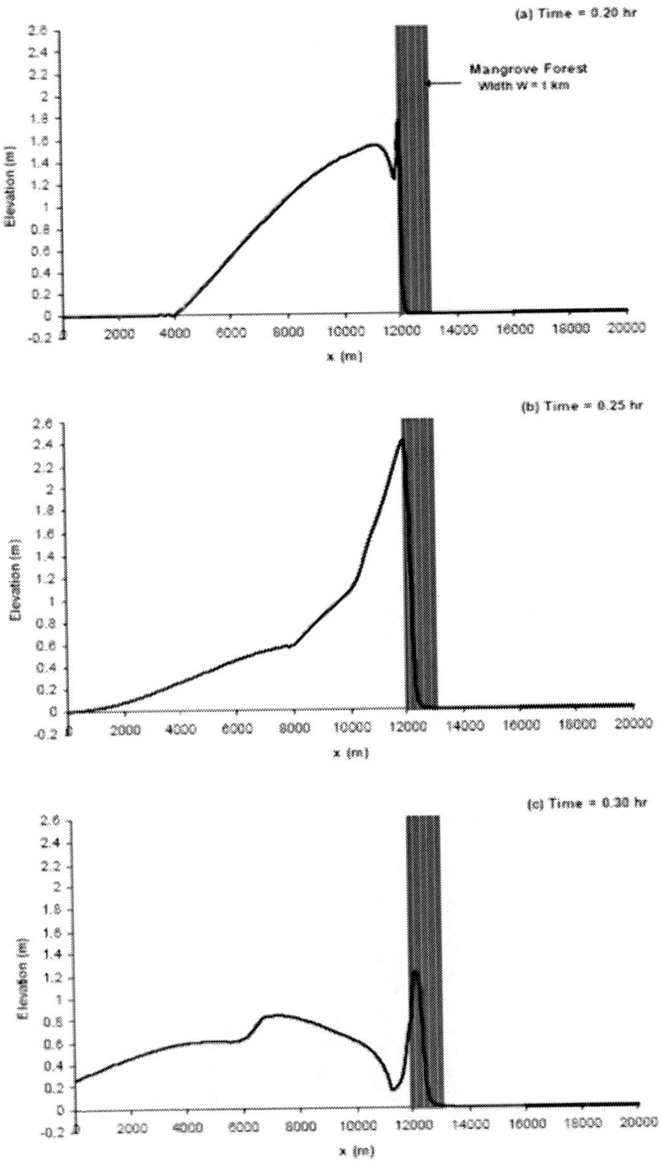

Figure 13. Waveforms at interval of 0.05 hr with n = 0.30 at 12 to 13 km, denoting the presence of mangrove forest.

Penang Case Study

To assess the role of mangrove forests in Penang in reducing tsunami runup and inundation, surveys of the mangrove trees were conducted in the mangrove forest located at Pantai Mas, Penang (Figure 14). Mangroves of the species *Avicennia officinalis* are observed to be abundant (Figure 15) but as we approach further towards land, the species *Bruguiera cylindrica* can be found (Teh, 2008). The width of the mangrove forest along this stretch of the beach in Penang is estimated to be in the range of 500 m to 1000 m. The measured characteristics of Pantai Mas mangrove trees are simplified for the simulations where three densities of the mangrove forest are considered. Two forest widths of 500 m and 1000 m will be considered in the simulation, the results of which arte briefly described below.

Figure 14. Location of Pantai Mas in Penang island.

Figure 15. Avicennia officinalis in Pantai Mas mangrove forest.

Forest Width 1000 M

As waves move through the forests, the heights and velocities will be reduced by the forests. The symbol r_η is used to refer to the fraction of wave height η remaining after the waves pass through the forests, with similar meaning r_u for velocity. Figures 16 and 17 provide the percentage of wave heights and velocities remaining after the waves pass through the mangrove forest in Penang with a width of 1000 m. For a wave period of 10 minutes, the elevation and velocity left behind the mangrove forest are 30 % for elevation and 20 % for velocity. For other wave lengths the percentage left behind are higher. Both r_η and r_u are lower for higher values of forest density.

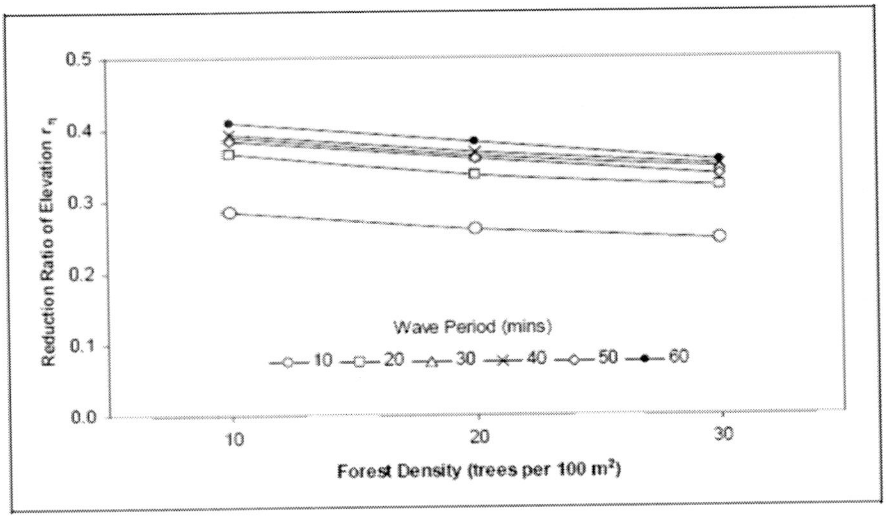

Figure 16. rη of a 1000 m wide Pantai Mas forest as a function of forest density for a range of wave periods.

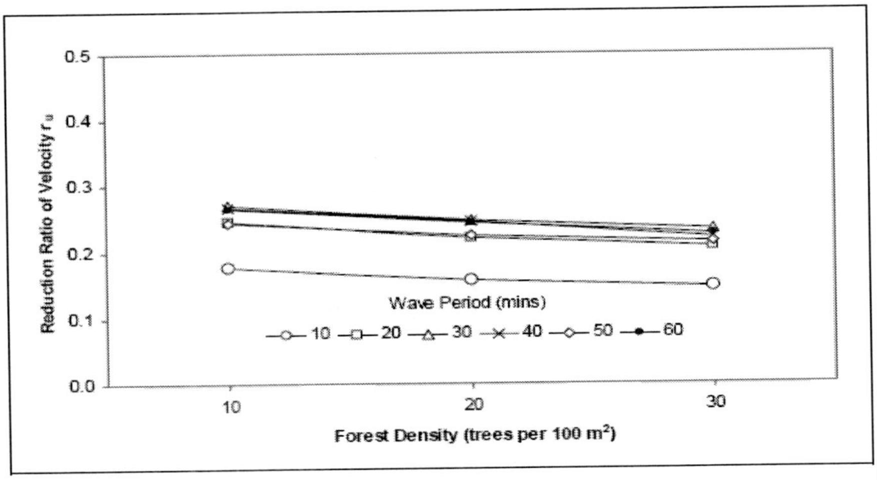

Figure 17. r_u of a 1000 m wide Pantai Mas forest as a function of forest density for a range of wave periods.

Forest Width 500 M

The percentage left behind after the waves pass through a forest with a smaller width of 500 m are higher (Figures 18 and 19) as compared with those after the waves pass through 1000 m of forests. For a wave period of 10 minutes, about 40 to 45 % of elevation and 35 % of velocity are left behind after the waves pass through forests of 500 m width. For the other wave periods considered, the percentage left behind the mangrove forest are higher, being in the range of 50 % to 60 % for wave heights and in the ranges from 35 % to 45 % for velocity. Both r_η and r_u are lower for higher values of forest density.

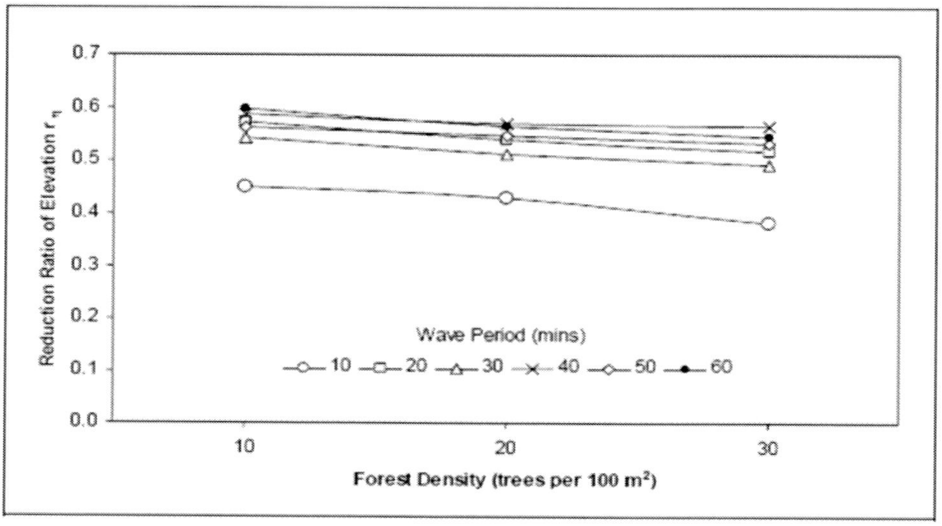

Figure 18: r_η of a 500 m wide Pantai Mas forest as a function of forest density for a range of wave periods.

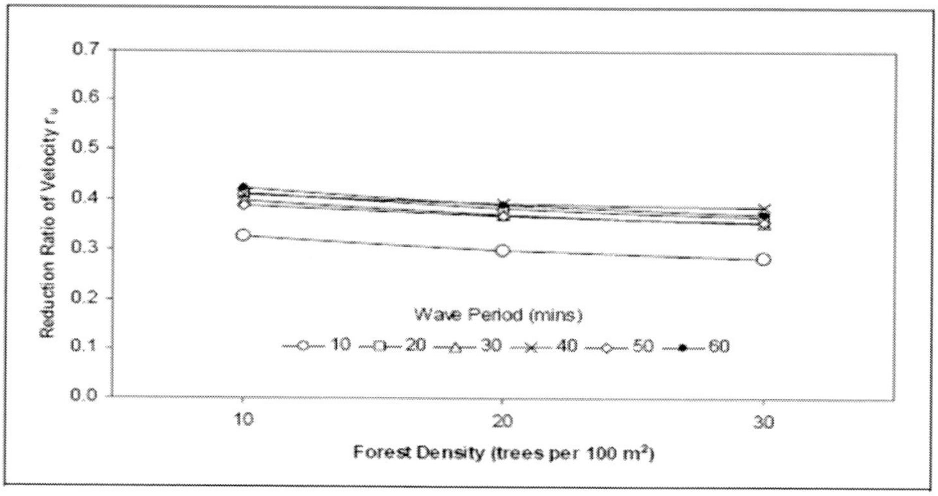

Figure 19. ru of a 500 m wide Pantai Mas forest as a function of forest density for a range of wave periods.

Sensitivity analyses are conducted for the mangrove species with prop roots such as *Rhizophora stylosa* that can be found in certain mangrove forests in Penang by varying the parameter values for wave periods, forest width and density. The analyses indicate that the reduction effect positively increases with increasing width and increasing density of the mangrove forest. Measurements of *Avicennia officinalis* in the mangrove forest in Pantai Mas, Penang are used to assess the forest potential in reducing runup wave heights and velocities. Considering a forest width of 1000 m, simulations indicated that the wave heights and velocities behind the forest can be potentially decreased to less than 40 % and 30 % of their initial values respectively. In the case of 500 m wide forest, the wave heights and velocities behind the forest can potentially be reduced to about 60 % and 45 % of the original values respectively. We conclude that the presence of mangrove forests might be able to reduce the runup wave heights and velocities to a certain extent. However, the level of the reduction depends on the characteristics of the mangrove forest, particularly forest widths and density. However, it should be noted that mangrove trees might not be able to withstand the force of a significantly high runup wave height, with more than 4 m wave height (Shuto, 1987).

Towards Tsunami Resilient Communities

The 2004 Sumatra-Andaman earthquake and the Indian Ocean tsunami have highlighted vulnerabilities to extreme natural disasters in the world coastal communities. Most of the damage occurred because neither a tsunami early warning system nor a simple communication network was put in place among the countries in the region. Public education, total risk management and coastal zone planning to mitigate tsunami hazard were also non-existent in the region. After the 2004 Indian Ocean tsunami, many countries have been working independently and collectively to develop early tsunami warning systems for the Indian Ocean region countries. USA has committed more than $50 millions over the next several years to deploy 29 new deep ocean sensor systems in the Pacific Ocean rim and Caribbean Sea.

Recently the USGS issued a report confirming the potential risk of tsunami sources along the entire Pacific subduction zones. It identified the Manila (Luzon) trench as a high risk zone, where the Eurasian plate is actively subducting eastward underneath the Luzon volcanic arc on the Philippine Sea plate. Two other medium risk subduction zones in the neighboring area are also identified. These subduction zones can rupture and generate large tsunamis in the future that will have devastating impacts on the countries in the South China Sea region. During the NUS-TMSI workshop on "Earthquake and Tsunami: From Source to Hazard" conducted in Singapore in 2007, participants of the workshop supported the idea of conducting an annual series of South China Sea Tsunami Workshop (SCSTW) to stimulate and collaborate research on regional tsunami hazard mitigation plan, total risk management and an early warning system in the South China Sea region (http://140.115.145.170/~scstw/; http://math.usm.my/scstw3/).

SCSTW3

The first South China Sea Tsunami Workshop 1 (SCSTW1) was held in the Academia Sinica, Taipei, Taiwan on 5-7 December 2007. The sponsors included the Academia Sinica, Cornell University and many other research universities and institutions. The second South China Sea Tsunami Workshop 2 (SCSTW2) was held in the Academic Center, Shanghai Jiao Tong University, Shanghai, China between 1 and 3 December 2008. The sponsors included the National Natural Science Foundation of China, Cornell University and Shanghai Jiao Tong University among others. Following the success of SCSTW1 in 2007 and SCSTW2 in 2008, a consensus has been reached to organize SCSTW3 from 3 to 5 November 2009 in Universiti Sains Malaysia (USM), in collaboration with Cornell University USA and Syiah Kuala University Aceh, Indonesia. Other co-organizers include the Malaysian Meteorological Department, Academy of Sciences Malaysia, National Oceanography Directorate and Mercy Malaysia. Themes of the workshop include: scientific, computational, technical and engineering aspects of tsunami, as well as social-cultural-economic implications and dimensions. The primary goal is the development of tsunami resilient communities worldwide. The objectives of SCSTW3 include:

1) To review and enhance all aspects of tsunami research, community preparedness as well as scenario development in the South China Sea region;
2) To review and enhance existing tsunami research, early warning systems and tsunami resilient community programs;
3) To develop implementation plan for sustained tsunami early warning system, total risk management and coastal hazard mitigation programs.

The Topics to be discussed and presented include:

- Engineering design and construction for tsunami impact reduction;
- Coastal zone management and mitigation plans;
- Effective use of tsunami warning systems and mitigation;
- Mechanism of earthquakes and tsunami and their prediction;
- Numerical simulations and physical modeling of tsunami evolutions;
- Tsunami community preparedness;
- Tsunami risk reduction;
- Tsunami rescue operation;
- Tsunami risk mapping and evacuation routes;
- Roles of NGOs;
- Roles of local governments and councils;
- Education towards tsunami resilience;
- Human resource program for tsunami education and outreach;
- Restoration program and city planning after earthquakes and tsunamis;
- Role of coastal vegetations and their recovery after tsunamis;
- Related subjects such as storm surge, typhoons and coastal flooding etc.

The collaboration among various research institutions and government agencies in Malaysia in conjunction with other similar international institutions serves to help develop coastal communities that are resilient to the impacts of future tsunamis in the regions. To promote international collaboration we provide a brief exposition of two institutions active in conducting and promoting tsunami research in Malaysia.

Academy of Sciences Malaysia

Soon after the 2004 Andaman tsunami, the Government of Malaysia appointed the Academy of Sciences Malaysia (ASM) to coordinate, manage and monitor the implementation of the Seismic and Tsunami Hazards and Risks Study beginning on 13th December 2005. The Study was undertaken on behalf of the Malaysian Ministry of Science, Technology and Innovation represented by its implementation agency, the Malaysian Meteorological Department. To undertake the study, two government agencies and three local universities were commissioned to implement the study, which was co-coordinated by the Academy of Sciences Malaysia. The study implementers were: The Minerals and Geoscience Department Malaysia (MGDM), The Malaysian Meteorological Department (MMD), Universiti Putra Malaysia (UPM), Universiti Sains Malaysia (USM) and Universiti Teknologi Malaysia (UTM). This study was successfully completed in December 2008. The Study's major Terms of Reference were as follows: (i) to evaluate seismic and tsunami risk; (ii) to develop country macrozonation map and city microzonation maps; (iii) to identify areas vulnerable to earthquakes and tsunamis; (iv) to evaluate seismic factors in planning and design of major structures; and (v) to assess and to enhance the adequacy of existing monitoring and data collection system.

MMD Established Mntews

To provide improved protection against future tsunami similar to the 2004 Andaman tsunami, the Ministry of Science Technology and Innovation (MOSTI) of Malaysia established the Malaysian National Tsunami Early Warning System (MNTEWS) at the Malaysian Meteorological Department (MMD) to provide early warning on tsunami generated in the Indian Ocean, South China Sea or the Pacific Ocean that may affect Malaysia. The main objective of the implementation of an effective MNTEWS is to detect earthquake events that might result in a destructive tsunami and to disseminate accurate and timely warnings so that appropriate actions can be taken to mitigate the adverse impacts of the resulting tsunami. The MNTEWS consists of three components namely: (1) Data and Information Collection, (2) Data Processing and Analysis, and (3) Dissemination of early Warning regarding tsunami.

MMD has installed a network of 14 seismic stations, six in Peninsular Malaysia, five in Sabah and three in Sarawak as shown in Figure 20. With this network, MMD maintains real-time monitoring of earthquake and tsunami occurrence in the region on a 24/7 basis throughout the year. To complement the Malaysian network, MMD also receives real-time seismic data from eight seismic stations in Indonesia through VSAT and from 26 seismic stations from other countries through internet. With this seismic network, MNTEWS is capable of detecting earthquakes within a few minutes from the time of its occurrence.

A total of 6 tide gauge stations have been installed at six selected risk areas, with three over the northwestern part of Peninsular Malaysia, one in the northeast region of Peninsular Malaysia and two in Sabah as shown in Figure 21. This network of tidal gauges serves to measure and monitor wave conditions and water elevations along the shores of Malaysia. These stations serve as the first-line monitoring system, as they detect abrupt rise of water levels. A tsunami detection algorithm monitors abrupt changes in water levels to detect possibility of tsunami waves. Upon detection of potential tsunamis, it switches the data logger from the normal mode (with one-hour data transmission) to the event mode or the tsunami mode (with one-minute data transmission) through INMARSAT. MMD also monitors water levels at 40 international sea level stations in the Indian Ocean and Pacific Ocean through the network operated by the Global Sea Level Observing System (GLOSS).

Two deep ocean buoys currently are deployed at two strategic locations to facilitate early detection of tsunamis. The first buoy was deployed near Rondo Island of Indonesia and the second buoy was installed near Layang-Layang Island in the South China Sea as shown in Figure 22. Measurement is recorded every one second, with the average of these 15 measurements being stored every 15 seconds. The data is then transmitted to the National Tsunami Early Warning Center every hour via INMARSAT under normal operating conditions. However, in the event of a potential tsunami, the data will be transmitted to the warning center every 15 seconds. MMD also monitors the sea levels from international buoy network in the Indian Ocean and the Pacific Ocean through the Global Sea Level Observing System (GLOSS). A network of four coastal cameras has also been installed for monitoring of sea conditions at selected coastal areas.

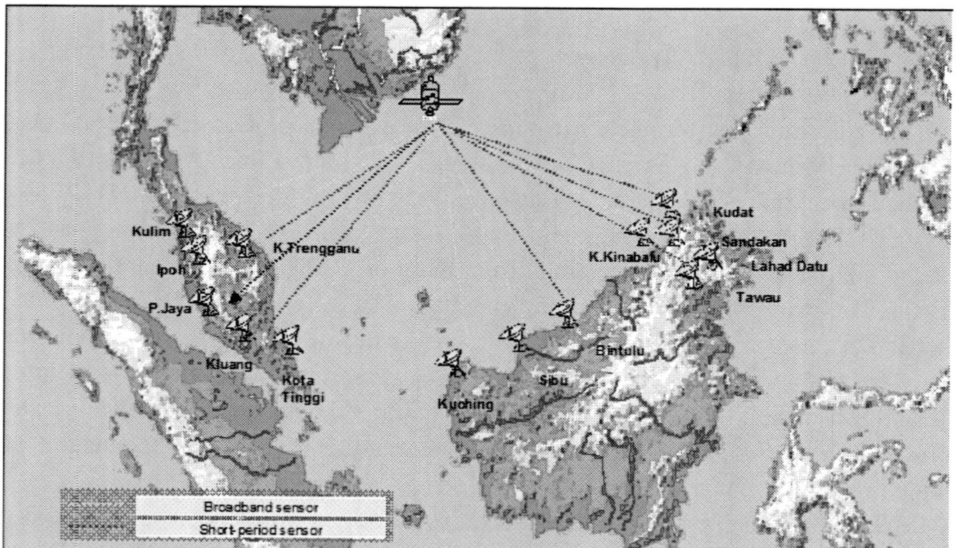

Figure 20. Malaysian National Seismic Network.

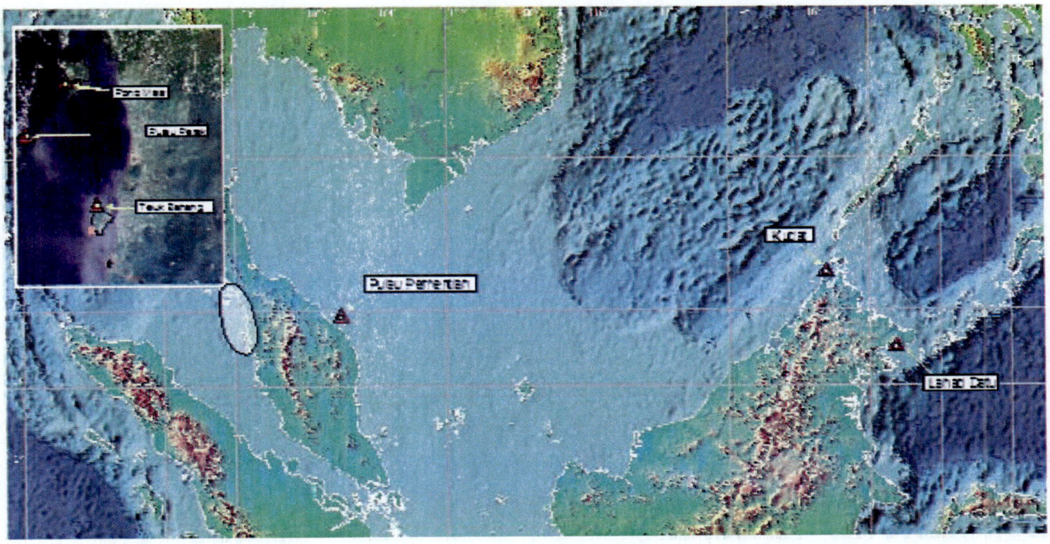

Figure 21. Tide gauge stations.

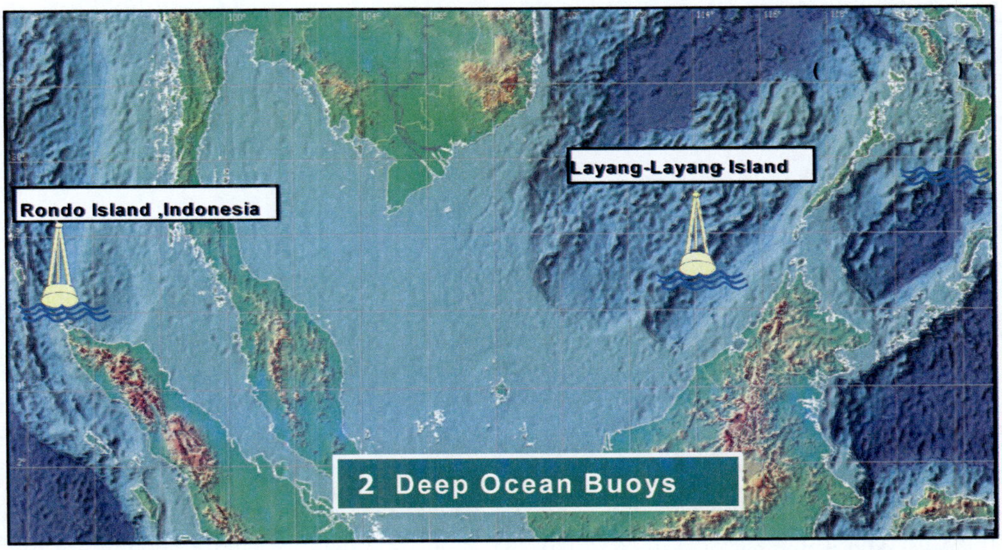

Figure 22. Deep ocean buoys.

MMD establishes a direct linkage through GTS to the Pacific Tsunami Warning Centre (PTWC) located in Hawaii and the Japan Meteorological Agency (JMA) in Tokyo. This linkage provides a direct communication means for receiving tsunami advisory services issued by these two centers for potential tsunamis that might occur in the Pacific Ocean, the South China Sea and the Indian Ocean. This information will assist in verifying the possibility of tsunami generation as a result of earthquake activities.

A large database of potential tsunami scenarios has been generated for the entire Indian Ocean, South China Sea, Sulu Sea and Celebes Sea using TUNAMI codes. This database enables the forecast of tsunami travel times and amplitudes at 50-meter depth offshore and at 1-meter depth along the coasts. The fault parameters are defined by the Scaling Law while other parameters such as strike, dip and rake angles are set to the most credible worst-case scenarios. Forecasts are then made available for Coastal Forecast Points covering all countries in the Indian Ocean, South China Sea, Sulu Sea and Celebes Sea in the event of any potential tsunami.

MMD has undertaken several measures to ensure efficient and timely dissemination of earthquake information and tsunami warnings. The advisory will be issued immediately upon verification of eminent occurrence of potential tsunamis along the Malaysian coasts. This advisory is disseminated to all relevant disaster management authorities, including the media and Non-Governmental Organizations (NGOs). This is performed via various channels such as SMS, hotline, fixed phone, telefax and internet. All earthquake and tsunami information and warnings are posted on the MMD website. In the event of a potential occurrence of a tsunami triggered by a strong earthquake, this information and warning will be disseminated to the public through local television stations via *Crawlers* and *News*. MMD has also implemented the Fixed Line Alert System (FLASH) to ensure that the warnings to evacuate will immediately reach the coastal communities at risk. Further, MMD has installed a siren network consisting of 13 sirens located along potentially affected coasts. This siren system can be activated remotely from the Tsunami Warning Center to warn communities at risk if a tsunami is eminent. To cater to multi-hazards, the existing monitoring and warning systems MNTEWS for earthquake and tsunami in the region is currently being enhanced through the addition of three seismic stations, 15 tide gauges stations, 14 coastal cameras and 10 sirens. Further, the INMARSAT communication link will be replaced by the VSAT communication link.

Tsunami Buoys in South China Sea

Recently, the Chinese government has awarded a contract to Science Application International Corporation (SAIC) to produce and deliver two SAIC Tsunami Buoy systems (STB) (http://140.115.145.170/~scstw/general/news0427.htm). The systems will be the basis of China's tsunami warning system network in the South China Sea region. The STB system consists of three subsystems: a surface communications buoy, a buoy mooring, and a bottom pressure recorder. The bottom pressure recorder includes a highly accurate sea floor pressure sensor that can detect earthquakes and sea level changes. Acoustic communications transmits the pressure data to the surface buoy which then relays the data by satellite communications to the Chinese Oceanographic Environmental Forecast Center for analysis.

Conclusion

This chapter begins with a brief account of the 26 December 2004 Andaman tsunami and its impacts on some beaches in Malaysia. It then presents field survey results on tsunami arrival times, wave runup heights and inundation distances on beaches in Penang and Kedah that are badly affected by this tsunami. The comprehensive compilation of this onsite survey results are then used to calibrate and validate a tsunami simulation model TUNA developed by the authors. A brief description of the mathematics and numerical scheme for TUNA are then provided. Simulation results indicate that TUNA performs satisfactorily and can indeed reproduce salient features of tsunami characteristics observed in beaches in Penang and

Kedah post the 2004 Andaman tsunami. The surveys conducted also provide some indications regarding the role of coastal mangrove forests in reducing the impact of tsunami. To understand quantitatively the role of mangrove in reducing the impact of tsunami, a numerical simulation model is developed and applied to a stretch of mangrove forests in Penang. A main theme of this chapter is the desire to develop coastal communities that are tsunami resilient. To help achieve this goal, a series of research workshops known as the South China Sea Tsunami Workshop (SCSTW) have been conducted in Taipei and Shanghai in 2007 and 2008 respectively. The third SCSTW3 will be held in Penang, Malaysia on 3 to 5 November 2009, in conjunction with the South China Sea Tsunami Working Group First Meeting on 6-7 November 2009, sponsored by UNESCO-IOC. This chapter ends with a brief description of the Malaysian National Early Warning System established at the Malaysian Meteorological Department. It is hoped that this chapter has provided a brief description of tsunami simulation research and mitigation program in Malaysia post 2004 Andaman tsunami with the hope that further collaboration with other research institutions could be enhanced by this chapter.

Acknowledgement

Financial support provided by Grants 1001/PMATHS/817024, 1001/PMATHS/817025, 1001/PMATHS/811093, 305/PMATHS/613131 and 203/PMATHS/671187 is gratefully acknowledged.

References

[1] Harada, K. and Imamura, F. (2005). Effects of Coastal Forest on Tsunami Hazard Mitigation-A Preliminary Approach. In K. Satake (Ed.), *Tsunamis: Case Studies and Recent Developments*, vol. 23, p. 279-292. Berlin: Springer.

[2] Hérbert, H., Schindelé, F, Altinok, Y., Alpar, B. and Gazioglu, C. (2005). Tsunami hazard in the Marmara Sea (Turkey): a numerical approach to discuss active faulting and impact on the Istanbul coastal areas. *Marine Geology* **215**, 23-43.

[3] Imamura, F., Shuto, N. and Goto, C. (1988). Numerical simulation of the transoceanic propagation of tsunamis. Proceedings of the 6th Congress Asian and Pacific Regional Division, *Int. Assoc. Hydraul. Res.* (IAHR), Japan, pp. 265-272.

[4] Intergovernmental Oceanographic Commission (IOC) (1997). Numerical Method of Tsunami Simulation with the Leap Frog Scheme, 1, Shallow Water Theory and Its Difference Scheme. In Manuals and Guides of the IOC, p. 12-19, *Intergovernmental Oceanogr. Comm.*, UNESCO, Paris.

[5] Koh, H.L. (1988). Finite element analysis of long period tidal propagation in a channel. In H. Niki and M. Kawahara, *Computational Methods in Flow Analysis*, p. 1262-1269. Okayama University of Science, Japan.

[6] Koh, H.L. (2004). *Environmental and Ecosystem Modeling* (Pemodelan Alam Sekitar dan Ekosistem), Universiti Sains Malaysia Publishers, 383 p.

[7] Koh, H.L., Teh, S.Y. and Izani, A.M.I. (2007). Tsunami Mitigation Management. Special Feature: Natural Disaster Management Technologies. The United Nations

Asian and Pacific Centre for Transfer of Technology (UN-APCTT) Nov-Dec 2007, *Asia Pacific Tech. Monitor* **24** (6), 47-54.

[8] Koh, H.L., Teh, S.Y., Liu, P.L.-F., Izani, A.M.I. and Lee, H.L. (2009). Simulation of Andaman 2004 Tsunami for Assessing Impact on Malaysia. *Journal of Asian Earth Sciences* (in press).

[9] Liu, P.L.-F., Woo, S.B. and Cho, Y.S. (1998). Computer Programs for Tsunami Propagation and Inundation, Cornell University, Sponsored by National Science Foundation, 104 p.

[10] Massel, S.R., Furukawa, K. and Brinkman, R.M. (1999). Surface wave propagation in mangrove forests. *Fluid Dynamics Research* **24** (4), 219-249.

[11] Mazda, Y., Wolanski, E., King, B., Sase, A., Ohtsuka, D. and Magi, M. (1997). Drag force due to vegetation in mangrove swamps. *Mangroves and Salt Marshes* **1,** 193-199.

[12] Okada, Y. (1985). Surface deformation due to shear and tensile faults in a half-space. *Bulletin of the Seismological Society of America* **75** (4), 1135-1154.

[13] Shuto, N. (1987). The effectiveness and limit of tsunami control forests. *Coastal Engineering in Japan* **30**, 143-153.

[14] Stein, S. and Okal, E.A. (2007). Ultralong period seismic study of the December 2004 Indian Ocean earthquake and implications for regional tectonics and the subduction process, B*ulletin of the Seismological Society of America* **97** (1A), pp. S279–S295.

[15] Sternberg, L., Teh, S.Y., Ewe, S., Miralles-Wilhelm, F. and DeAngelis, D. (2007). Competition between Hardwood Hammocks and Mangroves. *Ecosystems* **10** (4), 648-660.

[16] Teh, S.Y. (2008). *Modeling Evolution of Tsunami and Its Impact on Coastal Vegetation.* Ph.D. Thesis, Universiti Sains Malaysia, 197 p.

[17] Teh, S.Y., DeAngelis, D., Sternberg, L., Miralles-Wilhelm, F.R., Smith, T.J. and Koh, H.L. (2008a). A Simulation Model for Projecting Changes in Salinity Concentrations and Species Dominance in the Coastal Margin Habitats of the Everglades. *Ecological Modelling* **213** (2), 245-256.

[18] Teh, S.Y., Koh, H.L., Izani, A.M.I. and Lee, H.L. (2008b). Modeling Tsunami Runup by Moving Boundary. *Proceedings of the International Symposium on the Restoration Program from Giant Earthquakes and Tsunamis*, 22-24 January 2008, Phuket, Thailand, University of Tokyo, p. 301-306.

[19] Teh, S.Y., Koh, H.L., Liu, P.L.-F., Izani, A.M.I. and Lee, H.L. (2009). Analytical and Numerical Simulation of Tsunami Mitigation by Mangroves in Penang, Malaysia. *Journal of Asian Earth Sciences* (in press).

[20] Titov, V.V. (1997). *Numerical modeling of long wave runup*. Ph.D. Thesis, University of Southern California, Los Angeles, CA, 150 p.

[21] UKM (2006). The 26.12.04 *Tsunami Disaster in Malaysia: An Environmental, Socio-economic and Community Well-being Impact Study*. Ibrahim Komoo and Mazlan Othman (Eds.). Institut Alam Sekitar dan Pembangunan (LESTARI) and Akademi Sains Malaysia, 168 p.

[22] Yoon, S.B. (2002). Propagation of distant tsunamis over slowly varying topography. *Journal of Geophysical Research* **107** (C10), American Geophysical Union, 4.1-4.11.

[23] Yoon, S.B. and Liu, P.L.-F. (1992). Numerical simulation of a distant small-scale tsunami. In N. Saxena (Ed.), *Recent Advances in Marine Science and Technology, PACON92,* Pacific Congress on Marine Science and Technology, Kona, Hawaii, pp. 67-78.

In: Tsunamis: Causes, Characteristics, Warnings and Protection ISBN: 978-1-60876-360-3
Editors: N. Veitch and G. Jaffray, pp. 57-86

Chapter 3

2004 – Tsunami Characteristics of Wounds

Thavat Prasartritha
Center of Excellence in Orthopedics, Lerdsin Hospital Medical Services,Thailand

Abstract

Around 10:00 AM, 26 December 2004, Tsunami, the name that was unknown to most Thai people, had moved to the Andaman Sea having started from Phuket, Phang-Nga, Ranong, Krabi, Trang and Satun provinces. The main injuries of the survivors were aspiration and trauma. Trauma to parts of the body and extremities were sustained when the wave hit and swept back. Patterns of injury can be varied depending on the tidal velocity and size of particles that hit the body.

By the rolling and sweeping motion of the huge wave, victims were surged or heaved along an undetermined path and direction similar to a bullet being fired through surrounding targets with destructive potential. High impact velocity acting on a hard heavy object could directly cause permanent damage such as crushed, cavitation and tissue loss to any part of the body. Those lucky enough to survive were left with several injuries. The observed injury patterns and wounds are depended on the magnitude of the direct impact, indirect or secondary injury and condition of the landing surface. The variety of wounds seen in the Tsunami incident was enormous. Damage of skin and underlying soft tissues can range from self treated to life threatening condition which requires immediate resuscitation. The wounds had characteristics of multiple site involvement; any part of the body can be injured. Wound types categorized by the size and depth can be varied from area to area even in the same extremity. Extensive skin loss with severe contamination and multi-organ involvement are nearly similar to those of war wound injuries. All wounds can be graded by severity and contamination into *first degree* that involves only the covering skin and underlying soft tissues. *Second degree* has more involvement of skin and underlying soft tissue with retained contamination and skeletal involvement. In *severe degree,* there are extensive damages to the skin and underlying vital structures such as bone, joint, nerve and vessels. The small penetrating wound is the most dangerous with high incidence of sepsis and death. Fractures, dislocation and tendon injury were also common. Another characteristic finding was the early development of wound infection. Spreading of infection and sepsis were strongly related to the emergency aspect of the situation, which overwhelmed the available resource.

Care begins at the scene with immediate first aid and life saving measures by trained people. Patients should be taken to the hospital with resuscitative capacities and rescue pain control. Contaminated wound should be temporarily cleaned, covered and immobilized immediately. Hospital care begins with triage on arrival at the front gate. Primary wound dressing can be temporarily made by using tap water and then cover the wound with a sterile green towel while waiting for definite surgical intervention. For major life threatening wounds, the damage control surgery should be applied to minimize high rate of amputations and mortality. A wound with a small penetrating entrance should be considered dangerous. Aggressive debridement and thorough irrigation should be performed as soon as possible and the wound should be left open. Fasciotomies should be carried out to release pressure within the compartment. Broad spectrum antibiotics should be administered once cultures have been processed. Wounds should then be further re-explored and redebrided 24–48 hours after the initial procedures. Immediate stabilization of bone fractures with the least amount of stable splints to the most stable fixator is a basic principle of orthopedic care. Patients with more complex wounds and significant injuries should be stabilized and then immediately transported through a more advancing levels where definitive procedures can be effectively performed. Close surveillance and additional therapy for possible wound infection and early sepsis are supremely important. Skin or tissue reconstruction of high energy open wounds should be attended to as soon as possible once the wound bed is clean.

With 2004 tsunami, it is quite clear that the immediate Multiple Casualty Incident (MCI) must be handled at the time and place of impact with proper equipment, supplies and variety of experts in emergency management. The current review demonstrated that wounds had affected people and health care staffs. A better medical response can be definitively provided with a proper or well prepared planning and exercising. The mainstay of primary closure of wounds and increase use of antibiotics in civilian practice is not appropriate for any disaster incident. Initial wound assessment, appropriate multiple debridement, soft tissue and bone stabilization, further assessment and duration of the exposed wound are all of crucial factors that affect incidence of infection and healing.

2004 - Tsunami Characteristics of Wounds

Tsunami is a Japanese word that can be broken down to Tsu (harbour) and Nami (wave) [1,2]. It is a natural phenomenon where a series of waves are generated from a sudden or pulsating displacement caused by submarine landslides or submarine volcanic eruptions.

To date, the Tsunami-2004 is still the latest and deadliest tsunami that has occurred in the Indian Ocean [3,4].

2004 – Thailand Tsunami [5,18]

At 7:38 AM, 26 December 2004, an earthquake occurred in Phuket and the nearby provinces. At around 9.00 AM, many villagers and tourists were astonished by the declining sea water of over 1 km. This natural occurrence that had never happened before had attracted many tourists and local villagers to walk down and observe with curiosity and suspicion.

After 20 minutes, the first gigantic and subsequent dark waves instantly swept through and demolished everything along the coastlines with no exception. Victims were swallowed up while others were floating along with the current. Bodies were crushed against, trees, walls and various wreckages. When the waves subsided human remains could be seen at a boundless distance floating in the sea [5-9].

Then at around 11:00 AM., Tsunami, the name that was unknown to most Thai people was announced on the air across the Kingdom. Eventually details of the mass destruction and number of human casualties were televised nearly every minute throughout the country and worldwide. More than 10 countries were hit. In Thailand 6 provinces starting from Phuket, Phang-Nga, Ranong, Krabi, Trang and Satun were initially involved.

The worse hit area was Khao Lak in Phang-Nga (Figure 1). Initially, more than 400 people were hurt and sent to the nearby Takuapa and Phang-Nga hospitals. Takuapa, a 177-bed hospital, had a major incident disaster plan in place, which was prepared for only 80 traffic injury patients. During the first 24 hours of the disaster, more than 1,000 patients presented to the emergency room . Every inch of available space, regardless of the pathway, balcony, offices, and meeting rooms, were filled with Thai and foreign patients. The atmosphere was chaotic, depressed and mournful [10,11].

The second most damaged area was Phi Phi Island that is located about 1 hour by ship from the mainland of Krabi [12-14]. At 10:30 AM, the wave reached the island. About 600–700 people required medical attention. Initially evacuation was possible only by air. Following the repair of the damaged piers, the rest of the injured were transferred to the mainland by boats. By 12:00 noon, casualties were being treated at the Krabi Hospital. Triage was performed at the front gate of the hospital, while the dead were transferred to the nearby temples. By evening of that day, approximately 498 casualties had been treated at the hospital, with more arriving throughout the night by helicopter and boats from the outlying islands. The total number of deaths were more than 800. Six hundred (75%) came from Phi Phi Island.

The well-known beaches of Patong, Kamala, Kata, Karon and Naiyang in Phuket were besieged by the waves.

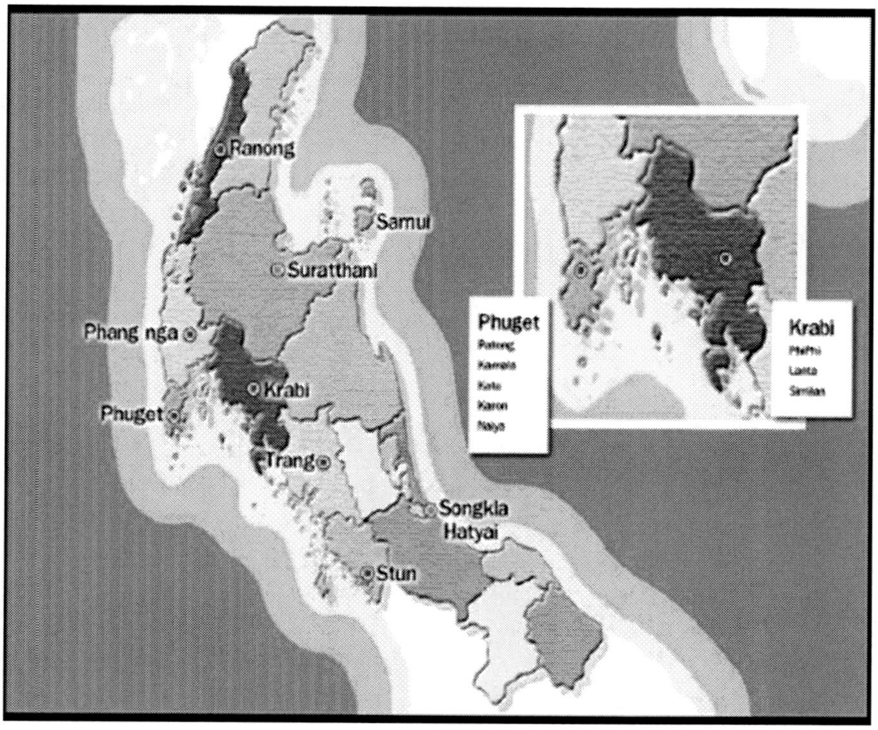

a)

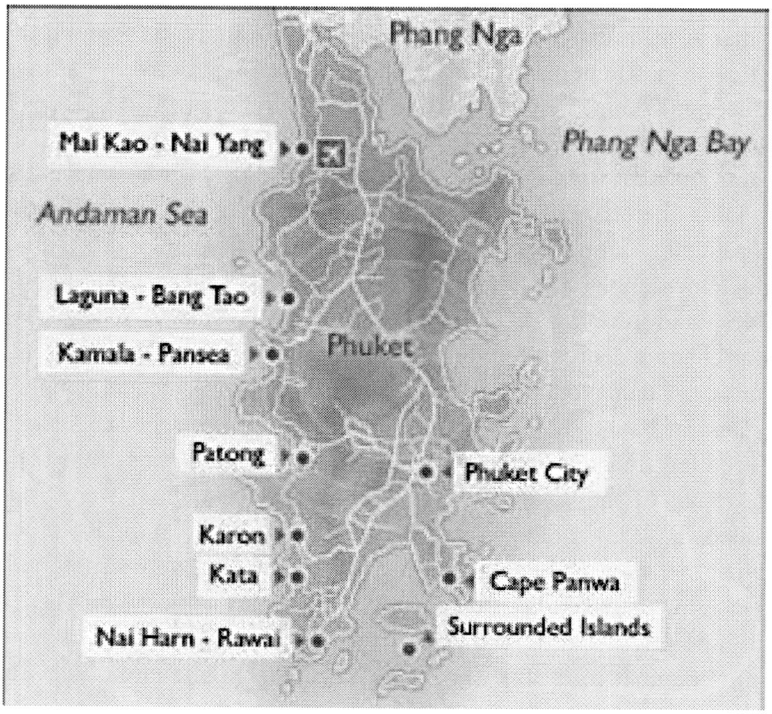

b)

Figure 1. (A) Map of six provinces hit by 2004 tsunami. (B) Map of beaches of Phuket Island.

The ocean water continually rose, forcing hundreds of people to evacuate and flee to the nearby city hall and mountains. Water had flooded buildings, rising up as high as the second or third floors of the hotels. Around noon, reports began pouring in about bodies being washed up along the beaches. Three public hospitals in Phuket were overwhelmed with a large number of patients. In Ranong, two public health centers were damaged. Two health volunteers lost their lives and two were severely injured. As Ranong is less attractive to tourists, the majority of the victims were natives and few Myanmar workers [6].

The six impacted provinces have a total of 10 hospitals with a capacity of 2,000 beds, 34 community hospitals and primary health care clinics. Under an emergency execution plan, the health professionals dedicated all their strength to assist the massive number of victims before the arrival of the backup teams. Apart from these, the nearby provinces; Suratthani, Chumporn, Nakhon Si Thammarat, Trang, Phattalung , Songkla and Hatyhai hospitals had urgently sent their emergency ambulances with rescue crews to the hardest hit areas. Along side of the health professionals, a lot of Thai and foreigners helped each other as much as possible. Some tourists acted as translators and some people offered food and water to complete strangers. Although the tsunami destroyed the lives of thousands, it revealed the kindness of millions of people [6,16].

The Tsunami paralyzed the communications between Phuket, Phang-Nga and Krabi, by damaging power lines and overcrowding the limited phone channels. Handheld radio equipment (walkie-talkies) were the only communicating means among working units. However, the limited supply of walkie-talkies was also a problem. By the report, the damaged area which included 6 provinces of South Andaman had extended as far as 41 districts, 277 sub districts and 329 villages [6,16].

Personal experience of the Tsunami Disaster by Dr. Porn Pongpanitanont the current (2009) Director of Takuapa Hospital, Khaolak, Phang-Nga.

It was around late morning of 26 December 2004, when Dr. Porn received the first phone call notifying him about the tsunami disaster. At that time, Dr. Porn was a level 9 Medic Officer, working in the government's Preventive Medicine unit of Phang Nga Provincial Health Office. The number of victims and the condition of the injured caused by the massive nature of the disaster were beyond his anticipation.

A major road accident involving trucks in 1992 was the most devastating incident he had encountered before the tsunami. But there were only 200 dead victims caused by that accident. This time, the real incident was totally and extremely different. There were so many wounded patients and dead bodies arriving at Phang Nga Hospital. Facilities were limited and the telecommunication system was not functioning. The surrounding atmosphere was confusing. Most of the victims were cold, hungry and suffered a strong desire to find their relatives.

The injured victims were categorized into 1-major severe injury that required immediate treatment 2 - average injury and 3 - minor injury. Temporary shelters providing foods, cloths and translation were set up. Some of the victims were immediately evacuated by ambulances and helicopters to Phuket International Airport and nearby hospitals. Although tremendous support from various organizations had arrived, there were still so many strained and exhausted situations in the first few days. Some patients and relatives as well as health care workers had to sleep on mats along the hallway of the hospital.

At first, dead bodies were sent to the hospital. Later the bodies were carried to Yan Yao temple for autopsy and identification. Local police officers were asked to record physical details and collecting evidences together with medical personnel.

Hospital staffs and support teams from various provinces worked towards the same goal- ”we would get everyone home safely”. Some volunteers had limitations: some could not work for a long period of time and some could stay only for 1-2 days. On the other hand, some volunteers stayed and helped for the whole month. Some even went so far as closing their companies in Bangkok temporarily in order to come and help, as well as donating money. The response for all sorts of unusual requests were swift and generous.

People wanted to claim the body of their relatives, based only on external characteristics of the corpse that they recognized, e.g. wearing jeans or wearing string on a wrist. But they had to wait until clear evidence was found after completing identification procedures, i.e. examination of fingerprint, dental history, dental x- ray and DNA analysis.

There was a case of one woman refusing to accept the body of her sister because of a ring on its finger. She did not recognize that ring on the photo, taken immediately when the body arrived at the hospital and was still in good shape, and was adamant that her sister did not have such a ring. Finally she accepted because of the finger print that matched with her sister’s ID card. She also admitted that both of them had not seen each other for six months.

In another case, the parents of a dead victim came to the hospital and claimed their son’s body. Everything was well matched, including DNA and chromosome XY result, except that the corpse itself had the physical characteristics of a woman. After in depth examination, we found that he had a sex change operation without telling his parents.

Working experience in dead victim identification was very useful in our operation. For instance, in checking DNA of people, sample of cell membrane could be easily obtained from blood checked at birth, hair on the comb, toothbrush (that had been used individually) or a swab from the mouth. In one case, a mother of a child victim had kept the baby’s umbilical cord for 14 years and brought it over for DNA check.

In working with teams from many nations, officers led by Khunying Pornthip Rojanasunan were sent to carry out autopsy and identification work since the beginning at Yan Yao temple. As the number of the victims was massive, the international team was responsible for foreign victims. Our role was to coordinate with the Director of Phang Nga Provincial Health Office to request for more support in developing autopsy and identification room, whereby air- conditioner as well as a Septic Anaerobic Filter must be installed.

The international team gave some advice regarding disease prevention and control, as well as environmental surveillance, since lots of bits and pieces of broken tiles were found at Yan Yao temple. The Royal Thai Police coordinated with the Pollution Control Department to measure airborne asbestos level before continuing our work.

From the situation, I was impressed with the kindness of both Thais and foreigners, who contributed their support in every way. Every one of us was working toward one goal, which was to enable everyone to return home safely. In the beginning, some officers were afraid to touch a corpse. But the gratitude they received from the victims’ relatives/families encouraged them to carry on the mission.

The role of Phang Nga Public Health Administration Center was to coordinate with various organizations, both from the government and private sectors, in providing medical treatment and rehabilitation service to injured victims. The Center also provided information, patient referral/transfer knowledge and disease prevention guidelines. Many volunteers involved in the whole process, starting from evacuating victims to hospitals, providing vaccine for Tetanus prevention, securing food and water sanitation, installing garbage disposal and waste water system. Operation team from Siriraj Hospital and community support were also greatly helpful.

After the emergency situation had ended, nursing teams still had to continue providing mental rehabilitation support for victims and officers involved in the operation.

As a recognition of our four staff members, who lost their lives during an operation in a mobile emergency ambulance that was parked in Khao Lak area, Phang Nga Hospital has organized a Tsunami Recall Ceremony on 26 December of every year. Plans are underway to build a monument of the ambulance remains, displaying a list of the dead victims and all the supported organizations. This is a reminder for our hospital team that, even though we have been thru the tragic disaster, we could still overcome it smoothly with the unity and support from everyone.

Injury and Wound Mechanics

The waves of the tsunami rushed into shore quite suddenly. As distraught tourists and locals fled for their lives, the huge wave crushed forth, engulfing every individual in its path.

The power of the Tsunami can be evidenced by :

a) *The wreckage seen everywhere* starting from the shore to more than 10 kilometers away (Figure 2). Pieces of substances were scattered all over the place. Boats and ships as well as cars and trucks were severely destroyed. Some were piled up in various unbelievable patterns.
b) *Number of missing and dead people [7]*. According to the Ministry of Interior's the tsunami disaster status reported on 5 September 2005, 5,395 were confirmed dead, while 2,817 were missing and 8,457 had been injured. More than 3,000 bodies were that of foreigners from 43 nationalities. A number of children lost their-lives while playing football along the beach. In the same time period more than 1,000 children were immediately orphaned.
c) *Patterns of wounds and associated injuries [10,11,15-17]* - The sustained injuries and wounds are absolutely different from those seen in the normal civilian setting. Due to the rolling and sweeping motion of the huge wave, victims were surged or swept along an undetermined path and direction similar to a bullet being fired through surrounding targets with destructive potential. Self-awareness and surviving skills are crucial in any crisis situation [18,19]. All contact with buildings, trees, rocks, woods or other debris and materials are potential hazards, particularly those hidden under the water. High impact velocity acting on a hard heavy object could directly cause permanent damage such as crushed, cavitation and tissue loss to any part of the body. While clinging around the trunk of a coconut tree, one local fisherman was hit on his back by the wave and sustained a blunt trauma to his lower abdomen. He underwent an exploration and colostomy and found that his sigmoid colon was nearly perforated (Figure 3). Some were knocked unconscious and drowned while others sustained blunt trauma to the chest, abdomen and other vital systems. Hypovolemic shock due to external and internal bleeding as well as

hypothermia and asphyxia can immediately kill an unfortunate patient. Those lucky enough to survive were left with several injuries. The observed injury patterns and wounds are depended on the magnitude of the direct impact, indirect or secondary injury and condition of the landing surface.

Figure 2A-G. Wreck produced by the Tsunami. Pieces of substances were scattered everywhere. Ships, cars, buildings, trees etc. were severely destroyed.

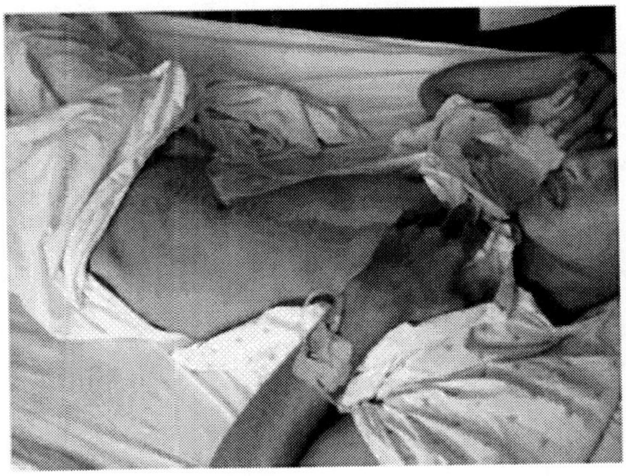

Figure 3. Patient hit by the tsunami while clinging around the trunk of a coconut tree. The picture shows colostomy bag and surgical scar.

The main injuries of the survivors were aspiration and trauma [9,10,13,15-17]. Mud, sand and dirt were seen present in the mouth and throat of the victims. Some patients had to clear sand out of their upper airway by producing a serial aggressive cough. Aspiration caused by sea water and contaminated soil can inevitably lead to asphyxia, acute respiratory distress syndrome (ARDS), pneumothorax , pulmonary congestion and edema. The resulting respiratory failure require immediate combination of specific antibiotics, aggressive and intensive respiratory care [11,15,20] (figure 4). A 17 – year – old girl was swept by the wave which carried her away to approximately 1 km. from her house. She developed polymicrobial lung and brain abscesses following the aspiration of not only salt water but also soil and particulate matter in a near-drowning experience. She gradually regained her motor power and can walk independently with appropriate course of intravenous and oral antibiotics [21].

Skin and its underlying soft tissues can be simply damaged or completely avulsed by direct friction from tangential force applied to the body surface [22]. Associated contusing, burning, cutting and bruising are common. Certain types of injury and wounding patterns required immediate aggressive treatments and were similar to war wounds. The result of multiple site and variation of tissue damages can be explained by the occurrence of energy exchanges, number and the type of particles involved. Extremities that have appeared to have had the flesh ripped from the body, possibly indicated the force by which the individual was struck. Apart from a direct hit, the sand and any dirt can act as multiple fragments such bullets penetrating and damaging along its trajectory path. Tissue damages confined to the inflicted track of the projectile are far more extensive with a high acting energy or force. The ensuring vacuum effect can also pull or suck the surrounding dirt into the wound [2,3].

Wound Characteristics : Wounds sustained from the Tsunami incident are greatly varied [16]. Damage to the skin and underlying soft tissues ranged from self treated to life threatening conditions which require immediate resuscitation. Characteristics of these wounds are multiple site involvement, where any part of the body can be injured. Variations in wound type and severity can occur throughout the body, and in some cases, on the same extremity. The majority of injuries appear to have occurred in the lower extremities. Manifestation of the damaged skin and its underlying tissues can indicate the aggressiveness and mechanism of the acting force.

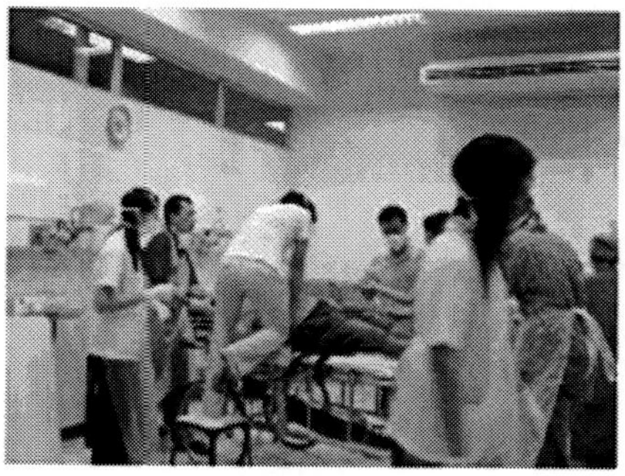

A

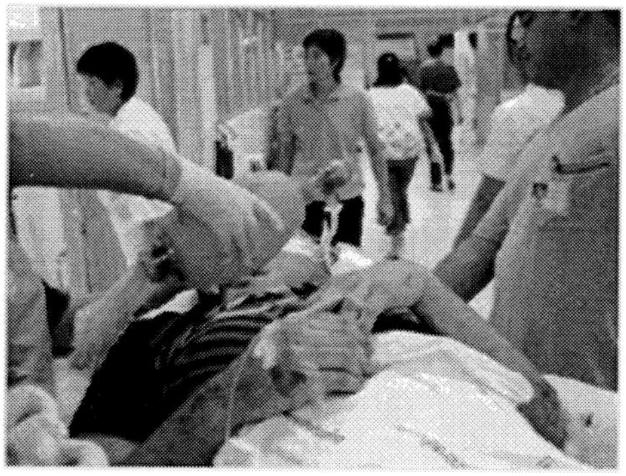

B

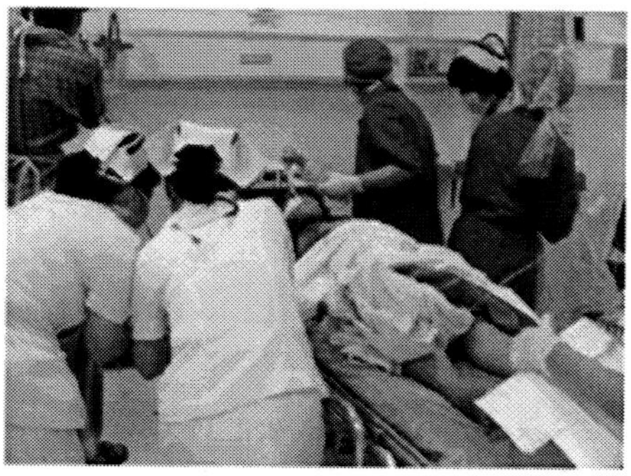

C

Figure 4. Seriously injured patients need immediate resuscitation.

All wounds can be graded by severity into contaminated and dirty wound [23-25] :

First degree - involvement only the covering skin and underlying soft tissues (figure 5a-h). Patterns can be presented as : simple contusion with ecchymosis, abrasion, skin loss, scratching (single or multiple), cutting or stabbing and a combination thereof. Contusions have various grading from mild to severe. Skin loss can be varied from none to involvement of the whole circumference of a limb. Wounds can be self-treated or need extensive debridement and reconstruction.

Second degree - more involvement of skin and underlying soft tissue with retained contamination and skeletal involvement. Tissue flap can be properly replaced with adequate coverage. The associated fracture or dislocation can be open or closed without vascular and neural injuries (figure 6). The wound has the potential of infection and sepsis.

Severe degree- extensive damages to the skin and underlying vital structures such as bone, joint, nerve and vessels (figure 7). The scope of wound can be varied from small penetrating to variable amount of skin loss or degloving injury. The wounds contained dirt, sand, mud and any kind of debris seen along the beach with possible multi organ involvement.bone and joints are open with severe contamination (figure 8). The real damage is far more extensive than initial appearance and needs further reconstruction. Early infection, sepsis, amputation even death are expected.

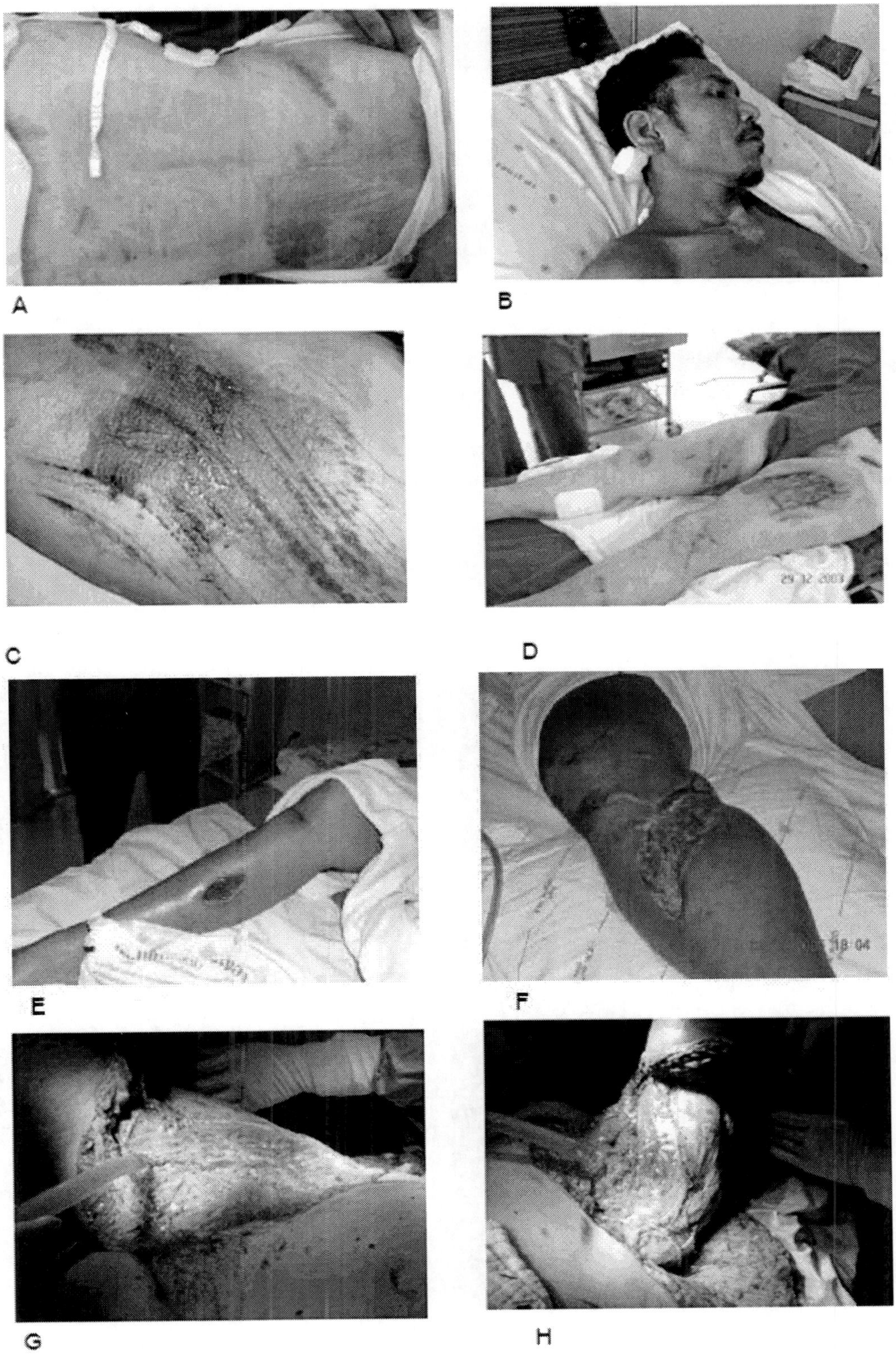

Figure 5. First degree wounds which vary from mild contusion to severe skin loss, all wounds have potential of infection and sepsis.

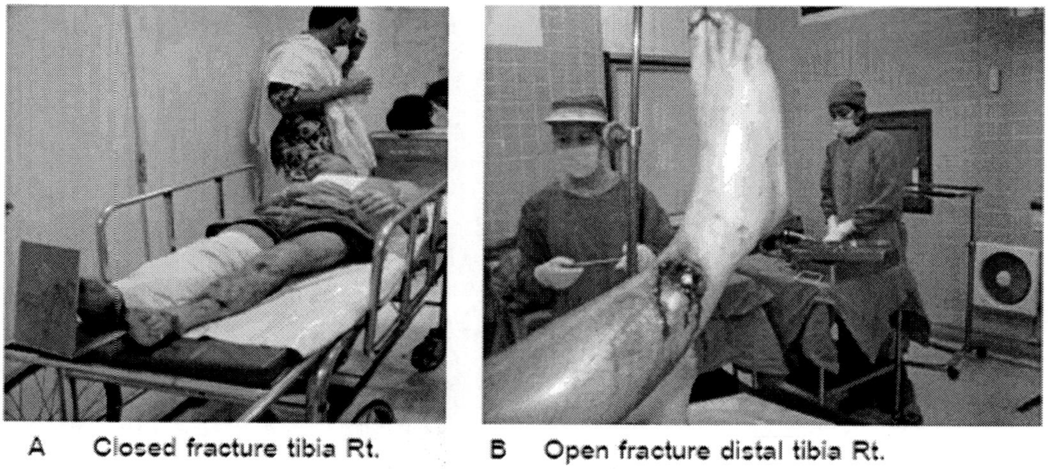

Figure 6. Second degree wounds with more involvement of covering skin and its underlying tissues. Fracture and joint injury can be closed or open with no neurovascular injuries.

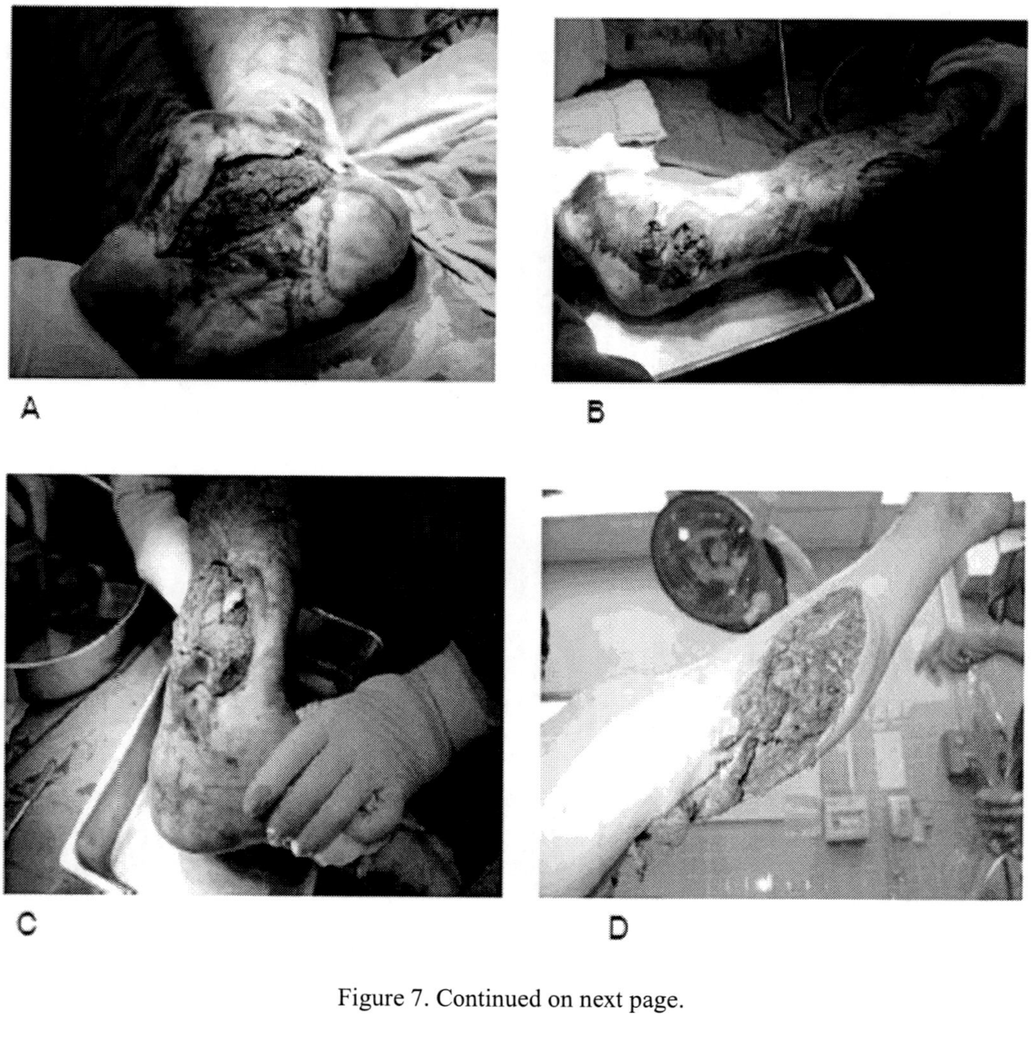

Figure 7. Continued on next page.

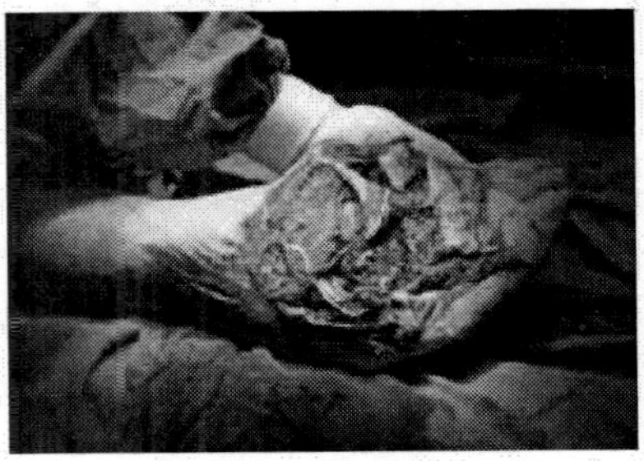

E

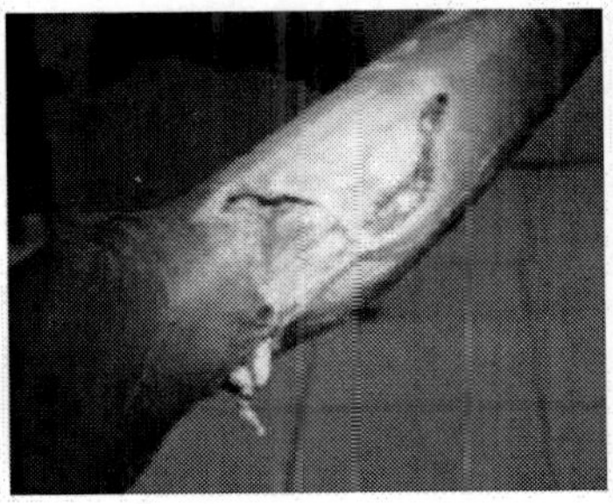

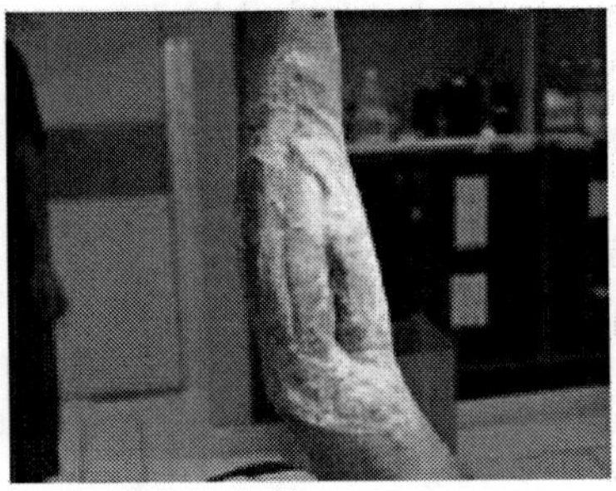

F G

Figure 7. Severe degree by extensive damages and contamination with involvement of bone, joint, nerve and vessels. Multi-organ involvement is common.

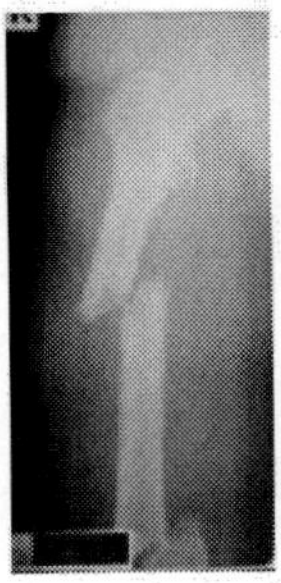

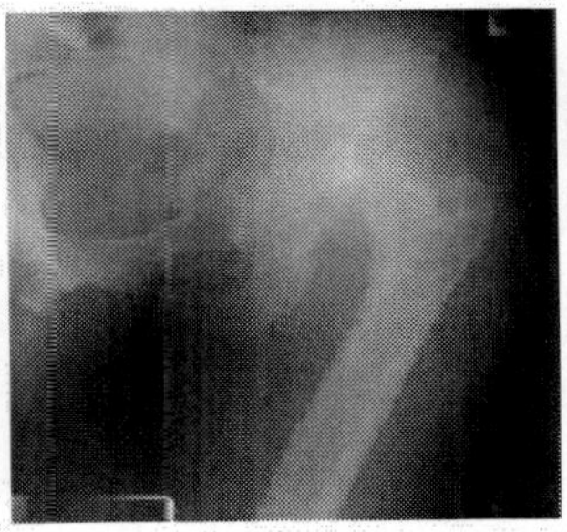

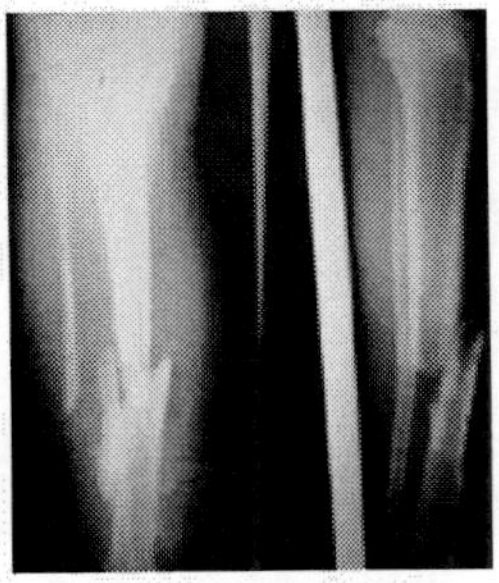

A B C

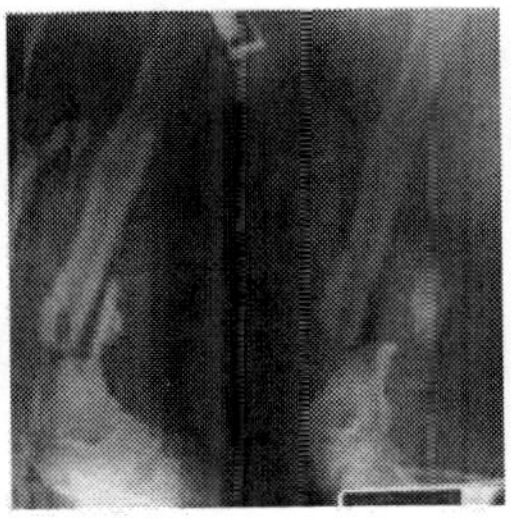

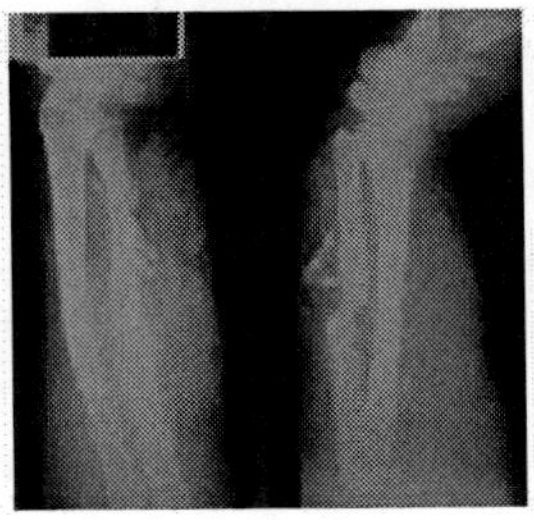

D E

Figure 8. Patterns of fractures. (A) Segmental fractures of femur. (B) Posterior hip fracture dislocation. (C) Comminuted fracture both bones leg. (D) Segmental fractures of distal humerus. (E) Open comminuted fractures of both bones forearm.

A wound known as necrotizing fasciitis is a wound that its underlying fascia was penetrated and inevitably caused invasive infection (Figure 9) [26-28]. A misnomer of flesh – eating bacteria is also commonly appreciated though germs do not actually eat the tissue. The disease is rare but causes severe destruction of tissues by releasing toxins. Generally, the rapid and progressive spreading of bacteria into deeper layers of soft tissues occur most commonly after surgical procedures. Some patients can be involved particularly those with diabetes and peripheral vascular disease. In the Tsunami, the occurring of the rare necrotizing infection not only indicates its extensiveness of the contamination but also gives evidence to the delay in diagnosis and treatment.

The magnitude of tissue injury is aggressive but difficult to ascertain adequately. Some cases were neglected due to the opening of wound being considered small and insignificant. One patient walked in with a minor wound but died within 24–48 hours. The small penetrating wound was undertreated. Later, it was found that the multi-organ failure from overwhelming sepsis occurred and ended this patient's life (Figure 10).

The contained devitalized tissues, contaminants and avascularized tendon, ligament and fascia absolutely act as a media or pabulum for bacteria growth. An aerobic environment within necrotic tissues also enhances the spreading of infection.

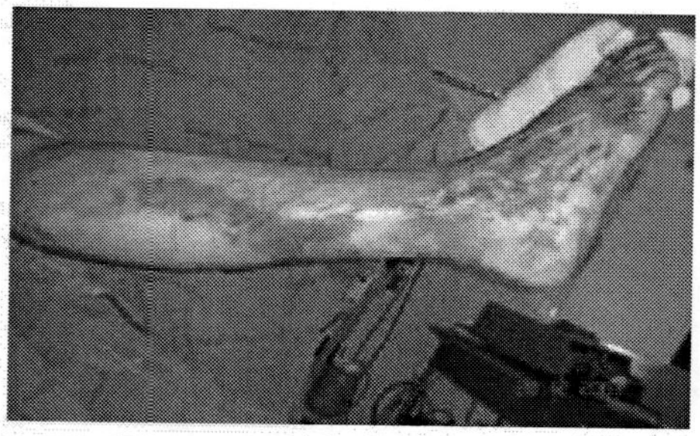

A1

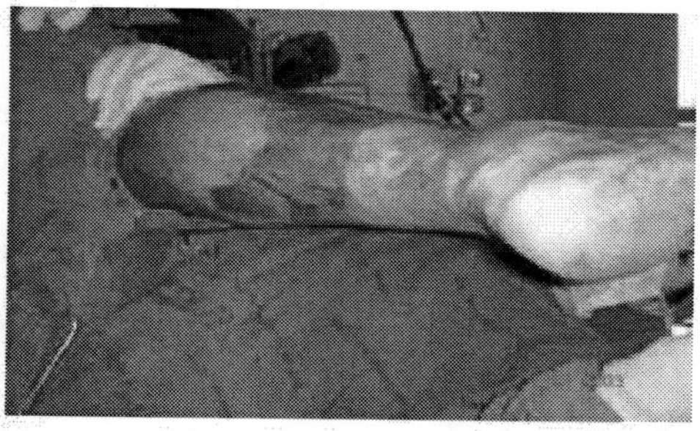

A2

A3

Figure 9. A1-2 - necrotizing fasciitis of calf, note the wide spreading of infection involving skin and its underlying fascia. A3 - following wide excision and removal of necrotic tissues, wound should be left open and the patient should be taken to ICU for close monitoring.

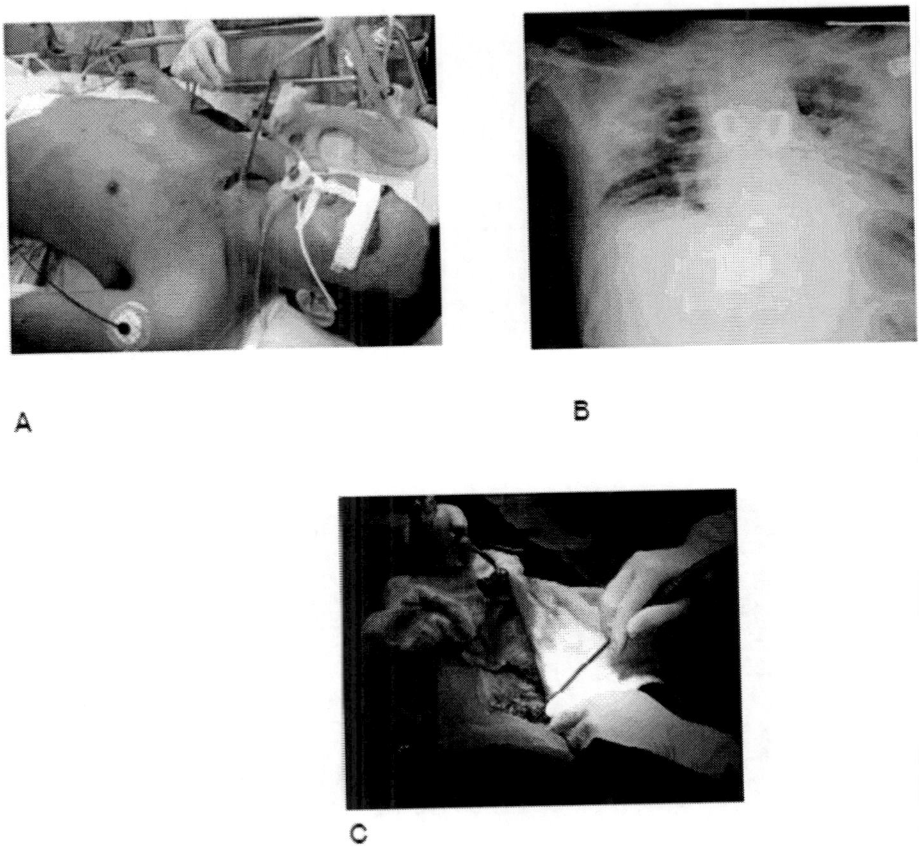

Figure 10. A - The local resident walked in with a mini wound at medial end of right clavicle and salted water pneumonia. The wound was dressed and he was sent home. Forty-eight hours later, he developed sepsis, renal shut down and coma. The photograph shows the extended wound and the gauze packing. B - The chest radiograph shows bilateral pulmonary infiltration. C - The infected wound was re-extended. Irrigation and excision of necrotic tissues were done and the wound was packed and left open. He died 72 hours following the re-exploration. His life could have been saved by aggressive antibiotics and early thorough cleansing of the wound.

Toxic shock causes organ and extremity dysfunction which may claim life and amputated limbs. Diagnosis should be confirmed by microsopic exam. Monitoring following aggressive debridement and antibiotics is required.

Another characteristic finding was the early development of wound infection (Figure 11). The evidence came from infection reports among Tsunami survivors [20,29-33]. Although most of the patients (94.9%) received antibiotics at the beginning, more than 70% had polymicrobial infection. Some of the bacterial infections were not commonly seen. The high rate of infection was strongly related to the lack of surgical resources such as limited operating rooms, equipment, number of experienced trauma care providers and fatigue after prolonged and continued work by surgeons, nurses and supporting personnel.

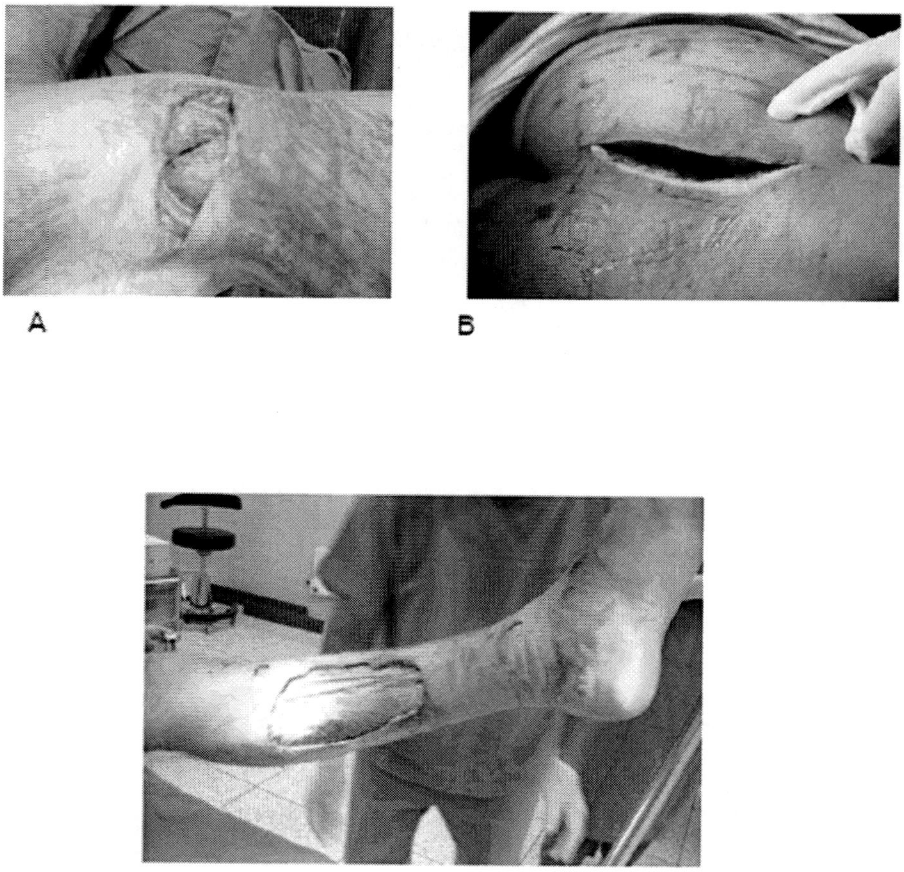

A B C

Figure 11. A - Infected penetrating wound. Note the severity of contamination and surrounding contusion. B – Photograph showing large penetrating wound. C - Wound with wide and deep spreading of infection along the leg.

Treatment

Tsunami is a major multiple casualty incident (MCI) occurring when a large number of injured people is generated within a short period of time [34]. Review of the Tsunami experience indicated that pre-hospital treatment was less controlled and more difficult. Transportation and evacuation from the scene were inefficient and slow (Figure 12). Medical personal competing in the primary assessments, were uncoordinated, records were incomplete, inaccurate and overlapped, which not only wasted time and resources but also raised ethical concerns [35]. Analgesia, though necessary was not provided adequately even in the hospital [36]. Despite communication being notified between nearby hospitals, numbers of patients were transferred without comprehensive resuscitation and stabilization. The main problems included those related to sand aspiration, asphyxia, wound covering and level of pain control.

A

B

C

D

Figure 12. Many patients and relatives waiting for air plane transportation, the process was delayed and inefficient.

Figure 13. ABC- triage was done at the front gate of Krabi hospital from daylight till late in the night. DEF- every inch of the hospital space was filled with uncountable patients and care providers.

All local hospitals were over whelmed by an uncountable number of hospital personnel and suffering people (Figure 13). They were ill prepared to handle the sudden influx of mass injured victims. The patient load was enormous and hospital resources had been fully mobilized. Although various experiences had been documented, the quality of trauma care had rarely been addressed [37,38].

Certain points deserve emphasis : In a typical MCI, about 10-15% of the survivors need immediate access to the hospital for life and limb saving measures [34,35]. There were a greater number of non critical patients and a limited number of qualified experts at each emergency scene. A majority of the less severe casualties had deviated attention from the care workers. Despite triage being conducted at the hospital's front gates, the lack of standardized systems and networking among local hospitals for referrals greatly decreased people's chances of survival.

A number of patients had died because of fatal wounds (Figure 14), associated injuries and inadequate treatment. Some of the more seriously injured patients were improperly evaluated and left unattended during their acute hospital stay. Part of the limited human resources were devoted to handle the dead (Figure 15). A number of deaths upon arrival or deaths by neglect and errors due to lack of experience were common and went without being recorded. There were also many gaps in the management of the mass fatalities [39].

There was a limited amount of information given by care works and physicians. Patient data and medical records were incomplete. The initial diagnosis and grading of wounds were subjected to serious individual variations and as such may be unreliable. Some of this data was sent with patients who were immediately transferred to other hospitals [16]. In the early stages of the pre-hospital period, no documentation and no morbidity reports were maintained.

Certain types of injuries and wounding patterns that required immediate aggressive treatments were similar to that of war wounds [40-43]. Wound management was delayed by its pathology poor working conditions, inexperience staff and fatigue due to prolonged operating hours. Although important and difficult, there was little attention focused on the role of analgesia which allows surgical procedures to be taken during the incident. The civilian setting and approach to the wounds which consisted of minimal surgical intervention, primary closure, and relying on common antibiotics were inappropriate. Although debridement was the most common procedure performed in the early phase, this could not lower the infection rate [29-33].

Indiscriminate arrival of relief supports and supplies had constituted a secondary problem on top of the first. As hospital capacity could not be expanded, reinforcement of personnel who did not deal with trauma on a daily basis created a helpless scenario. There were too many external people and unprepared volunteers. Unnecessary supplies and service that were delivered without assessment and management had posed problems and put strains on the communities [44] (Figure 16).

Apart from the resulting deformities (Figure 17), there were significant mental health problems found in adults and children in Tsunami – affected areas [45-46]. Although psychological support to reduce traumatic distress is of great value to normalize quality of life, there was almost no such early assessment or services being conducted.

Recommendation and Conclusion

It is important to assess what already has transpired or is being done to collect all key data and information [47]. Lessons learned must be converted into disaster preparedness plans. Enhancing and exercising the well established protocol are essential. Immediate health needs, service gaps to be accurately filled and resources requirement can be possibly achieved with sufficient information particularly for the early assessment phase [44-47].

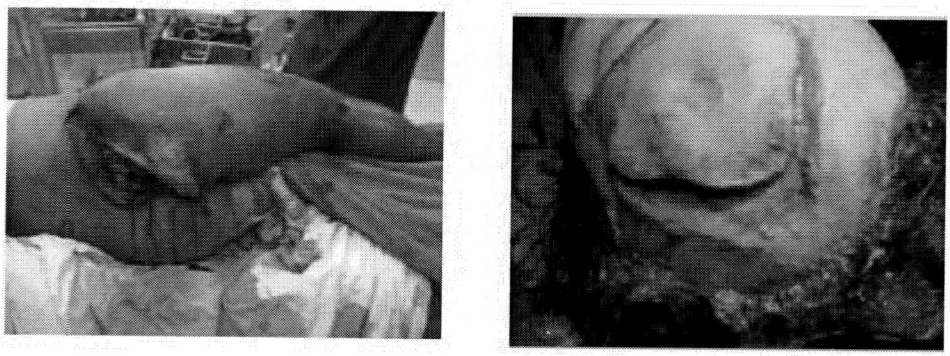

A B

Figure 14. Fatal wounds.

A B

Figure 15. A- dead bodies were carried to the nearest temple. B- processing of dead victim identification.

A

Figure 16. Continued on next page.

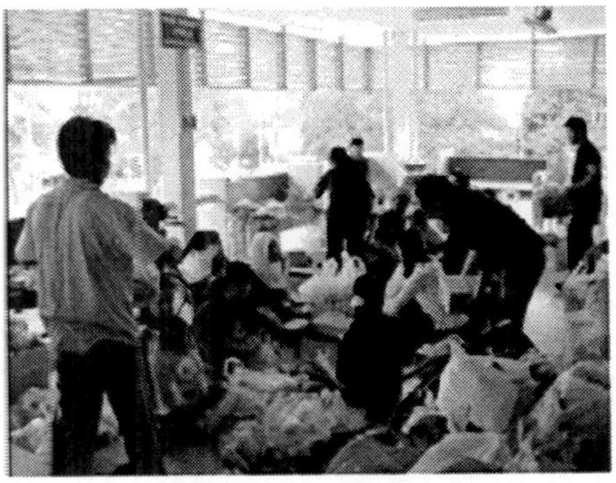

B

C

Figure 16. A - Portable shelters of local people who loss their homes. B,C - Piles of unnecessary supplies.

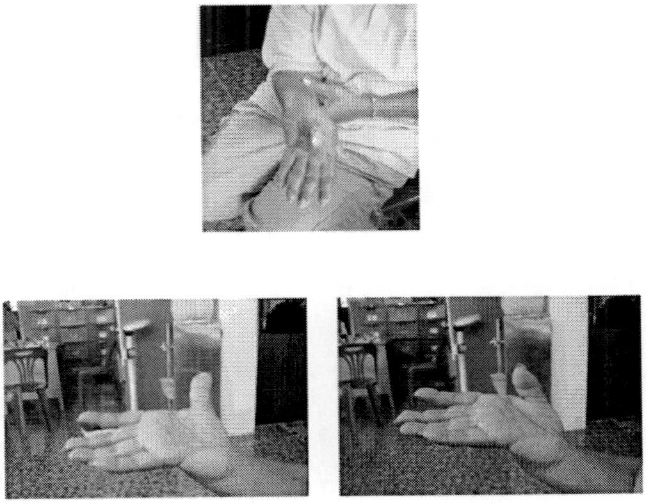

A - Thai female 32 years old sustained skin loss and tendon injuries of her right hand. She underwent more than 9 surgical procedures including skin graft replaced over her dorsum of right hand. Currently she can only move her thumb.

Figure 17. Continued on next page.

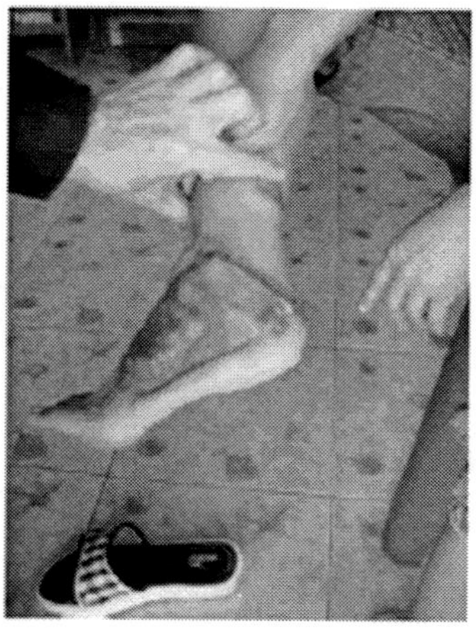

B - Thai female 42 years old sustained open wound with skin loss over her medial surface of right ankle and foot. The wound healed by multiple debridements and skin graft. She can now walk but with a limp. Her ankle cannot be fully dorsiflexed.

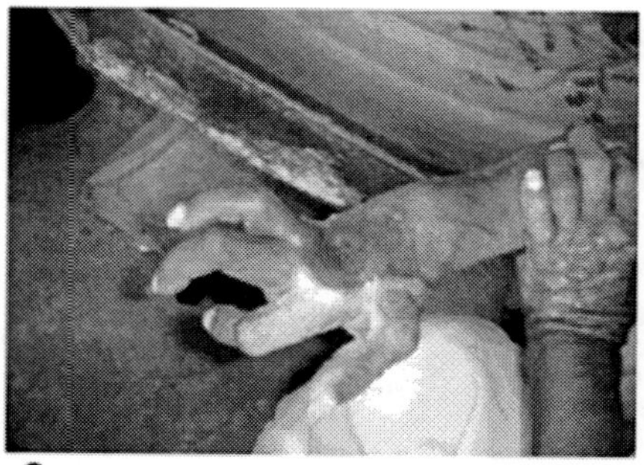

C₁

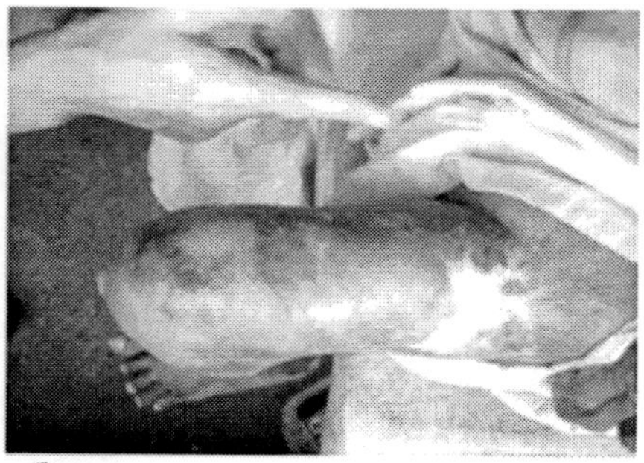

C₂

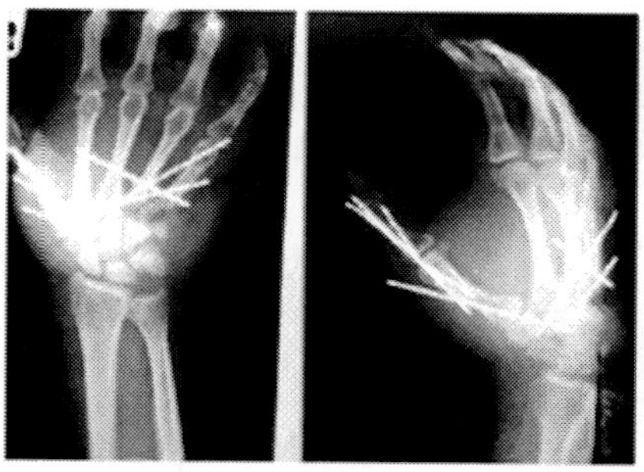

C₃

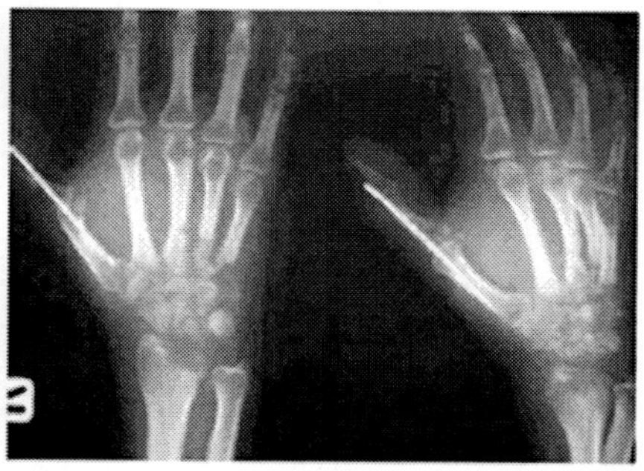

C₄

C- Thai male aged 75 years old sustained open fracture with skin loss over dorsal surface of his right hand and open wound of his left thigh. Metacarpal bones were fixed by several Kirschner wires. The wound infected and the 5th digit was amputated. Through uncountable surgical procedures, the wounds healed with contracture and malalignment of the digits.

Figure 17. Deformities following treatment.

Objective needs, assessments from psychosocial distress should also be integrated into disaster preparedness plans. Leadership with a firm knowledge to do the right things is needed in a large scale disaster like the Tsunami of 2004. In a situation with a limited resources, well trained and experienced leaders are needed to harmonize strong coordination and cooperation. An effective core group must consist of well experienced command physicians and well trained in charged nurses [34-35].

Care begins at the scene with immediate first aid and life saving measures by trained people [48-51]. Patients should be taken to the hospital with resuscitative capacities and rescue pain control. Clearing the Air way to provide breathing and ensuring circulation are the priority. Contaminated wounds should be temporarily cleaned, covered and immobilized immediately.

Initial trauma approach by military doctrine with symptomatic pain relief should be conducted [52-54]. There are various options relating to pre-hospital analgesia though the supporting evidence is limited [55]. Obtaining vital signs upon arrival including pain scale measurements are routine. Patients' demographic data and co morbidities should be quickly addressed prior to surgical initiating management. Primary wound dressing can be temporarily cared for by using tap water and then covering the wound with sterile green towel while waiting for definite surgical intervention. Some life saving procedures can be immediately done without imaging.

Initial evaluation from head to toe by direct and thorough inspection of wounds and associated injuries should immediately begin once the patient is stabilized. Comparing between limbs should be accomplished by palpation along the axes and prominence of bony marks. Unconscious and unstable patients need special attention.

Damage control concept used in battlefield surgery has been recently advocated to prevent and treat life-threatening conditions [56]. Initial surgery to control hemorrhage, resuscitation–rewarming, re-exploration and definite repair by rapid transportation through several advanced levels of medical care should be undertaken as soon as possible. The technique has been extended for extremity [57], which is characterized by bone stabilization to minimize blood loss and infection rate followed by medical stabilization and secondary definitive orthopaedic management. Blood loss and occult bleeding from any part of the body should be identified and controlled before radiographs can be taken.

Extent and contamination of wounds should be rapidly graded [58]. Referring to the well known Gustilo Andersion [24], open fracture classification has been extended for more practical uses. Severity of the damaged extremity can be better scored by grading of covering tissues of skin and fascia, bone and joint involvement as well as its underlying vital tissues such as muscles, nerve and vessels [25]. In the last several decades, the determination of outcome of an injured limb can be basically classified based on the nature and severity of the wound. Although these are not good predictors of ultimate functional outcomes following successful reconstruction, the provided objective criteria do assist surgeons during the initial state of treatment [59-60].

Broad spectrum antibiotics should be administered within minutes of wound discovery followed by therapeutic antibiotics once microorganisms have been identified [61-62]. Aggressive debridement and thorough irrigation should be performed as soon as possible and the wound should be left open [54,63-65]. Tourniquet should not be used as a routine, particularly for distal extremity injuries. Removal of the definite devitalized tissues and embedded particles remains a challenging problem. The surrounding skin should be shaved, scrubbed and rinsed by large volumes of water. Extent of grossly damaged, degloved skin, exposed ligament and non vital tissues can be carefully determined by its consistency, bleeding, color and contraction of muscle. Fasciotomies should be carried out to release pressure within the compartment. Toothbrushes can be perfectly used for cleaning the sandy wounds. In case of unavailable sterile saline, bottle water is an acceptable substitute [66]. There was no significant increase in the risk of infection when wounds were irrigated with tap water. Part of the survived flap can be replaced and stitched after removal of subcutaneous fat or taken as a skin graft. Thorough debridement and hemostasis, as well as adequate drainage should be completed before application of compressive dressings or sterile green towel.

Wounds with small penetrating entrance should be considered dangerous [16,41,49]. The smaller the opening, the more serious an injury should be anticipated. Enlargement of wound and excision of all contaminants should be accomplished as soon as possible. Patients should be taken to the ICU for continued monitoring and further resuscitation. Leading common causes of late mortality should also be closely watched.

Wounds are then further re-explored and re-debrided 24-48 hours after the initial procedures. In war wounds, duration of the exposed wound has been shown to affect the outcomes [49,54,64]. The best result still comes from proper initial management and subsequent repeated debridement to obtain a healthy wound bed. Temporary coverage by a Vacuum Assisted Closure (VAC) system, often used in US military hospitals to treat extremity war wounds is becoming popular [67-73].The technique can be effectively used in the Tsunami to prepare the wound for secondary closure. Delay in treatment whether in the ER or tertiary facilities would cause serious complications which are difficult to treat and can be considered as global economic burden. Coverage of the injured limb within 7-15 days had high saving and low infection rates. In cases of exposed wounds, a muscle flap is an ideal option for delivery of antibiotics and bone healing [64,65,74,75].

Choice of limb salvage or amputation can be better justified with a more practical grading system. In a case of traumatic amputation, planning of flap coverage should be done at the first incision. Currently, encouraging results with excellent pain relief and function are expected by recent development of advance technology, dressing materials and increasing use of prophylactic and specific antibiotics [76-77]. Various special physicians must collaborate and co-operate from the beginning. For instance, plastic and orthopedic surgeons should sufficiently engage in the debridement and reconstructive process particularly for the coverage and healing of the exposed limbs [63,74,75].

Immediate stabilization of bone fractures with very minimal stable splints to the most stable fixators is a basic principle of orthopedic care [76]. Each method may be used efficiently and effectively by a clear understanding of their benefits, pitfalls and complications. Discrepancy in the treatment modalities may be related to surgeons' preference and experience.

Theoretically, the prevention of respiratory and wound complications needs to begin immediately after primary treatment [78-79]. Missed injuries may generally be presented with early sepsis and unexplained organ failure. Secondary assessments should be routinely done at the ward and intensive care units for further hemorrhage, asphyxia, wound and pulmonary infection.

Another critical success factors of war wound management is the rapid and efficient evacuation [46,50-51]. Those with more complex wounds and significant injuries should be removed from the scene to a local hospital through advancing levels where definitive procedures or amputation can be effectively performed. It should be noted that any faster and or more efficient transportation still can not in any way save a victim who has no response to verbal stimuli.

Medico-legal threat is possible. Information is a key resource in dealing with medical malpractice claims of negligence - a conduct in rendering care that falls short of a standard and results in injury [79]. Pictures and imaging evidences of the damaged extremity should be clear enough for victim identification. All on duty physicians should protect their patients from harm. Surgeons need to take affirmative steps to develop and conduct appropriate surgical procedures. An understanding and use of current guidelines as well as medical documentation and communication with suffering people may avoid litigation.

Medical records and patient's data are often lost and incomplete [16]. Electronic medical record systems and electronic linkage from the primary site to other level of care facilities should be funded and developed. This information is essential and can be converted into strategic plans and effective actions for defining better procedures and systems.

With the 2004 tsunami, it is quite clear that the immediate MCI must be handled at the time and place of impact. Proper equipment, supplies and variety of experts in emergency management is needed to respond effectively. The current paper demonstrated that wounds had affected people and health care staffs. Numerous critical issues were detected and emphasized.

The Tsunami had taken the lives of many doctors, nurses and health care workers. It also swept away hospitals, clinics and cut critical supply lines. Safety concerns should also be paid to those who voluntarily deliver care services. A better medical response can be definitively provided with a proper or well prepared plan and training exercise. Although health workers had contributed a great deal of disaster response, a more coordinated and facilitated manner can be achieved. The lessons elaborated can be incorporated to what is needed and how we can do better in the future. The mainstay of primary closure of wounds and increase use of antibiotics in civilian practice is not appropriate for any disaster incident. Duration of the exposed wound is one of the crucial factors affecting incidence of infection and healing. Hopefully the appearing contents may stimulate physicians' interest and encourage multidisciplinary team effort to provide wound care from the beginning until its healing has been established.

Acknowledgements

The author is indebted to all the directors: Surat Thani Hospital; Phitsanu Manichot, M.D. and Deputy Director, Sirisak Janwanitsataporn, M.D.; Takuapa Hospital, Porn Pongpanitanont, M.D.; Phang-nga Hospital, Samran Tanapai, M.D.; Songklanagarind University Hospital, Associate Professor Sumet Peeravud, M.D.Dean, Faculty of Medicine ,Prince of Songkla University ; Krabi Hospital, NiphonPopatanachi, M.D.; Ranong Hospital, Pornlert Jitpratoom, M.D.; Trang Hospital, Jaruat Jampa, M.D. for their kind support of all data and information. Mr. Chainarong Samingchairoj, Mr. Prapat Tantisalidchai for their assistance with the English language edition. Miss Prapa Geennikul for providing references. No benefits in any form have been received.

References

[1] Runyon J. The history of tsunamis [online]. 2006 [cite 2006 November 20]. Available from: URL http://www.edhelper.com/Reading Comprehension_38_12.html.

[2] Management Training Programme (DMTP). Introduction to hazards[online]. 1997. [cited 2006 November 20]. Available from: URL http://www.undmtp.org/english/hazards/ 22 hazards.pdf.

[3] Wikipedia. 2004 Indian Ocean earthquake [online]. 2006 [cited 2006 November 21]. Available from: URL: http://en.wikipedia.org/wiki/2004_Indian_Ocean_earthquake.

[4] Department of Disaster Prevention and Mitigation. The earthquake and tsunami situation in the western coast of Thailand [online]. 2006 [cited 2006 November 21]. Available from: URL: http://www.thaiembdc.org/whatsnew /tidalwaves2004 /index.html.

[5] World Health Organization. Tsunami and health: situation report No.31. January 29, 2005[online]. [cited 2006 November 21]. Available from: URL: http://www.who.int/hac/crises/international/asia tsunami/sitrep/31/en/index.html

[6] Kunaratanapruk, S. and Somboonsook, B. (2006). Editor. *A record on the public health experiences of the Tsunami disaster.* Bangkok: Thunwa Printing.

[7] Ministry of Public Health of Thailand. (2005). *Thailand health aspects and the management of the TSUNAMI disaster.* Bangkok, Ministry of Public Health of Thailand.

[8] Ministry of Interior. Tsunami disaster situation September 5, 2005 [online]. 2006 [cited 2006 November 16]. Available from: URL: http://61.19.54.131/tsunami/overall.php?pack=report_10.

[9] Wattanawaitunechai, C. and. Peacock, S.J. (2005). Tsunami in Thailand—disaster management in a district hospital. *N Engl J Med,* **352**(10), 962–4.

[10] Watcharong, C. and Chuckpaiwong ,B. (2005). Orthopaedic trauma following tsunami: experience from Phang Nga, Thailand. *J Orthop Surg (Hong Kong)*, **13**(1), 1–2.

[11] Kateruttanakul, P. and Paovilai, W. (2005). Respiratory complication of tsunami victims in Phuket and Phang-Nga. *J Med Assoc Thai,* **88**(6),754–8.

[12] Ammartyothin, S. and Ashkenasi, I. (2006). Medical response of a physician and two nurses to the mass-casualty event resulting in the Phi Phi Islands from the tsunami. *Prehospital Disaster Med,* **21**(3),212–4.

[13] Johnson, L.J. and Travis, A.R. (2006). Trauma response to the Asian tsunami: Krabi Hospital, southern Thailand. *Emerg Med Australas,* **18**(2),196–8.

[14] Johnson, L.J. and Travis, A.R. (2006). Trimodal death and the injuries of survivors in Krabi Province, Thailand, post-tsunami. *ANZ J Surg,* **76**(5),288–9.

[15] Yorsaengrat ,W. and Chungpaibulpatana, J. (2006). Respiratory complication of tsunami disaster victims in Vachira Phuket Hospital. *J Med Assoc Thai,* **89**(4),518–21.

[16] Prasartritha, T. and Tungsiripat, R. (2008). The revisit of 2004 tsunami in Thailand: characteristics of wounds. *Int Wound J,* **5**(1),8-19.

[17] Limchawalit, K. and Suchato, C. (2005). Images in clinical medicine. Tsunami sinusitis. *N Engl J Med,* **352**(25),e23.

[18] Lord, S. and Davis, P. (2005). Drowning near drowning and immersion syndrome. J R *Army Med Corps,* **151**(4),250-5.

[19] Cotes, L. and Hargarten, S. (2006). Hennes H. Recommendations for water safety and drowning prevention for travelers. *J Travel Med,***13**(1),21-34.

[20] Kongsaengdao, S. and Bunnag, S. (2005). Treatment of survivors after the tsunami. *N Engl J Med,***352**(25),2654-5.

[21] Kao, A.Y. and Munandar, R. (2005). Case records of the Massachusetts General Hospital. Case 19-2005. A 17-year-old girl with respiratory distress and hemiparesis after surviving a tsunami. *N Engl J Med,***352**(25),2628-36.

[22] Kudsk, K.A. and Sheldon, G.F. (1981). Degloving injuries of the extremities and torso. *J Trauma,***21**(10),835-9.

[23] Williams, N.A. and Leaper, D.J. (1998). Infection. In Leaper, D.J. and Harding. K.G.(Eds.), 24 *Wounds Biology and Management* (71-87). Oxford [U.K.], New York : Oxford University Press.

[24] Gustilo, R.B. and Mendoza, R.M. (1984) Williams DN. Problems in the management of type III (severe) open fractures: a new classification of type III open fractures. *J Trauma,***24**(8),742-6.

[25] Rajasekaran, S. and Naresh Babu J. (2006). A score for predicting salvage and outcome in Gustilo type-IIIA and type-IIIB open tibial fractures. *J Bone Joint Surg Br,***88**(10), 1351-60.

[26] Carter, P.S. and Banwell, P.E. (2004). Necrotising fasciitis: a new management algorithm based on Clinical classification. *Int Wound J,***1**(3),189–98.

[27] Kihiczak, G.G. and Schwartz, R.A. (2006). Necrotizing fasciitis: a deadly infection. *J Eur Acad Dermatol Venereol,***20**,365–9.

[28] Cainzos, M. and Gonzalez-Rodriguez, F.J. (2007). Necrotizing soft tissue infections. *Curr Opin Crit Care,***13**(3),433-9.

[29] Hiransuthikul, N. and Tantisiriwat, W. (2005). Skin and soft-tissue infections among tsunami survivors in southern Thailand. *Clin Infect Dis,***41**(10),e93–6.

[30] Maegele, M. and Gregor, S. (2006). One year ago not business as usual: wound management, infection and psychoemotional control during tertiary medical care following the 2004 Tsunami disaster in southeast Asia. *Crit Care,***109**(2),R50.

[31] Lim, P.L. Wound infections in tsunami survivors: a commentary. *Ann Acad Med Singapore,* **34**(9),582–5.

[32] Nieminen, T. and Vaara M. (2005). Burkholderia pseudomallei infections in Finnish tourists injured by the December 2004 tsunami in Thailand. *Euro Surveill* ,**10**(3), E050303.4.

[33] Uckay, I. and Sax, H. (2008). Multi-resistant infections in repatriated patients after natural disasters : lessons learned from the 2004 tsunami for hospital infection control. *J Hosp* 25 *Infect,***68**(1),1-8

[34] Hirshberg, A. and Holcomb, J.B. (2001). Hospital trauma care in multiple-casualty incidents: a critical view. *Ann Emerg Med,***37**(6),647–52.

[35] Borra, A. and Perez, L.J. (2005). Panel 2.5: mass-casualty management and hospital care. *Prehospital Disaster Med,***20**(6),412–3.

[36] Charuluxananan, S. and Bunburaphong, P. (2006). Anesthesia for Indian Ocean tsunami- affected patients at a southern Thailand provincial hospital. *Acta Anaesthesiol Scand*, **50**(3),320–3.

[37] Kapila, M. and McGarry, N. (2005). Health aspects of the Tsunami disaster in Asia. *Prehosp Disaster Med,***20**(6),368-77.

[38] Wahlstrom, M. (2005). Overview of the Tsunami disaster. *Prehosp Disaster Med,* **20**(6),378-81.

[39] Kunaratanapruk, S. and Phupat, T. (2006). Editors. *Lesson learned from dead bodies management in Tsunami disaster.* Bangkok: Thunwa Printing,

[40] Geiger, S. and McCormick, F. (2008). War wounds: lessons learned from Operation Iraqi Freedom. *Plast Reconstr Surg,* **122**(1),146-53.

[41] Rautio, J. and Paavolainen, P. (1988). Afghan war wounded : experience with 200 cases. J Trauma,**28**(4),523-5.

[42] Mazurek, M.T. and Ficke, J.R. (2006). The scope of wounds encountered in casualties from the global war on terrorism: from the battlefield to the tertiary treatment facility. *J Am Acad Orthop Surg,* **14**(10 Spec No.),S18-23.

[43] Cooper, G.J. and Ryan, J.M. (1990). Interaction of penetrating missiles with tissues: some common misapprehensions and implications for wound management. *Br J Surg,* **77**(6), 606-10.

[44] Smith, J. and Fink, S. (2005). Session 1.4: health services delivery: a critical review of 26 experience. *Prehosp Disaster Med,* **20**(6),389-92.

[45] Van Griensven F. and Chakkraband, M.L. (2006). Mental health problems among adults in tsunami-affected areas in southern Thailand. *JAMA,* **296**(5),537–48.

[46] Thienkrua, W. and Cardozo, B.L. (2006). Symptoms of posttraumatic stress disorder and depression among children in tsunami-affected areas in southern Thailand. *JAMA,* **296**(5),549–59.

[47] Aldis, W. and Rockenschaub, G. (2005). Panel 2.1: assessing impact and needs. *Prehospital Disaster Med,* **20**(6),396–8.

[48] Bagg, M.R. and Covey, D.C. (2006). Levels of medical care in the global war on terrorism. *J Am Acad Orthop Surg,* **14**(10 Spec No.),S7-9.

[49] Hamdan, T.A. (2006). Missile injuries of the limbs: an Iraqi perspective. *J Am Acad Orthop Surg,* **14**(10 Spec No.),S32-6.

[50] Willett, K.M. and Dorrell, H. (1990). ABC of major trauma. Management of limb injuries. *BMJ,* **301**(6745), 229–33.

[51] Cayten, C.G. (1996). Prehospital Management, Triage, and Transportation. In Ivatury, R.R. and Cayten, C.G. Editors. *The Text book of Penetrating Trauma.* (153-69). Baltimore : Williams and Wilkins.

[52] Covey, D.C. (2006). Combat orthopaedics: a view from the trenches. *J Am Acad Orthop Surg,* **14**(10 Suppl),S10–7.

[53] Tenuta, J.J. (2006). From the Battlefields to the States: the Road to Recovery. The Role of Landstuhl Regional Medical Center in US Military Casualty Care. *J Am Acad Orthop Surg,* **14**(10Suppl),S45–7.

[54] Pollak, A.N. (2006). Timing of debridement of open fractures. *J Am Acad Orthop Surg,* **14**(10 Suppl),S48–51.

[55] Borland, M.L. and Jacobs, I. (2002). Options in prehospital analgesia. *Emerg Med* 27 *(Fremantle),* **14**(1),77-84.

[56] Shapiro, M.B. and Jenkins, D.H. (2000). Damage control: collective review. *J Trauma,* **49**(5),969-78.

[57] Hildebrand, F. and Giannoudis, P. (2004). Damage control: extremities. *Injury,* **35**(7), 678–89.

[58] Coupland, R.M. (1992). The Red Cross classification of war wounds: the E.X.C.F.V.M. scoring system. *World J Surg,*16(5),910–7.

[59] Ly, T.V. and Travison, T.G. (2008). Ability of lower-extremity injury severity scores to predict functional outcome after limb salvage. *J Bone Joint Surg Am,***90**(8),1738-43.

[60] O'Toole, R.V. and Castillo, R.C. (2008). Determinants of patient satisfaction after severe lower-extremity injuries. J Bone Joint Surg Am,**90**(6),1206-11.

[61] Jones, M.E. and Karlowsky, J.A. (2003) Epidemiology and antibiotic susceptibility of bacteria causing skin and soft tissue infections in the USA and Europe: a guide to appropriate antimicrobial therapy. *Int J Antimicrob Agents,***22**(4),406-19.

[62] Holtom, P.D. (2006). *Antibiotic prophylaxis: current recommendations. J Am Acad Orthop Surg,***14**(10 Suppl),S98-100.

[63] Murray, D.S. (1990) Wordsley Hospital, Stourbridge, West Midlands, UK. Skin loss of the lower limb. *Injury,***21**(5),309-10.

[64] Haury, B. and Rodeheaver, G. (1978). Debridement: an essential component of traumatic wound care. *Am J Surg,***135**(2),238-42.

[65] Andersen, R.C. and Frisch, H.M. (2006). Definitive treatment of combat casualties at military medical centers. *J Am Acad Orthop Surg,***14**(10 Spec No.),S24-31.

[66] Whaley, S. (2004). Tap water or normal saline for cleansing traumatic wounds? *Br J Community Nurs,* **9**(11),471–8.

[67] Herscovici, D. and Sanders, R.W. (2003). Vacuum-assisted wound closure (VAC therapy) for the management of patients with high-energy soft tissue injuries. *J Orthop Trauma,*28 **17**(10):683-8.

[68] Preston, G. (2008). An overview of topical negative pressure therapy in wound care. *Nurs Stand,***23**(7),62-4, 66, 68.

[69] Expert Working Group. (2008). Vacuum assisted closure: recommendations for use. A consensus document. *Int Wound J,***5** Suppl 4,iii-19.

[70] Vikatmaa, P. and Juutilainen, V. (2008). Negative pressure wound therapy: a systematic review on effectiveness and safety. *Eur J Vasc Endovasc Surg,***36**(4),438-48.

[71] Ubbink, D.T. and Westerbos, S.J. (2008). A systematic review of topical negative pressure therapy for acute and chronic wounds. *Br J Surg,***95**(6),685-92.

[72] Murray, C.K. and Hsu, J.R. (2008). Prevention and management of infections associated with combat-related extremity injuries. *J Trauma,***64**(3 Suppl),S239-51.

[73] Gregor, S. and Maegele, M. (2008). Negative pressure wound therapy: a vacuum of evidence?. *Arch Surg,***143**(2),189-96.

[74] Levin, L.S. (2008). Principles of definitive soft tissue coverage with flaps. J Orthop Trauma,**22**(10 Suppl),S161-6.

[75] Yazar, S. and Lin, C.H. (2004). One-stage reconstruction of composite bone and soft tissue defects in traumatic lower extremities. *Plast Reconstr Surg,***114**(6),1457-66.

[76] Mazurek, M.T. and Burgess, A.R. (2006). Moderator's summary : stabilization of long bones. *J Am Acad Orthop Surg,***14**(10 Suppl),S113-117.

[77] Puzas, J.E. and Bukata, S.V. (2006). Accelerated fracture healing. *J Am Acad Orthop Surg,* **14**(10 Suppl),S145-151.

[78] Kauvar, D.S. and Lefering, R. (2006). Impact of hemorrhage on trauma outcome: an overview of epidemiology, clinical presentations, and therapeutic considerations. *J Trauma,***60**(6 Suppl),S3-11.

[79] Berlly, M. and Shem, K. (2007). Respiratory management during the first five days after spinal cord injury. *J Spinal Cord Med,* **30**(4),309-18.

[80] Bal BS. (2009). An introduction to medical malpractice in the United States. *Clin Orthop Relat Res,* **467**(2),339-47.

In: Tsunamis: Causes, Characteristics, Warnings and Protection ISBN: 978-1-60876-360-3
Editors: N. Veitch and G. Jaffray, pp. 87-111

Chapter 4

APPLICATION OF COASTAL FOREST IN TSUNAMI DISASTER MITIGATION

***Rabindra Osti*[a] *and Dinar Istiyanto*[b]**
International Centre for Water Hazard and Risk Management – ICHARM
Minamihara 1-6, 305-8516 Tsukuba, Ibaraki, Japan

Abstract

Tsunamis and storm surges have killed more than one million people, and some three billion people live with a high risk of these disasters that are becoming more frequent and devastating worldwide. Effective mitigation of such disaster is possible via healthy coastal forests, which can reduce the energy of the wave. Many independent studies are conducted to evaluate the control functions of coastal forest against tsunami and several recommendations are proposed to improve the mitigative effects of such natural shield. However, in many instances the ideas diverse in their view points. In order to synthesise the research findings and to propose a best practicable and optimal solution, a comparative study was conducted among available results and recommendations. Especial focus was given to the Tsunami 2004 in Indian Ocean. This paper reviews previously produced numerical and experimental results and compares them with field observation. The relationship between the degree of damage reduction and associated parameters especially width, size and density of coastal vegetation are discussed.

Keywords: Tsunami, coastal forest, disaster mitigation, implementation, sustainability

1. Introduction

The role of Mangrove forest is immense in coastal community development and maintaining coastal environment. Wide, elongated, dense and mature mangrove forests growing along the coasts can help to reduce the devastating impact of tsunamis and coastal storms by reducing their wave energies. Additionally, Mangrove forests provide variety of services to the coastal ecology and to the societies. These services include coastal erosion

[a] E-mail address: rabiosti@gmail.com, osti55@pwri.go.jp.
[b] E-mail address: pakdinar@yahoo.com, dinar55@pwri.go.jp. Tel: +81-29-879-6809, Fax: +81-29-879-6709.

prevention, protection of coral reefs from siltation, pollutant control, timber production, production of food and traditional medicines, and shelters for some indigenous people as well as for variety of flora and fauna.

Tsunami 2004 in Indian Ocean, which was responsible for the loss of about 225,000 lives and million worth of property, was one of the worst events in the history of natural disasters worldwide. Although the magnitude of tsunami waves was high all along the affected coasts, human losses and the amount of damages to inland property and built infrastructures were less in places where healthy Mangrove or coastal forests existed such as in Andaman and Nicobar Islands and some parts of Tamilnadu state in India (Selvam, 2005). Many other places, where the coastal forest had been significantly degraded or converted into other land uses, suffered high losses. The importance of Mangrove forests to mitigating tsunami disaster has always been highlighted, and now many scientific studies are carried out to understand the characteristics of coastal forest against tsunami disasters from different prospects.

The control function of coastal forest against tsunami 2004 in Indian Ocean is widely discussed in many literatures *e.g.* Osti et al., 2008; Selvam, 2005; Dahdouh-Guebas *et al.*, 2005; IUCN, 2005; Upadhyay *et al.*, 2002; Padma, 2004; Williams, 2005, etc. A role of Mangrove forest in mitigating tsunami disasters, especially in the case of tsunami 2004, was studied by Selvam (2005) based on satellite and field data. Similarly, reduction of tsunami flow pressure by increasing the density and width of planted zone was simulated, numerically or by laboratory experiments, by several researchers [*e.g.* Harada and Imamura (2000), Hiraishi and Harada (2003), Harada and Kawata (2004), Dinar *et al.* (2006)], which show the importance of these two factors in tsunami flow reduction by coastal forest. Kathiresan and Rajendran (2005) used linear regressions to identify the value added by the Mangrove forest to reduce the per-capita mortality due to Tsunami 2004 in India. However, weakness on their regression analysis was pointed out by Kerr *et al.* (2006), suggesting more reliable studies to verify the findings. There are a few more studies (*e.g.* Mazda *et al.* 2006; Wolanski 2007), which suggest that the wave attenuation function of certain type of Mangroves is highest in certain conditions and not necessarily for any magnitude of waves. Results of these studies have emphasized the need of further studies to search for maximizing Mangrove functions against tsunami and storm surge of different physical characteristics. Although these individual research help to understand the basic functions of Mangrove forest, an applicability of these results is yet to be proven in a variety of circumstances. While analysing the control functions, the most important thing is to identify the correlation among different parameters *e.g.* i) physical characteristics of tsunami wave, ii) local topography, iii) characteristics of Mangrove forest and, iv) built-in environment. The physical components of Mangrove forest include a) normal height, shape and size of trees, b) inner plants, c) stiffness of individual tree, d) width and length of the forest, e) seasonal variations, f) locations and orientation of trees, etc. In addition, tsunami and Mangrove related research should highlight the environmental, societal and economical concerns of the region. Community forest and community based disaster management approaches are gaining high recognition in developing countries (Osti, 2004) and such a know-how can be used to develop coastal forests for the welfare of coastal communities and for the sustainable development of the region. A multidisciplinary approach is required to address the current gaps. This paper highlights the importance of Mangrove forest against tsunami disasters, and fosters the way forward for its improvement.

2. Tsunami and Coastal Forest: Problems and Prospects

Tsunami is one of the deadliest natural hazards and its mitigation is much challenging because of its complex development process and physical characteristics. Tsunami is usually triggered by different natural processes such as earthquake, sub-marine landslides, sub-arial landslides and hydrometeorological conditions. Record shows that more than one million people were already killed by tsunami in many parts of the world, placing East Pacific at the top of highly affected regions (Table 1).

Table 1. Historical fatal tsunami events

Year/Tsunami Name/Locations	Deaths	Year/Tsunami Name/Locations	Deaths
2009 Samoa Islands	100-200	1933 Sanriku, Japan	3,008
2007 Solomon Islands	52	1929 Newfoundland, Canada(SML)	29
2006 Java, all over Indonesia	800	1923 Sagami bay, Kanto, Japan	2,144
2004 Indian Ocean tsunami, Asia	~225,000	1908 Messina, Italy	~100,000
1999 Thatta and Badin, Pakistan	400	1906 Ecuador and Colombia	1,500
1998 Papua New Guinea	2,200	1899 Bada Sea, Indonesia	3,620
1997 Different locations	400	1896 Sanriku, Japan	26,360
1996 Minahassa Penisula	24	1888 Ritter Island in (LS)	~3,000
1996 Biak, Irian Java	161	1883 Krakatoa, Indonesia (Vol)	36,500
1996 North coast of Peru	12	1868 Arica, Northern Chile	25,674
1995 Jalisco, Mexico	1	1868 Hawaii	81
1994 Eastern Java, Indonesia	223	1854 Nankaido, Japan	3,000
1994 Mindoro Island	70	1952 Paramushir Island (former Soviet Union)	3,000
1993 Okushiri Island, Japan	200	1792 Mt Unzen, Kyushu, Japan	9,745
1993 Different locations	59	1782 South China (China Taiwan)	~40,000
1992 Flores Island and Babi Island	1953	1771, Ryukyu Trench, Japan	13,486
1992 Nicaragua	170	1755, Lisbon, Portugal	~60,000
1983 Western part of Japan	104	1746 Lima, Peru	3,800
1979 Irian and Lomblem Indonesia	639	1707 Tokaido-Nankaido, Japan	4,900
1979 Nice, France	23	1703 Tokai-Kashima, Japan	5,233
1979 San Juan Island, Colombia	250	1703 Awa, Japan	~100,000
1976 Cotabato city, Philippines	8,000	1692 Jamaica	3,000
1975 Hawaiian tsunami, Hawai	2	1674 Banda Sea, Indonesia	2,243
1964 Prince William Sound, USA	130	1611 Sanriku Japan	5,000
1960 Chilean tsunami, many parts	2,290	1606 Bristol Channel, UK (ME)	2,000
1960 Hilo Hawaii	61	1605 Nankaido, Japan	5,000
1958 Lituya Bay, Alaska, USA	3	1570 Chile	2,000
1946 Nankaido, Japan	1,997	497 BC Potidaea, Greece	Many
1946 Aleutian tsunami, Hilo Hawaii	165	1410 BC, Santorini, Greece	~100,000
1945 Arabian Sea, Makran Coast	~4,000	5000 BC North Atlantic (SML)	Many
		Total Deaths > 1,000,000	

Vol-Volcano, ME-Meteorological Extremities, SML-Sub-marine Landslide, LS-Landslide
(**source**: different literatures)

The trend of destruction in human lives, environment and built environment due to tsunami had increased significantly during the period 1700 to 1930 (Figure 1). In this period, countries like Japan, Italy and Portugal were severely affected (Table 1). Thanks to the strong economic growth and the development of advance technologies, developed countries have considerably improved the counter systems against tsunami in their territories.

However developing countries are still lacking of resources to cope with problems and have been facing extreme threats of tsunami and storm surge disasters. Damage due to Tsunami 2004 is an active example. In addition to unpredictable natural events, the tendency of destruction is further exacerbated by other human factors such as rapid increase in population density (Figure 1), urbanisation, industrialisation and coastal deforestation. Additionally, disaster awareness among coastal residents is vital because less aware people who can not properly prepare for the events are the forefront victims of the disasters. Level of disaster awareness is significantly low in developing countries.

There are several countermeasures and especially structural measures are being widely adopted in both developed and developing countries. This has reduced the aesthetic values of coastal areas, imposed adverse environmental impact, and inhibited people and nature from interacting with sea. This development yielded some benefits to the community but has also replaced the traditional assets, therefore bringing much complexity. It is recognised that the nature itself makes it possible to mitigate the impact of tsunami by growing coastal forests, balancing the coastal environment in a proportional manner. Therefore, development and preservation of coastal forest should be considered as an eco-friendly solution against tsunami and storm surge disasters.

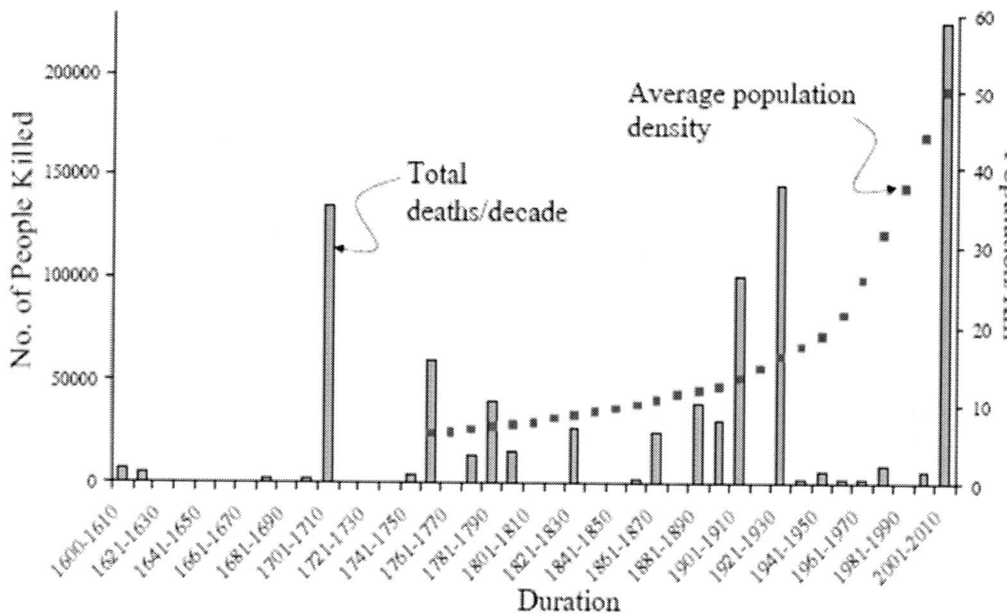

Figure 1. An increasing trend of global population density and number of people killed by tsunamis in recent decades.

There are less than 15 million hectares of Mangrove forests (Philippe *et al.*, 2005) over 60,000 square miles in the warm water of tropical oceans all over the world. According to

FAO's recent survey report (Wilkie and Fortuna, 2003), about 41% of world's Mangrove forest lies in Indonesia, Brazil, Nigeria and Australia, and 60% of total Mangrove area is found in just ten countries. Despite the fact that Mangrove forests offer multifaceted benefits to the coastal communities and environment, significant amount of Mangrove forests have been destroyed over the last few decades. In the last 20 years, the world lost almost 50% of its Mangrove forests, making them one of the most endangered landscapes (MIC, 2006). Figure 2 shows the rate of Mangrove forest destruction in each continent. The coastal forest has significantly depleted in Asia, North and South America (Wilkie and Fortuna, 2003). However in Africa, the second largest continent for coastal forest resources, the rate of forest destruction is considerably low. The primary causes of coastal forest destruction in all continents are climate variation, urbanisation, industrialisation, aquaculture development, mining, tourism, coastal area protection works and agriculture, which include but are not limited to: a) crude oil and other pollutants supply/accumulation b) prolonged flooding from artificial dikes or causeways, c) charcoal and timber industries, d) development of hotel, infrastructures, recreational facilities, e) land encroachment, e) excavation and uprooting for mining, f) rapidly expanding shrimp aquaculture industry, g) environmental stresses including remarkably high salinity and, h) overexploitation of forest for firewood and building materials.

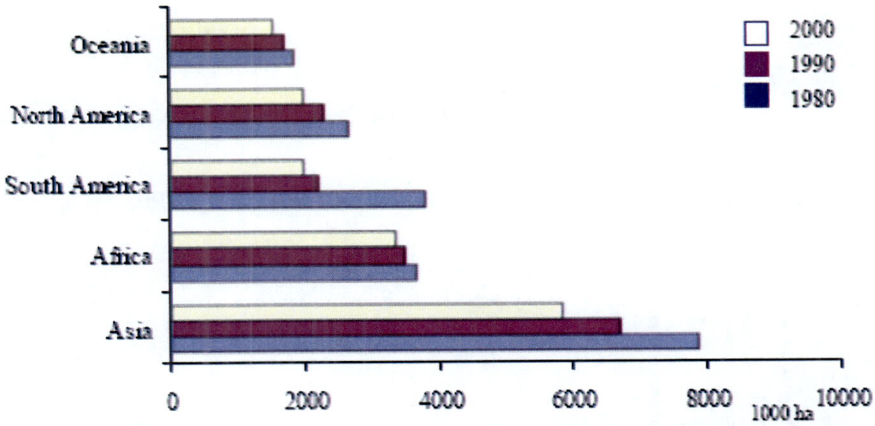

Trend of Destruction (%)	Asia	Africa	South America	N & C America	Oceania
1980-1990	-14.9	-5.2	-42.1	-13.1	-8.0
1990-2000	-12.8		-3.4	-10.4	-14.3 -
1980-2000	-25.8	-8.4	-48.1		-25.5 -

Figure 2. Global trend of Mangrove forest destruction (analysed by authors).

These changes appear along with the increasing trend of coastal population (Fig. 3) and associated coastal area development. Fifty percent of the world's population currently live within sixty kilometers of the coast, which at present is more than 3 billion people (WB, 2007). United Nations Environmental Programme's technical report (UNEP, 2007) says that an average population density in the coastal zones was 77 people/Km^2 in 1990 and this value increased to 87 people/Km^2 in 2000. It has been predicted that in 2010, 2025 and 2050, population density will increase to 99 people/Km^2, 115 people/Km^2 and 134 people/Km^2 respectively. On the other hand, although global population increase is not directly related to coastal forest destruction, it is indirectly responsible for consuming forest products and being a source of migration. The destruction of coastal forest and the simultaneous increment in coastal population have exposed the communities to high tsunami and storm surge risks and led to adverse environmental impact. It is also a fact that the devastating tsunami and storm surges are among the other causes of massive Mangrove forest destruction.

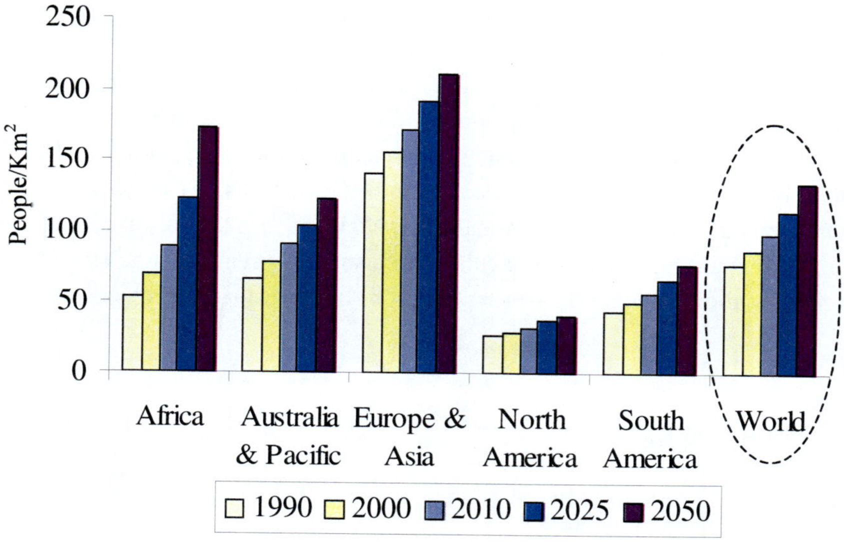

Figure 3. Increase in coastal population density by continent (UNEP, 2007).

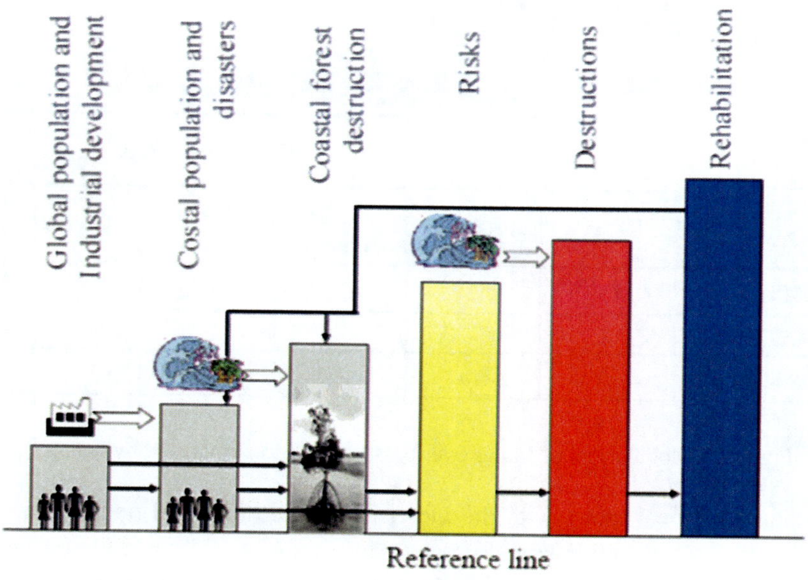

Figure 4. Deforestation and co-related risk rising process in tsunami disasters.

Moreover there is a further threat of deforestation in the name of reconstruction (Kazmin, 2005; MAP, 2005; Barbier, 2006) and there is always a chance of illegal felling of trees and encroachment of forest land. These consequences have made rehabilitated or existing coastal population more vulnerable to potential tsunami disasters. The dynamic cycle of forest destruction and associated tsunami risk raising process is illustrated in Figure 4 in the form of driver, pressure, state, impact and response. The breakthrough of the vicious loop illustrated in Figure 4 is only possible by prescribing appropriate strategies and their implementation to preserve and develop coastal forest.

3. The General Role of Coastal Forest in the Reduction of Tsunami Disaster and Important Factors in Their Interaction

In general, the role of coastal forest in the reduction of tsunami disaster includes the following functions: (Shuto, 1987; Tanaka *et al.*, 2007)

1) Trapping effect, *i.e.* stop driftwood (fallen trees, etc.), debris (destroyed houses, etc.) and other floatages (*e.g.* boat) ,
2) Energy dissipation effect, *i.e.* reduces water flow velocity, flow pressures and inundation water depth,
3) Soft-landing effect, *i.e.* provide a life-saving means by catching persons carried-off by tsunamis and enable them to land on tree branches,
4) Escaping effect, *i.e.* provide “a way” of escape by climbing trees,
5) Collects wind-blown sands and raises dunes, which act as natural barrier against tsunami.

Interaction between tsunami and coastal forest are influenced by tsunami inland flow characteristics and forest characteristics. Factors involved in the tsunami and coastal forest interaction are summarized in Table 2.

The ultimate impact of coastal forest on tsunami risk reduction can be checked in reference to the observed tsunami inundation depth in the inland area. Therefore tsunami inundation depth, which can be used to calculate tsunami in-land velocity and forces, is very important parameter for planning and designing tsunami defense infrastructures. In order to understand the tsunami-forest interaction, both forest and tsunami characteristic should be evaluated. For an example, the effect of tsunami wave period or tsunami wave length on the effectiveness of tsunami flow reduction by coastal forest has been described by using numerical simulation (Harada and Kawata, 2004). It indicates that the flow reduction effect is decreased along with the increase of wave period. However, since tsunami is a very long wave comparing to the common width of coastal forest, effects of length of tsunami wave on the reduction is generally negligible. Direction of tsunami wave propagation in reference to the alignment of costal forest is another factor to be accounted while analyzing the effect of coastal forest on tsunami reduction. Higher degree of reduction can be expected when the tsunami wave approach in a perpendicular direction to the thickest part of forest width.

On the other hand, the forest characteristics depend on a single tree capacity as well as the capacity of a unit forest as written in Table 2. The capacity of a single tree to stand against tsunami forces determines the survivability of the tree. This capacity is usually represented by its breaking limit under certain tsunami flow depth. In addition to breaking limit or stiffness of the tree, the ultimate reduction of tsunami force depends on other characteristics of the tree such as canopy, height, etc. Besides the capacity of individual or group of trees, ground soil should be strong enough to withhold the trees during active tsunami event. Although the characteristics of individual tree are important, the total effectiveness mostly depends on the capacity of a unit forest and more importantly on forest density, height and width, local topography, ground cover, etc.

The factors, especially the number of trees, trunk diameter, vertical structure composition (portion of roots, trunk and canopy) and horizontal arrangement, are the main variables that influence forest density level. Like in any other kinds of water waves, tsunamis can be deformed due to shoaling or refraction during its propagation to the coastal area and eventually deceased on shore when the above mentioned conditions are satisfied by coastal forest (Yeh, 1991).

At present, in the modeling of tsunami and coastal forest interaction, drag force is consider as the main hydraulic force of tsunami that working onto the tree bodies (Hiraishi and Harada, 2003; Harada and Kawata, 2004; Tanaka *et al.*, 2007, 2008; Yanagisawa *et al.*, 2008). The inertial force due to the velocity acceleration is much smaller than the drag force, thus inertial force is usually neglected (Harada and Kawata, 2004, Yanagisawa *et al.*, 2008) in modeling practices.

Effects of coastal forest on tsunami impact reduction are basically due to the reduction in inundation depth, flow velocity and applying forces. Reduction of these parameters means reduction in inundation area and the impact. Reduction of flow velocity also means to provide considerable span of time for evacuation before tsunami reaches the residential area.

Table 2. Factors involved in the tsunami and coastal forest interaction

Tsunami Characteristics	Forest Characteristics	Expected Effect
Inundation depth -Wave force Period/wave length Wave direction Total volume	Single tree capacity: - Type of vegetation (Height, canopy size, trunk diameter) - Breaking moment capacity	Flow velocity reduction Inundation depth reduction Impact force reduction
	Ground slope, soil type, ground cover etc.	
	Unit forest: - Forest width - Tree arrangement - Forest density -- Trees number -- Trunk diameter -- Vertical structure Tree height Root-trunk-canopy composition	

4. Survival Capacity of a Coastal Forest against Tsunami

Coastal forest works against tsunami force as long as it acts in a unit system. However the effective resistance decreases along with the decrease in the number of survived trees (Shuto, 1987; Dinar *et al.*, 2006; Tanaka *et al.*, 2007, 2008). The damage patterns of the trees due to tsunami forces can be classified into (1) cut-off or washed away *i.e.* only the roots and bottom part of trunk remain, (2) uprooted and washed away *i.e.* no parts of the tree is remained, (3) cut-off or uprooted but not washed away; the collapse parts of tree is still at the place, (4) tilted or inclined (Shuto, 1987; Tanaka *et al.*, 2007). The survived trees are only those, which can stand stiffly against tsunami. Once the tree has damaged, it has no sufficient capacity to reduce tsunami forces because destroyed trees can hardly provide resistance against tsunami flow that has high and long wave period (Yanagisawa *et al.*, 2008). Trunk size and the strength of ground soil against erosion or scouring due to wave action are considered as the main factors for the survival of the trees. There are few studies conducted on the surveillance of the coastal forest, which are summarized below.

Minimum Trunk Diameter

A rough criterion for quick judgment of survivability of coastal forest against tsunami is provided by Shuto (1987). Figure 5 by Shuto (1987) shows damage situation of pine trees forest in terms of trunk diameter and tsunami height above ground surface. The minimum trunk diameter to stand against tsunami with no expectation of damage should be approximated by (Curve II in Figure5)

$$d = 0.37H^3 \tag{1}$$

Only when soils at the fringe or at sparse place of a forest are severely scoured due to the concentrated water flow, trees may be tilted or turned over. Whereas, the smallest diameter of a tree able to stop floatages (without necessarily reduce the tsunami energy) should be determined from the following relation (Curve I in Figure-5)

$$\left.\begin{array}{ll} H = 4.65 & \text{for } d<10 \\ d = 0.1H^3 & \text{for } d>10 \end{array}\right\} \tag{2}$$

In the above, d is the diameter at the breast height (considered to be 1.2m from the ground) of a tree in cm and H is the inundation depth in m.

Figure 5 also gives indication that in case of tsunami inundation higher than 4.65m, most of the trees were cut down or turn over with no resistance effect against tsunami. It is indicated too that trees with trunk diameter greater than 10cm survived against tsunami below 4.65m.

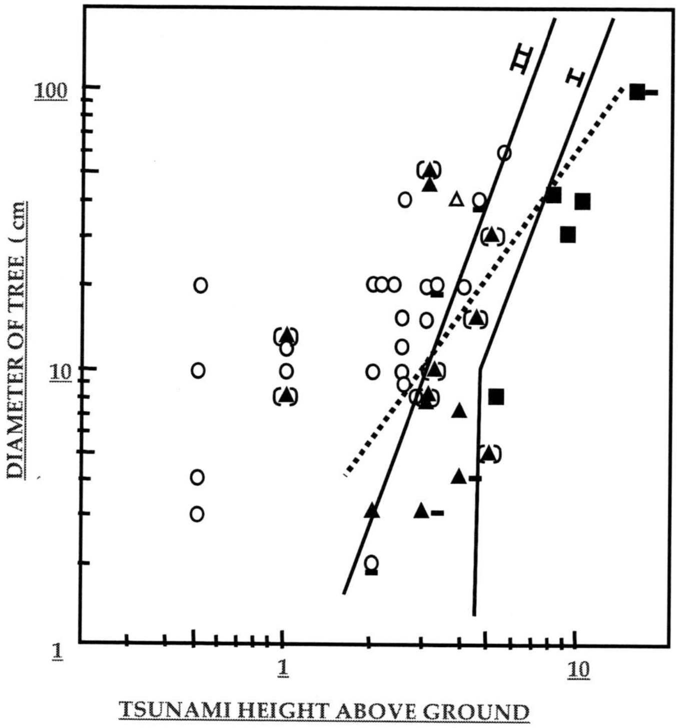

Notes:
O – no damage to tree with the effect of stopping floatages
= – no damage to tree with the effect of stopping floatages when a tsunami behaves as a standing wave
● – damage to some of the trees with the effect of stopping floatages
■ – cut down of the tree and no effect
△ – reduction of the current velocity and inundation depth with no damage in the forest
▲ – reduction of tsunami energy behind the forest with the damage to the forest
Underline – dense undergrowth
Bracket– damage to trees in poor condition
Horizontal bar – actual tsunami height being bigger than the values indicated in this figure

Figure 5. Degree of damage to tree in terms trunk diameter and tsunami height above ground surface (Source: Shuto, 1987).

Wave Thrust and Tree Breaking Moment

Many researchers (*e.g.* Tanaka *et al.*, 2008; Matsutomi, 2008) have identified that bending failure *i.e.* the failure due to bending moment acting on trees during tsunami, is the main cause of the tree damage. If the bending moment imposed by tsunami is greater than the maximum resistance to bending moment of the tree, the failure mechanism is anticipated. The center rotation of bending moment of vegetation with prop roots or aerial roots, *e.g.* Rhizopora sp. or Pandanus sp., is usually at the joint between the main trunk and the prop roots. Whereas, it is in between breast height of the trunk and very close to the ground for Bruguiera sp., Casuarina equisetifolia, or pine trees.

The wave-induced bending moment acting on the tree should be calculated by the following correlations (Yanagisawa *et al.*, 2008)

$$M_d = 0.5\,F\,(h - H_R) \tag{3}$$

$$F = 0.5\,C_D\,\rho\,A_0\,u^2 \tag{4}$$

In the Equations (3) and (4), M_d is the bending moment (in Nm), h is the tsunami inundation depth (m), F is the hydraulic drag force acting on the tree, C_D is the drag coefficient, ρ is water density, A_0 is the vertical projection area of inundated part of the tree, and u is tsunami velocity. Tsunami velocity is assumed to be vertically uniform (depth-averaged) and the tree stem is represented by a cylinder. Whereas H_R is the height of the prop roots (m) above the ground in case of vegetation with prop roots or aerial roots (Yanagisawa *et al.*, 2008), there is still no clear reference of H_R selection for the hard trunk trees, *e.g.* Casuarina and pine trees. Determination of C_D should consider vertical structure of the tree and the expected tsunami inundation height as well. If only the main trunk of non-prop-roots tree (or plus small bottom parts of canopy) are inundated, C_D may be considered as a single cylinder and defined according to Shore Protection Manual 1984 (Yanagisawa *et al.*, 2008):

$$C_D = \begin{cases} 1.2 & \text{for} \quad Re \leq 2 \times 10^5 \\ 1.2 - 0.5\left(\frac{Re}{3\times10^5} - \frac{2}{3}\right) & \text{for } 2\times10^5 \leq Re \leq 5\times10^5 \\ 0.7 & \text{for} \quad Re \geq 5\times10^5 \end{cases} \tag{5}$$

$$Re = \frac{|u| \times d}{\nu} \tag{6}$$

$$u = \sqrt{gh} \tag{7}$$

where Re is the Reynolds number, ν is the kinematic viscosity, g is acceleration of gravity, h is tsunami inundation depth and d is representative trunk diameter. Usually, d is diameter at the breast height (sometimes written as *dbh*) for trees without prop-roots and at 30cm above the joint of prop-roots and main trunk. For more general cases in which tsunami inundates most parts of the trees, including prop-roots, trunk and canopy, Tanaka *et al.* (2007) proposed $C_{d\text{-all}}$ for the integral drag coefficient of all inundated parts of the tree, instead of $C_D A_0$, which is calculated as follow:

$$C_{d\text{-all}} = \alpha\beta \times C_d \tag{8}$$

$$\alpha\beta = \frac{1}{h}\int_0^h \alpha(z)\beta(z) = \frac{1}{n}\sum_{i=1}^{n} \alpha_i \times \beta = \frac{1}{n}\sum_{i=1}^{n} \frac{dA_i}{dA_{1.2}} \times \beta \tag{9}$$

$$dA_i = \sum_{m=1}^{m=i_{max}} dA_{im} \tag{10}$$

$$dA_{1.2} = \frac{d}{100} \times z_t \tag{11}$$

In Equations (8) to (10), h(m) is tsunami flow depth, α is branches effect on C_d, β is leaves effect on C_d, dA_i(m^2) is the summation of vertical projection area of all tree part at the layer i^{th}, $dA_{1.2}$(m^2) is the vertical projection area of the trunk at 1.2m above the ground, z_t is the determined thickness of partial layer at the relevant position. Additional drag by leaves was taken as constant $\beta = 1.25$ (in leaf-bearing layers) or $\beta = 1$ (in leafless layers). The breaking moments of coastal trees have not been well investigated. However, Tanaka *et al.* (2008) tested several representative coastal vegetations in Sri Lanka, *i.e.* Pandanus odoratissimus, Scaevola sericea, Lumnitzera racemosa, Rhizopora mucronata, Casuarina equisetifolia, and Avicenia marina. The vegetations were classified according to their elasticity and habitat, *i.e.* coastal vegetations (including Pandanus odoratissimus), flexible trees (including Rhizopora mucronata) and trees with hard trunk (including Casuarina equisetifolia). Based on those tests, they derived moment breaking equations for the tested conditions. The breaking moment equations are respectively written as M_{GP1}, M_{GP2} and M_{GP3R} as follow,

$$M_{GP} = 4.45 \text{ x } dbh^{2.62} \tag{12}$$

$$M_{GP2} = 20.52 \text{ x } dbh^{2.83} \tag{13}$$

$$M_{GP3R} = 4.9 \text{ x } (1.5dbh)^3 \tag{14}$$

where dbh is the diameter (cm) of a tree at the breast height.

For the above mentioned coastal trees, Tanaka *et al.* (2006) have further calculated and drawn a graph that correlates trunk diameter with its related tsunami height that is expected to cause tree breaking. Figure 6 shows that Pandanus odoratissimus whose diameter is less than 20cm will start to be broken under about 3m tsunami inundation. Rhizopora apiculata-type with trunk diameter less than 20cm will no longer stand against tsunami when the inundation depth is more than 4.5m. Whereas, Casuarina equisetifolia with diameter about 20cm will be broken when tsunami inundation reach about 6m. Application of Figure 6 is limited to tree conditions that were used to derive the correlation, especially the local correlation between trunk diameter and tree height.

By re-plotting data in Figure6 into the logarithmic template of graph in Figure5 we obtain a new graph in Figure7. The latter graph correlates required minimum trunk diameter to stand against certain tsunami inundation depth. Curve I (dotted line) and Curve II (strip and dotted line) were drawn according to Equation (2) and Equation (1) of Figure5 respectively. Curve I and Curve II were drawn based on pine trees data. Curve I is the minimum trunk diameter that can stop the floatage but have no effect on tsunami energy reduction, whereas Curve II is the minimum trunk diameter that will not be destroyed by its related tsunami inundation depth and may reduce tsunami flow.

It can be seen from Figure 7 that Excoecaria agallocha and Pandanus odoratissimus (R=1m) have breaking moment capacity less than pine tree, whereas oppositely Rhizopora apiculata-type and Casuarina equisetifolia have higher breaking moment capacity than pine tree. Lumnitzera racemosa and Pandanus odoratissimus (R=2m) have almost similar breaking moment capacity with pine tree although for wave greater than 3m, Pandanus odoratissimus tend to have less capacity.

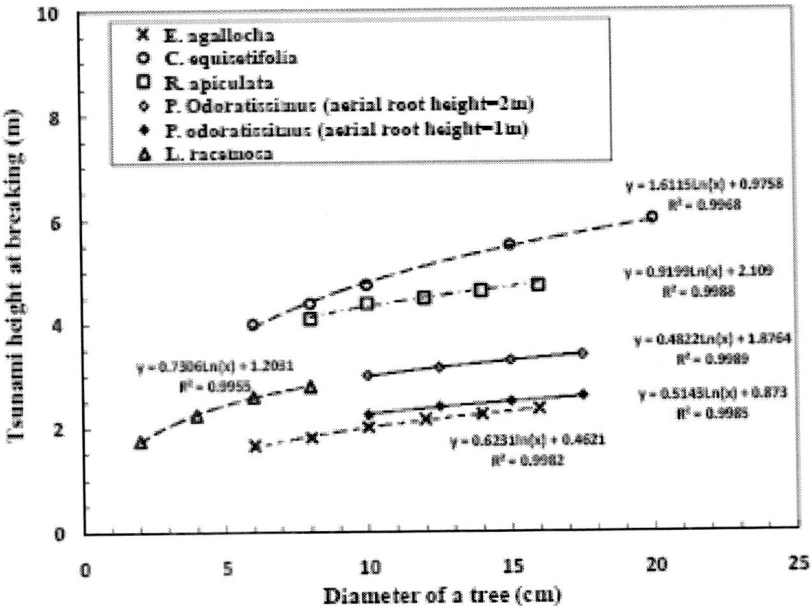

Figure 6. Relation between diameter at breast height and Tsunami height at breaking of trees (redrawn from Tanaka et al., 2006).

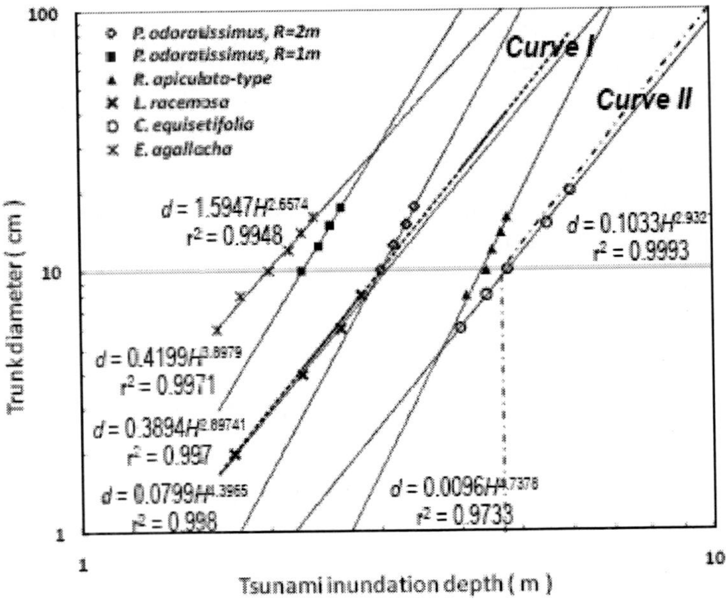

Figure 7. Trunk diameter at breaking and its related tsunami inundation depth.

For each of vegetation, an r^2-trend line was extended. It is seen that parallel trend lines were shown for Pandanus odoratissimus (R=1m), Pandanus odoratissimus (R=2m) and Rhizopora apiculata-type, whereas the other parallel trend is for pine tree, Excoecaria agallocha and Casuarina equisetifolia. The first group is for trees with prop-root, whereas the second group is for hard trunk trees without prop-root. The first group has smaller range of trunk diameter in compared to the second group.

5. Effect of Forest Density on the Reduction of Tsunami Flow

Once the trees have survived against tsunami force, their effect on tsunami flow depends on the combination of forest density and forest width in the direction of tsunami flow. Forest density is a combination function of individual tree size (including root, trunk, branches and leaves) and number of tree per unit area (Shuto, 1987; Harada and Imamura, 2000; Tanaka *et al.*, 2007, 2008). The higher the product of these two factors, the higher the density of coastal forest. We may call this density as horizontal density. All research results, numerical as well as laboratory experiment, show the increase of flow reduction by the increase of forest density (Hiraishi and Harada, 2003; Harada and Kawata, 2004; Dinar *et al.*, 2006; Tanaka *et al.*, 2007, 2008).When we consider the depth of tsunami inundation, the vertical structure of the tree, *i.e.* the composition of tree's root and canopy, which influence the vertical distribution of density, became additional important factor affecting the effectiveness of tsunami flow reduction.

Forest Density for Low-Inundation

In case of tsunami flow depth less than the height of canopy bottom, forest density calculation can be simply based on representative trunk diameter. The term of "summed diameter" that was introduced by Shuto (1987) to evaluate the effectiveness of pine tree forest in the reduction of tsunami energy may be used in this case. Shuto (1987) considered that the hydraulic resistance of a forest is given by evaluating the hydraulic resistance of a tree and summing it up over the number of trees in the direction of water flow. Considering that the resistance of a tree is proportional to the product of the projection area dH and the square of flow velocity v^2 (velocity v is assumed to be equal to gH; g is gravitational acceleration; H is tsunami inundation depth), and n is number of trees along the direction of water flow, Shuto (1987) proposed the following expression of tsunami energy reduction,

$$d\,H\,v^2\,n \simeq d\,n\,H^2 \tag{15}$$

In the above expression, H^2 is representative characteristics of tsunami at the site and dn, which is called as "summed diameter", is a major component of the resistance of a forest.

Based on the above approach, Shuto (1987) provided graph on effect of, and damage to, tsunami control forest in terms of tsunami height and summed diameter of trees as shown in Figure8.

According to Figure8, it can be concluded that forest with dn less than 30 had no effect on tsunami flow reduction. The higher the value of dn, the higher its reduction effect on tsunami flow.

For several species of mangrove, which have considerable volume of prop-root or aerial root, *e.g.* rhizopora apiculata or rhizopora mucronata, these roots resistance effect should be well considered. Effect of this root shall be considered as a friction coefficient against the flow (Yanagisawa, 2008) or as a hydraulic resistance related to the determined drag force coefficient (Tanaka *et al.*, 2006, 2007, 2008).

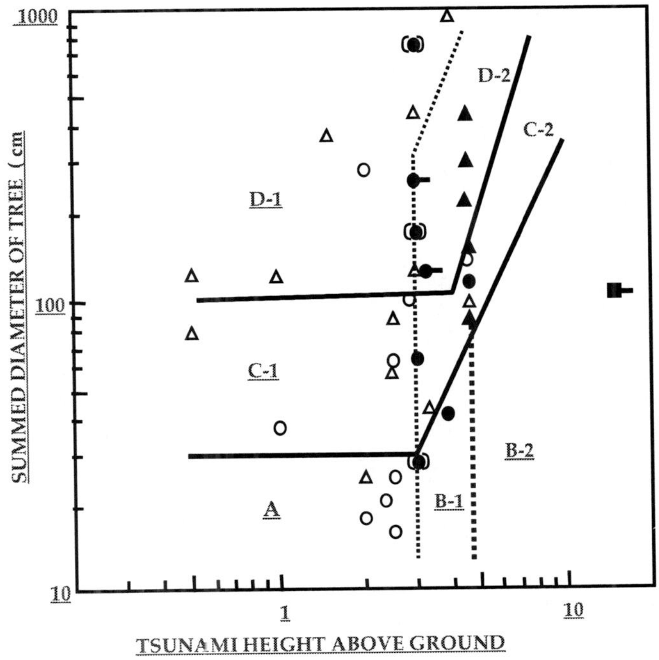

Notes:
O – no damage to tree with the effect of stopping floatages
= – no damage to tree with the effect of stopping floatages when a tsunami behaves as a standing wave
● – damage to some of the trees with the effect of stopping floatages
■ – cut down of the tree and no effect
△ – reduction of the current velocity and inundation depth with no damage in the forest
▲ – reduction of tsunami energy behind the forest with the damage to the forest
Underline – dense undergrowth
Bracket– damage to trees in poor condition
Horizontal bar – actual tsunami height being bigger than the values indicated in this figure

Figure 8. Effect of, and damage to, tsunami control forest in terms of tsunami height and summed diameter of trees [Source: Shuto (1987)].

Forest Density for High-Inundation

When tsunami flow interacts with most parts of the tree, flow resistance caused by all tree faculties should be included. Due to the unique vertical structure of each type of tree, their effect to tsunami flow also unique in terms of different tsunami flow depth. Tanaka *et al.* (2007) analyzed that vertical structure of the tree significantly affects total drag coefficient, $C_{d\text{-}all}$, working against tsunami flow. Every age period of the tree poses unique total drag coeffcient in terms of tsunami inundation depth.

The values of $C_{d\text{-}all}$ were calculated according to the expressions formulated by Tanaka *et al.* (2007) in Equations (8) to (11). Figure9(a) gives an example of calculated $C_{d\text{-}all}$ in terms of tsunami height for several representative trees grown at Sri Lanka and Thailand. The product of $\alpha(z)\beta(z)$ in terms of tsunami height for the related representative vegetations is shown in Figure9(b). Regarding the correlation between tree diameter and tree distance, Tanaka *et al.* (2007) used the data from field investigation to draw the correlation for representative vegetations. Then, by using this, they generated the value of vegetation thickness in unit area, dN_u, in term of tsunami height. See Figure10(a) and Figure10(b) respectively.

However, it should be considered that the above discussed vertical structures are specific to field investigations situation after Indian Ocean tsunami 2004 at Sri Lanka and Thailand. The utilization of graphs in Figure9 and Figure10(b) are limited to the trees with relevant types and dimension.

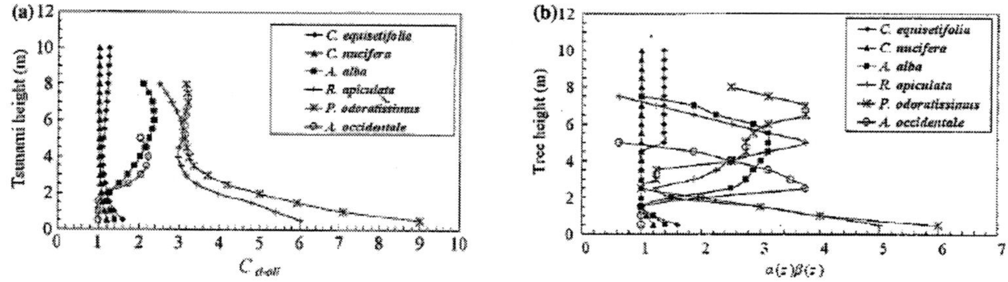

Figure 9. (a) Cd-all values of each representative tree in term of tsunami height. (b) Vertical distribution of branches and leaves effects, $\alpha(z)\beta(z)$ (Source: Tanaka et al., 2007).

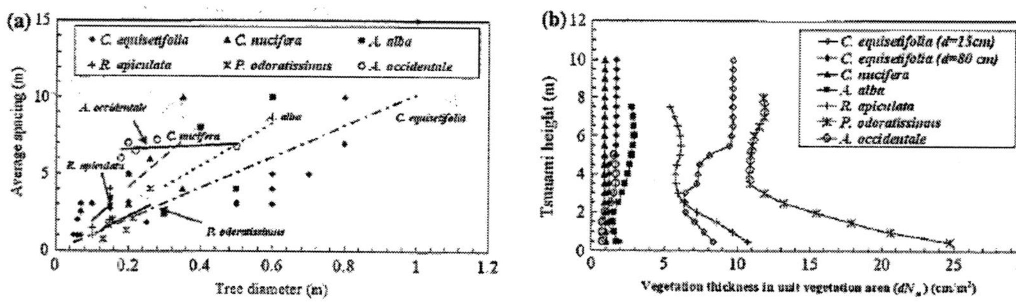

Figure 10. (a) Relationship between trunk diameter and the average space between trees, (b) relationship between dNu and the tsunami height (Source: Tanaka et al., 2007).

In the design process, information of number of tree per unit area in terms of trunk diameter is usually more practice to get the value of forest density and su mmed diameter (*dn*) in one run. For this purpose, we may replot data from Figure10(a) to be a graph (Figure11) that correlates trunk diameter and number of tree in unit area after tranpose the average spacing data to be the number of trees in unit area. It is known from Figure11 that in average Casuarina equisetifolia and Avicenia alba are possible to have greater trunk diameter than the other tree types. It can be seen too that for a single stand forest, the number of tree per unit area is sensitive to the change in trunk diameter.

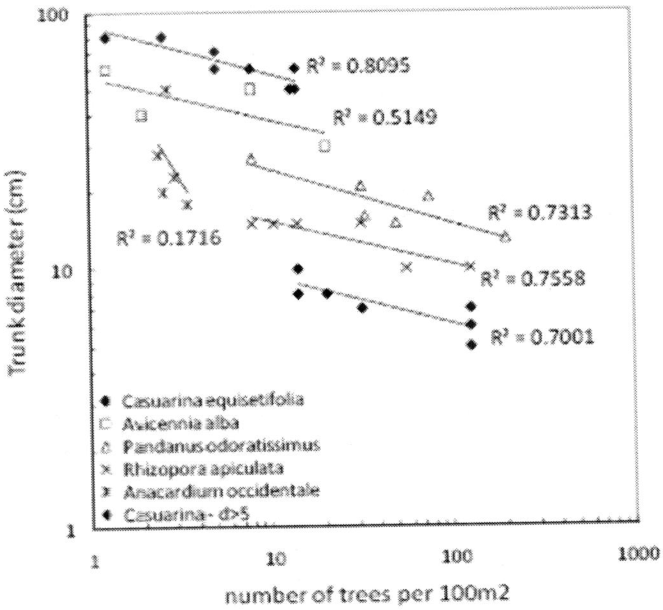

Figure 11. Number of tree per 100m2 in term of trunk diameter for several representative trees in Sri Lanka (re-calculated and re-plotted from Tanaka et al., 2007).

The Importance of Variation in the Forest Components

Naturally, a unit forest consists of various types of vegetations. The dominant species usually grows higher than the others and occupies most of the forest area. The spacing between the dominant trees is influenced by the dimension and shape of canopy (branches and leaves). The bigger the dimension of canopy, the longer the distance between trees (Yokozawa and Hara, 1995; O'Brien *et al.*, 1995; Zuhaidi, 2009). The distances available between trees provide space for other species to grow up within the forest. Although these complementary species will not flourish as the primary species, they fill up the remaining space and increase the density of the forest. These complementary species may be types of trees or bushes.

Shuto (1987) used the term of undergrowth for this complementary vegetation. Although no quantitative description on the effect of undergrowth on tsunami flow, Figures 5 and 8 show that forest with dense undergrowth exhibited no or less damage under tsunami attack and performed reduction on the current velocity. In Kalutara, Sri Lanka, two layers of vegetation forest, *i.e.* Pandanus odoratissimus and Casuarina equisetifolia, was indicated to give effective protection against tsunami attack to the land behind the forest (Tanaka *et al.*, 2007). The dense vertical structure of Pandanus (5m height) with many prop roots as well as branches and leaves reduce the flow effectively, while the strong and high Casuarina (25m) with bigger trunk diameter stop floatages and reduce flow velocity at the upper space.

In this regard, forest stands variation should be considered in the design of newly developed tsunami-protective coastal forest to provide maximum protection effects.

Effect of Trees Arrangement in the Forest

Matsutomi (2008) carried out laboratory experiments to investigate the effect of trees arrangement on the reduction of water flow force behind the forest. Four models of trees arrangement were each subjected to a 9,140 cm^3/s uniform flow discharge in a 30cm-width channel. The experimental results showed indication that trees arrangement has effect on the reduction of water flow force. Models arrangement in the channel and the experiment results is shown in Figure12.

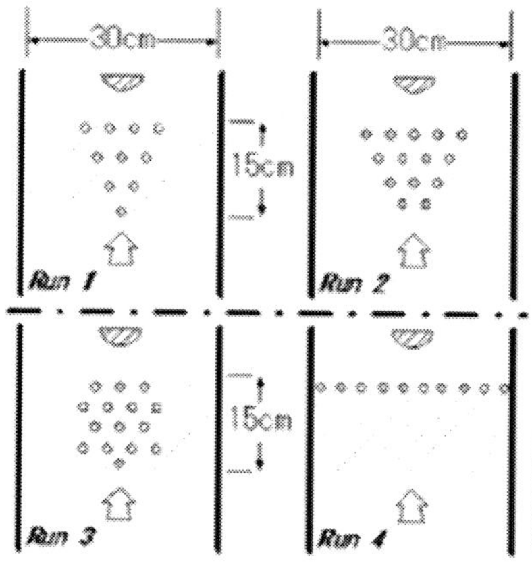

Notes:
= point of force measurement
○ = trees position in an arrangement
= direction of incoming flow

Run No.	Water depth *h* (cm)	Current velocity *u* (cm/s)	Drag *D* (gf)
0 (No trees)	7.99	38.1	66.9 (0.656 N)
1	8.24	37.0	37.1 (0.364 N)
2	8.21	37.1	46.3 (0.454 N)
3	8.29	36.8	36.0 (0.353 N)
4	8.00	38.1	48.9 (0.479 N)

Figure 12. Model arrangement in the experiment channel and the results of experiment (Source: redrawn from Matsutomi, 2008).

As these experiment results give only initial indication, further research is necessary before more firm reference is available about the effects of tree arrangement on the reduction of water flow force.

The forest ecology science suggests that there are certain allometric relationships among tree faculties (Asano, 2007; Yokozawa and Hara, 1995), *e.g.* between tree height and trunk diameter (Dauda *et al.*, 2004), trunk diameter and canopy diameter (Zuhaidi, 2009), which affect the composition of tree structure and number of tree within certain area. These allometric relationships are unique for each type of vegetation at an instantaneous period of time (Yokozawa and Hara, 1995). Good understanding on these relationships enable the design of optimum composition and arrangement of forest stand towards effective tsunami-protective coastal forest. Since canopy diameter is usually greater than trunk diameter, the distance between trees is determined by the diameter of canopy. If the allometric relationship between canopy diameter and trunk diameter can be formulated, various combinations of canopy diameter, trees distance and trunk diameter can be compared to find the maximum forest density of single stand by assuming that forest density is a product of number of trees and trunk diameter. At present, however, knowledge about the allometric relationships of coastal trees and vegetations is not well established yet.

6. Effect of Forest Width on the Reduction of Tsunami Flow

The tendency of forest width effects on tsunami flow reduction was initially shown by Shuto (1987) for the case of mainly pine trees forest of Japan. He found that at width of 20m or less (in the direction of tsunami), forest has no capacity of reducing tsunami flow. As long as tsunami flow depth is less than 3m, forests of more than 20m width are capable of reducing tsunami flow. The flow reduction increases with the increase of forest width. However when tsunami flow depth is greater than 4.65m, coastal forest width has virtually no effect on tsunami flow reduction although it may work as floatage stopper effectively. Graph in Figure13 shows the effectiveness of the width of pine tree forest on reduction of tsunami energy (Shuto, 1987).

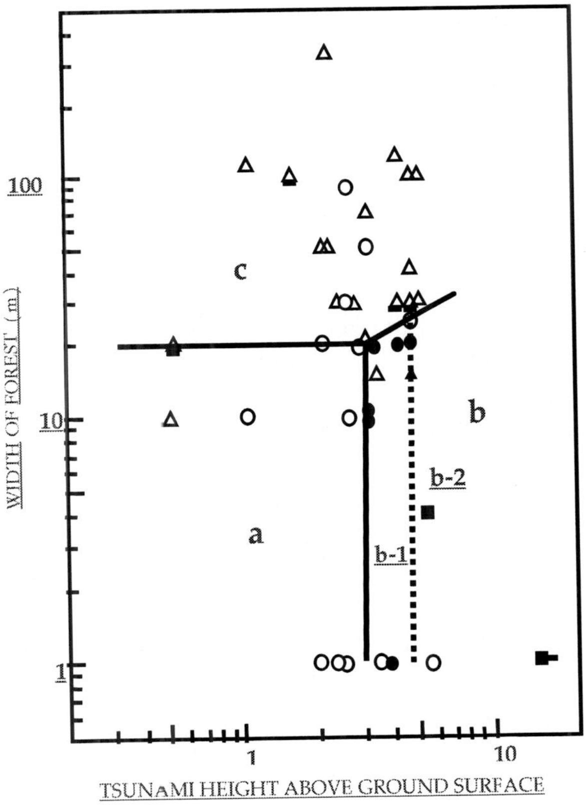

Notes:
O – no damage to tree with the effect of stopping floatages
= – no damage to tree with the effect of stopping floatages when a tsunami behaves as a standing wave
● – damage to some of the trees with the effect of stopping floatages
■ – cut down of the tree and no effect
△ – reduction of the current velocity and inundation depth with no damage in the forest
▲ – reduction of tsunami energy behind the forest with the damage to the forest
Underline – dense undergrowth
Bracket– damage to trees in poor condition
Horizontal bar – actual tsunami height being bigger than the values indicated in this figure

Figure 13. The effectiveness of the width of pine tree forest on reduction of tsunami energy (Source: Shuto, 1987).

Several laboratory experiments as well as numerical simulation have been carried out by many researchers to investigate the forest width effects on tsunami flow.

Numerical simulation results of Harada and Kawata (2004) show that the effect of forest width to reduce tsunami flow is far greater than the effect of forest density. Within the same forest width of 50m, forest density of 10(trees/100m^2), 30(trees/100m^2), and 50(trees/100m^2) gave inundation rate behind the forest of 0.83, 0.82 and 0.81 respectively. But 400m forest width with forest density of 10(trees/100m^2), 30(trees/100m^2), and 50(trees/100m^2) gave inundation rate behind the forest of 0.23, 0.18 and 0.15 respectively. It is shown that the inundation rate reduction were highly different between two different forest width conditions, whereas the rate changes only slightly following the variation of forest density. Here, inundation rate is inundation elevation behind coastal forest, which is non-dimensionalized by inundation elevation at the same point without coastal forest.

Numerical simulation results by Tanaka *et al.* (2007) and Harada and Kawata (2004) also show that doubling the forest width did not halve the inundation depth. For example for the case of Pandanus odoratissimus, 100m and 200m widths of forest reduced water elevation by 17% and 30% respectively.

7. Effects of Forest Ground Topography

Slope of the coastal area poses significant influence on the tsunami flow at the beach. This effect should be considered in the calculation of tsunami flow on the beach in addition to the effect caused by coastal vegetation. In general, the effect of coastal slope on tsunami flow is treated as in the wave runup calculation. In the numerical modeling of tsunami runup on the beach slope, the slope gradient, the bed surface roughness and surface similarity (Iribarren number) of the wave are considered as important variables.

According to the numerical simulation results by Tanaka (2008), the effect of forest topography to tsunami flow reduction could be greater than the effect of coastal forest. Effect of coastal forest at a steep slope on tsunami flow is not very significant as if it was at a gentler slope. In other words, coastal forest will effectively reduce tsunami flow at a gentle slope. Tanaka (2008), as an example, shows tsunami inundation reduction rate, in terms of coastal forest width and beach slope variations, for the vegetations of Pandanus odoratissimus. This graph is for the condition of 5m tsunami wave height and forest conditions as follow: tree height : 6m, diameter at breast height: 0.155m, density of trees: 0.403 trees/m^2 (spacing: 1.69m).

8. Implementation

Although tsunamis are mostly the product of earthquakes resulting from the displacement of plates in the process of energy releases, such events can recur over the time and may cause devastating tsunami in the region. In order to protect coastal communities from potential tsunami hazards, sufficient preparation is required. Since Mangrove forest is natural solution and has multifaceted benefit to the environment and people, it should be promoted in an appropriate manner.

Tsunami is not dangerous if the wave magnitude can be reduced to the acceptable limit and the provision of massive structural measures for that purpose may not be appropriate due to a variety of reasons. Sometimes structural interventions yield additional threat to the coastal people and environment therefore they are not the ultimate solutions. However, structural counter measures can be considered as a secondary or supplementary measure. The non-structural measures such as flood hazard map, warning and evacuations are inevitable to secure people's lives. Figure 14 shows the methods of maintaining permissible risk for increasing tsunami magnitude, wherein the size of tsunami, *e.g.* small, moderate or large, and potential damage are expressed in relative terms. However the definition of size and damage could vary from one purpose to another. Forests destroyed by tsunami events must be rehabilitated according to scientific knowledge. In so doing, traditional values of coastal forest for nature and people should not be ignored. Mangrove ecosystems have been well managed by indigenous populations for centuries, and Mangrove forest still acts as a shelter for them around the world. Therefore, societal, environmental and economical coverage of Mangrove forest development should be considered while optimizing the benefits.

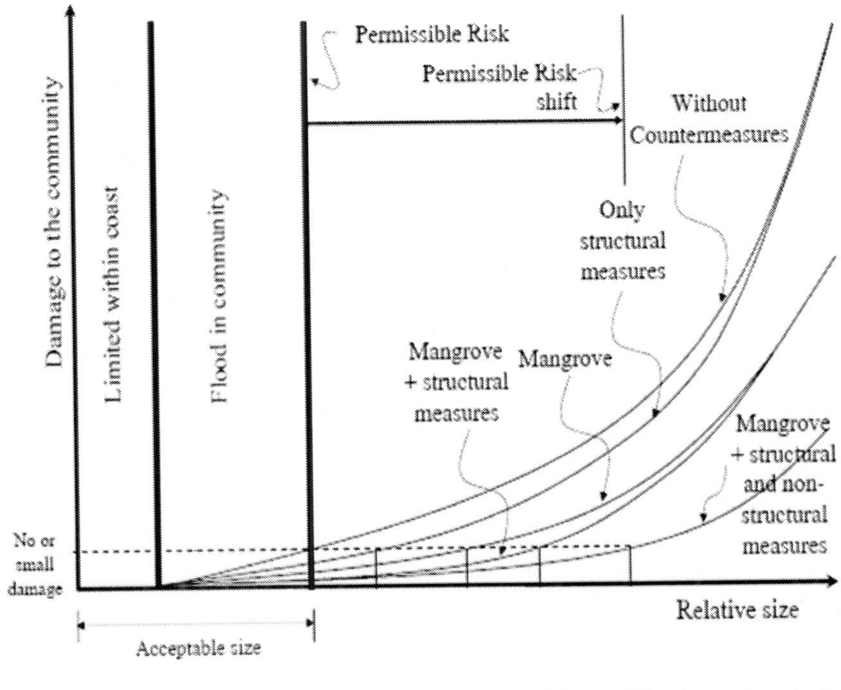

Figure 14. Tsunami risk reduction strategy with different possible combinations of control measures.

There are several national and international organisations working in the sectors of tsunami disaster mitigation. Some of them are focusing on Mangrove forest development, while some are on habitat reconstruction or poverty eradication. The ultimate goal of tsunami disaster mitigation can be achieved by effectively coordinating these all activities in an integrated manner. At the grass root level, this is possible by adopting community based disaster management practices. As a part, a concept of community forest can be promoted as an entry to the holistic development of coastal community. Community based development approach has already gained full recognition as a means for sustainable development. There is a clear perception that tsunami risk management cannot be treated in isolation. Rather, it should be part of community development. In addition, community based tsunami mitigation could be the best way to elimin ate overlapping and duplication in development approach. However there are still some points to be clarified, such as how the idea can be integrated into community based development as well as into sustainable poverty eradication plan. Tsunami hazard mapping can be a tool to guide this development process; therefore, it is essential to launch a campaign aimed to allow communities to understand their vulnerabilities, strategies, activities and the role they could play in managing flood risks without relying on external entities. Flood hazard map, which is found very effective in facilitating safe and smooth evacuation of people in emergency (Osti *et al.* 2008), could be a fundamental tool for community based tsunami disaster management. The community based approach will not only correct the defects of top-down approach, but also encourage the stakeholder's participation in a holistic manner.

9. Conclusions

Currently, the Earth is the home of some 6.5 billion people, and many of them are residing in coastal areas, exposed to the direct risk of tsunami and storm surges. In the last few decades, almost one million people were killed, and million worth of property was destroyed by tsunami events in different parts of the world. Unless disrupted, nature has made itself possible to mitigate the impact of such disasters by growing Mangrove forests along the coastal areas, which can considerably reduce the energy of tsunami waves entering into the land ward direction. Besides these advantages, coastal forests hold several additional benefits to people and nature. Unfortunately, these natural shields are declining day by day as a consequence of natural processes and increasing global population's demand on land, food and forest products. Nowadays, it becomes difficult to recover these natural resources because of changing environment and unfavourable human interactions. Therefore, special attention has to be given to develop and preserve these valuable resources. In order to maximise the benefits and minimise the risk of tsunami, practicable approach should be adopted, for which interdisciplinary and multidisciplinary studies are prerequisite. Community based tsunami risk management and community forest approach should be promoted to promote the holistic development of coastal communities. Tsunami hazard mapping practices can be promoted to facilitate the development process. In fact, tsunami hazard maps are very useful to endorse appropriate land use planning and in the meantime to bring awareness as well as to promote safe and smooth evacuation in emergencies.

References

Asano, T. (2006) 'Mitigation effects of mangrove forests against tsunami attack'. *Proceeding of the 30th International Conference on Coastal Engineering*, Vol.2, pp. 1541-1552

Barbier, B. E. (2006) 'Natural barriers to natural disasters: replanting Mangroves after the tsunami'. *Frontiers in Ecology and the Environment.* **4** (3). pp. 124–131.

Dahdouh-Guebas, F. *et al.* (2005) 'How effective were Mangroves as a defence against the recent tsunami?'. *Current Biology.* **15** (12). pp. 443–447.

Dauda, T.O., Ojo, L.O. and Nokce, S.K. (2004) 'Unexplained Relationships of Height-Diameter of Three Tree Species in A Tropical Forest'. *Global Nest: the Int. Journal* **6**(3), pp. 196-204

Dinar C.I., Utomo, K.S., Suranto and Jauzi, M.Z. (2006) 'The Influence of Rhyzopora-Shrub on Tsunami Propagation at The Beach'. *Proc. of Nat. Seminar on Tsunami, JICA-CDRC (in Indonesian language)*, Indonesia, pp.311-322.

Harada, K. and Imamura, F. (2000) 'Experimental Study on the Resistance by Mangrove under the Unsteady Flow'. *Proc. of the 1st APACE Conf.*, Vol.2, Dalian, China, pp.975-984.

Harada, K. and Kawata, Y. (2004) 'Study on the Effect of Coastal Forest to Tsunami Reduction'. *Annual of Disas. Prev. Res. Inst., Kyoto Univ.*, No. 47 C.

Hiraishi, T. (2005) 'Greenbelt technique for tsunami disaster reduction'. *Proceedings of APEC-EqTAP Seminar on 'Earthquake and Tsunami Disaster Reduction'*, Jakarta, Indonesia. pp. 1-6.

Hiraishi, T. and Harada, K. (2003) 'Greenbelt Tsunami Prevention in South-Pacific Region'. *Report of the Port and Airport Research Institute*, Vol. 42(2).

IUCN (The International Union for the Conservation of Nature and Natural Resources) (2005) *Early Observations of Tsunami Effects on Mangroves and Coastal Forests*. Gland. Switzerland.

Kathiresan, K., N. Rajendran (2005) 'Coastal Mangrove forests mitigated tsunami'. *Estuarine, Coastal and Shelf Science*. 65. pp. 601-606.

Kazmin, A. (2005) 'Disputes Over Land Threaten New Start'. *The Financial Times Limited.* December.

Kerr, A. M., A. H. Baird and S. J. Campbell (2006) 'Comments on ''Coastal Mangrove forests mitigated tsunami'' by K. Kathiresan and N. Rajendran [Estuar. Coast. Shelf Sci. 65 (2005) 601e606]'. *Estuarine, Coastal and Shelf Science*. **67**. pp. 539-541.

MAP (The Mangrove Action Project) (2005) *Public Forum to Counter Forest Encroachment.* The Mangrove Action Project News. 155th Edition.

Matsutomi, H. (2008) 'Use of Coastal Forest for Tsunami Disaster Mitigation'. *Lecture material of Comprehensive Tsunami Dis. Prev. Training Course, Workshop on Coastal Forest for Tsunami Disaster Mitigation*, ICHARM, Tsukuba, Japan.

Mazda, Y., M. Magi, Y. Ikeda, T. Kurokawa and T. Asano (2006) 'Wave reduction in a Mangrove forest dominated by Sonneratia sp'. *Wetlands Ecology and Management.***14** (4). pp. 365-378.

MIC (Mangrove Information Centre) (2006) *About Mangroves*. Official Brochure of MIC. Dutch Caribbean.

MSNBC News (2005) *Mangrove Forests Seen as Life-savers in Tsunami.* Reuters. Jan. 24, 2005.

O'Brien, S.T., Hubbell, S.P., Spiro, P., Condit, R. and Foster, R.B. (1995) 'Diameter, Height, Crown, and Age Relationships in EightNeotropical Tree Species'. *Ecology* **76**(6), pp. 1926-1939.

Osti, R. (2004) 'Forms of Community-participation and agencies role for the implementation of water-induced disaster management: protecting and enhancing the poor'. *Disaster Prevention and Management*, vol. 13 (2). pp. 6-13.

Osti, R., T. Shigenobu, T Toshikazu (2008) 'Flood hazard mapping in developing countries; a prospect and problems'. *Disaster Prevention and Management.* **17**(1), pp. 104-113.

Osti, R., T. Shigenobu, T. Toshikazu (2008) 'An overview on the importance of coastal forest in tsunami and storm surge disaster mitigation'. *Disasters*. **32**(4), pp. 88-99.

Padma, T. V. (2004) 'Mangrove Forests can Reduce Impact of Tsunamis'. *SciDev.Net*. **30** December 2004.

Philippe, M., P. Holmgren, F. Achard, H. Eva, H. Stibig and A. Branthomme (2005) 'Tropical forest cover change in the 1990s and options for future monitoring'. *Philosophical Transactions of Royal Society. B* **360**. pp. 373–384.

Selvam, V. (2005) *Impact Assessment for Mangrove and Shelterbelt Plantation*. Tsunami for Tamil Nadu Forestry Project. M.S. Swaminathan Research Foundation. New Delhi, India.

Shuto, N. (1987) 'The Effectiveness and Limit of Tsunami Control Forest'. *Coastal Eng. In Japan* **30**(1), pp.143-153.

Tanaka, N., Nandasena, N.A.K, Jinadasa, K.B.S.N., Sasaki, Y., Tanimoto, K. and Mowjood, M.I.M. (2008) 'Developing effective vegetation bioshield for tsunami protection'. *Journal of Civil Engineering and Environmental System* **26** (2), pp.163-180.

Tanaka, N., Sasaki, Y., Mowjood, M.I.M., Jinadasa, K.B.S. and Homchuen, S. (2007) 'Coastal vegetation structures and their functions in tsunami protection: experience of the recent Indian Ocean tsunami'. *Landscape Ecol. Eng.* **3**, pp.33-45.

Tanaka, N., Takemura, T., Sasaki, Y. and Mowjood, M.I.M. (2006) 'Species different of the breaking condition of coastal vegetation in Sri Lanka by tsunami and the difference of tsunami arrival time behind the vegetation'. *Ann. J. of Coastal Eng., JACE (in Japanese)*, Japan, Vol.53, pp.281-285

Tanaka, N. [2008] "Effects and Limitations of Coastal Vegetation in Tsunami Protection: Points for Mitigation and Future Planning," *Lecture material of Comprehensive Tsunami Dis. Prev. Training Course, Workshop on Coastal Forest for Tsunami Disaster Mitigation*, ICHARM, Tsukuba, Japan.

UNEP (United Nations Environment Programme) (July 2007) 'Physical Alteration and Destruction of Habitats'. July, http://www.gpa.unep.org/content.html?id=199andln=6.

United States Geological Survey-USGS (2009) 'Surviving a Tsunami—Lessons from Chile, Hawaii, and Japan, http://pubs.usgs.gov/circ/c1187/#tree.

Upadhyay, V. P., R. Ranjan and J. S. Singh (2002) 'Human–Mangrove conflicts: The way out'. *Current Science*. **83** (11). pp. 2328-2336.

WB (World Bank) (July 2007) 'Coastal and marine management'. July, http://go.worldbank.org/FWQVNO6O80.

Wilkie, M. L. and S. Fortuna (2003) *Status and Trends in Mangrove Area Extent Worldwide, Forest Resources Assessment. Working Paper* No. **63**. Forest Resources Division. FAO. Rome.

Williams, N. (2005) 'Tsunami insight to Mangrove value'. *Current Biology*.**15** (3). pp.73-79.

Wolanski, E. (2007) Estuarine Ecohydrology. Amsterdam, Elsevier.

Yanagisawa, H., Koshimura, S., Goto, K., Miyagi, T., Imamura, F., Ruangrassamee, A. and Tanavud, C. (2008) 'The reduction effects of manrove forest on a tsunami based on field surveys at Pakarang Cape, Thailand and numerical analysis'. *Estuarine, Coastal and Shelf Science Journal* **81**, pp.27-37.

Yeh, H.H. (1991) 'Tsunamis Bore Runup' in *Tsunami Hazard: A Practical Guide for Tsunami Hazard Reductiont,* ed. E.N. Bernard (Kluwer Academic Publisher, The Netherlands), pp.209-220.

Yokozawa, M. and Hara, T. (1995) "Foliage Profile, Size Structure and Stem Diameter-Plant Height Relationship in Crowded Plant Populations," *Annals of Botany* **76**, pp.271-285.

Zuhaidi, Y.A. (2009) 'Local Growth Model in Modelling the Crown Diameter of Plantation-Grown Dryobalanops Aromatica'. *The Journal of Tropical Forest Science* **21**(1), pp. 66–71.

In: Tsunamis: Causes, Characteristics, Warnings and Protection ISBN: 978-1-60876-360-3
Editors: N. Veitch and G. Jaffray, pp. 113-129

Chapter 5

COASTAL PROTECTION MEASURES FOR TSUNAMI DISASTER REDUCTION

***Emel Irtem*[1,*], *M. Sedat Kabdasli*[2] *and Nuray Gedik*[1]**
[1]* Balikesir University, Department of Civil Engineering, Balikesir, Turkey
[2] Istanbul Technical University, Civil Engineering Faculty, Maslak, Istanbul, Turkey

Abstract

Tsunami may be generated by earthquake triggered movement of the sea bottom, landslides and collapses. It has caused great impacts on human life and coastal environments, including massive loss of human life, devastation of coastal ecosystems and settlements, and damage to infrastructure and facilities. In this chapter, tsunami protection works will be investigated as hard structures likes tsunami breakwaters, sea walls etc. and natural barriers including coastal forest etc. except soft approaches (education, awareness, evacuation, etc.). Although hard structures may provide highly effective protection, they may have high cost and may also cause large amount of negative environmental impact on the coastal areas. That is why, natural coastal barriers which have lower environmental impact and higher additional natural value can be considered as a protection measure against tsunami effects.

Tsunami damage occurs mostly in the nearshore zone and in the coastal area behind the coastline because of the tsunami hydrodynamics during the run-up period. Therefore, tsunami run-up height was also investigated. Tsunami run-up heights for impermeable and permeable (not armored and armored) beaches were examined and empirical equations suggested. Following, coastal protective measures, to reduce tsunami damage on coastal areas were analyzed as hard structures and natural barriers. Furthermore, the effects of coastal forests on tsunami run-up heights were discussed and empirical equations suggested.

Introduction

Socioeconomic life rapidly expanding all around the world results in increasing population on coastal region and 50% of world's population still live on these regions. It is estimated that this population on coastal regions will rise up to 75% as of year 2020 (Reeve,

2004). Accordingly, most of industrial and commercial activities take place on these regions. However, such coastal regions are extremely vulnerable to natural and coastal disasters caused by diverse factors. One of the most destructive coastal disasters is tsunamis caused by water body movements based on ground movements, sea bottom landslides, slides and collapses generally triggered by earthquakes. These tsunamis, which are actually a kind of long wave, may be highly destructive based on social economic activities on coastal areas. For example, tsunami triggered by earthquake took place in the Indian Ocean on 24 December 2004, affected 11 countries and caused 250.000 people death. Since tsunami caused deaths are significantly more than number of deaths resulted directly from this earthquake itself, this situation has drawn global attention to tsunamis and people all over the world gained awareness about high risks of tsunami disasters.

The most interesting fact known about tsunamis is that although they have a very small height and wave periods which can be considered as infinite on offshore conditions and therefore, has really no particular effect, their run-up height increases extremely on coastal areas and thus, has a remarkable negative effect on region beyond coastline. For this reason, studies intended for reducing significant tsunami risks mainly focus on determining tsunami run-up heights and taking preventive or eliminative measures against tsunami effects on coastal and continental regions. Especially, it is acknowledged that determining maximum run-up height of tsunamis is "essential" in terms of reducing losses caused by tsunami disasters (Liu et al, 1995). Tsunami waves can be modeled as solitary waves (Lie et al. 1991; Synolakis, 1986, 1987; Hall and Watts 1953; Al-Banaa and Liu, 2007, etc) or N waves (Tadepalli and Synolakis 1994, 1996).

It will be highly important to estimate run-up height of tsunamis while designing coastal structure because tsunamis are mainly effective on coastlines around their run-up zones. In this chapter, general information about run-up heights is presented with relevant literature. An experimental study on run-up heights of tsunami on permeable coasts (armored and not armored) has been discussed. Considering these experimental data, some empirical formulas are suggested for permeable coastal regions. Following, coastal protective measures, to reduce tsunami damage on coastal areas, are analyzed under two titles as hard structures and natural barriers. Additionally, another experimental research on effects of coastal forests on tsunami run-up heights is discusses and empirical equations are suggested.

Tsunami Run-Up Height

General

The most significant characteristic of tsunamis is the occurrence of strong water body movements landward from coastlines. As such in all gravity waves, a tsunami stops its wave movement after reaching a particular depth, starts water body movement and this movement proceeds to a particular point beyond coastline based on geographical characteristics of this coastline, and then, water mass begins drawing back from this point. The horizontal distance between still water level (SWL) and peak point of running up water is called as run-up distance (L_R) while vertical distance is called as run-up height (R)(Figure 1).

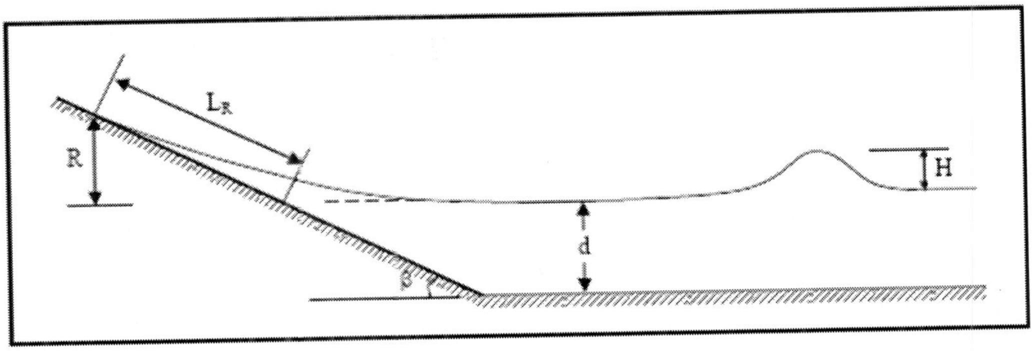

Figure 1. Run-up distance and run-up height.

Tsunami run-up heights depend on various factors such as beach slopes and surface roughness. As an extreme condition, it can easily be said that this run-up height can be very close to zero on a plain land while run-up distance can be kilometers long. Generally, the studies are primarily focused on determining run-up height based on morphological characteristics of coastal areas and that such coastal areas mostly have a beach slope of 1/100 – 1/1.

Estimation of Tsunami Run-Up Height

Theoretic and experimental researches have been made in order to determine run-up heights of long waves. Pedersen and Gjevik, 1983; Synolakis, 1987; Synolakis and Skjelbreia, 1993; Kanoglu and Synolakis, 1998; Liu et al, 1991; Maiti and Sen, 1999; Pelinovsky, 1999; Carrier, 2003 provided information and analytical solutions about run-up up of non-linear waves on impermeable beachs. In these approaches, the run-up problem are examined by solving basic equations for specific initial and boundary conditions or using empirical equations. Some experimental studies made on run-up of solitary waves are given in Hall and Watts, 1953; Pedersen and Gjevik, 1983; Synolakis, 1987; Teng et al., 2000; Li and Raichlen, 2001; Shankar and Jayaratne, 2003; Gedik et al., 2005.

Basic parameters effecting run-up height are water depth, wave height and beach slope and whether this wave breaks or not. In Hall and Watts (1953) which is the first experimental research in this context, it is indicated that run-up height (R) on impermeable beaches for non-breaking solitary waves only depend on water depth (d) and wave height (H) as shown in Eq. 1.

$$\frac{R}{d}=3.1\left(\frac{H}{d}\right)^{1.15} \tag{1}$$

The beach slope is used by Synolakis (1987) in an equation giving run-up height of non-breaking solitary waves on impermeable beaches. In this equation, β is the angle between sea bottom and horizontal axis and the Eq. 2 is known as run-up law in literature.

$$\frac{R}{d} = 2.831\sqrt{\cot\beta}\left(\frac{H}{d}\right)^{5/4} \tag{2}$$

Besides above mentioned parameters, run-up height is also affected by roughness and viscosity. Teng et al. (2000) analyzed maximum run-up height of non-breaking solitary waves both on rough and smooth beaches as experimentally. It has been found that viscosity and roughness are highly effective on mild slopes (< 10^0) and that maximum run-up height is reduced more than 50% when compared with cases where viscosity and roughness are neglected.

Synolakis (1987); Kobayashi and Katjadi (1994); Li and Raichlen (2003); Hughes (2004); Hsiao et al. (2008); Madsen and Fuhrman (2008), suggested equations in relation with run-up heights of breaking solitary waves. In above mentioned studies, run-up height is generally analyzed on smooth beaches and confirmed with experimental models.

For non-breaking tsunamis, the effect of not armored beaches and armored beaches on tsunami's run-up height has been experimentally analyzed by Gedik et al (2005). These experiments are carried out in a wave channel which is 22.5 m length, 1.00 m width and 0.50 m depth and covered with glass walls. The not armored beach having a slope of 1:5 is built up of natural sand with a specific gravity of 2.63 and a diameter of 0.35 mm. First of all, armor units with a diameter of 10 mm, a specific gravity of 2.318 a thickness of 0.05 m are spread over not armored beach. Then, experiments have been conducted on obtained armored beach. Thereafter, this process has been repeated with different armor units having a diameter of 13.8 mm and a specific gravity of 2.289 (Gedik et al. 2005).

Run-up heights (R) on impermeable beach have been measured, (R/d) has been formed, and these results have been compared with dimensionless run-up height parameters obtained from Eq. 2 which is also known as run-up law. It is shown a good agreement between the results (Figure 2).

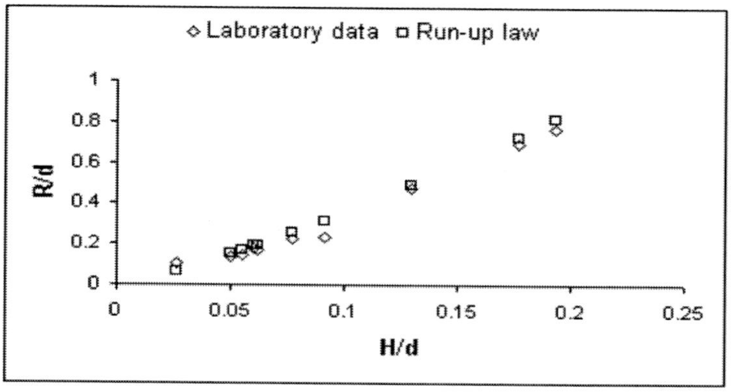

Figure 2. Comparison of dimensionless run-up heights parameters obtained from laboratory data and run-up law for impermeable beach (Gedik et al., 2005).

In order to obtain run-up heights on not armored and armored beaches, these experiment results are evaluated and below mentioned empirical equations considered sand and armor unit characteristics are suggested (Eqs. 3-4):

$$\frac{R}{d}=4.10^{-4}\left(H\frac{G_{sp}}{D}\cot\beta\right)^{0.921} \qquad \text{for not armored} \qquad (3)$$

$$\frac{R}{d}=5.10^{-3}\left(H\frac{G_{st}}{D_{n50}}\cot\beta\right)^{0.9539} \qquad \text{for armored} \qquad (4)$$

where, (R/d) is dimensionless run-up height, (D) is sand diameter, $G_{sp} = \gamma s / \gamma w$ is specific gravity of sand, (γ_s) sand density, (γ_w) water density, and $G_{st} = \gamma_{st} / \gamma_w$ is specific gravity of armor units, (γ_{st}) armor unit density.

R/d, calculated in run-up law (Eq. 2), and R/d, obtained from equations (Eqs. 3-4) suggested for armored and not armored beaches, are compared in Figure 3. It is found out that armor units reduce run up height around 50%. This result is agree with Teng et al. (2000)'s research on effect of roughness on run-up heights.

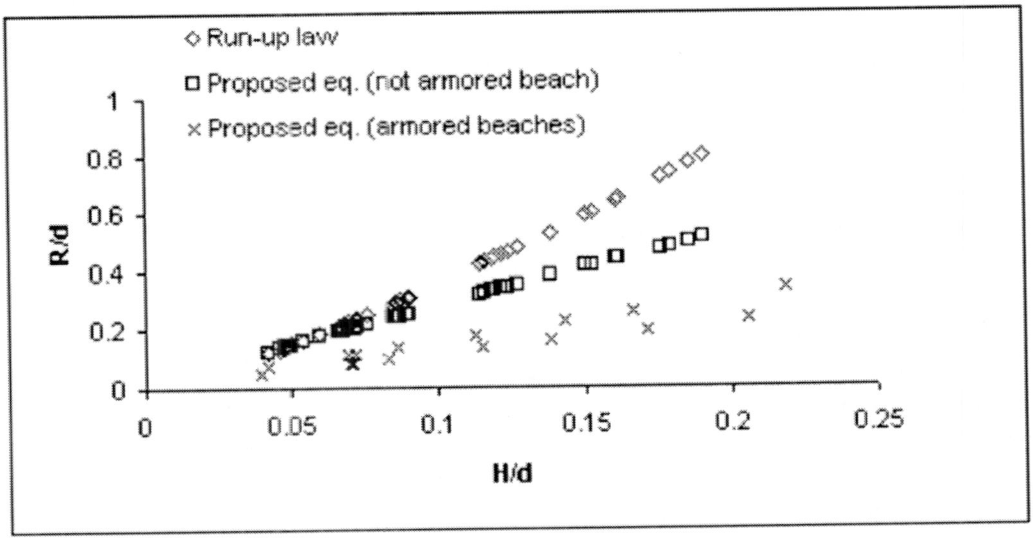

Figure 3. Comparison of dimensionless run-up heights parameters obtained from laboratory data and run-up law for permeable beach (Gedik et al., 2005).

Coastal Protection Measures against Tsunami

While run-up process, tsunami causes extreme damage to human life and environment as well as structures located onshore region. For this reason, humans have developed various measures in order to reduce tsunami effect. Generally, these measures can be classified in two groups as hard structures and natural barriers. Hard structures may not be suitable for all coastal regions because their construction costs are significantly high, their dimensions are large and they have negative effects on environment. In this situation, natural coastal barriers (coastal forests, mangroves etc.) are preferred. However, when natural barriers are not satisfactory in reducing damages caused by tsunami waves, hard structures are applied despite of their negative effects on environment.

Hard Structures

Hard structures are compositions of two or more structures such as caisson, composite, rubble mound, submerged breakwater, inclined or vertical seawalls and groins which are traditionally used in coastal engineering against storm surge effects. The general characteristics of these structures are same but dimensioning criteria are different for tsunami and presently, various studies are being carried out to improve above mentioned criteria. As known, although classical coastal structures are highly effective against storm surges with significant wave steepness, they may not be adequate against long waves due to their classical dimensioning criteria.

The most important problem in using hard structures against tsunami disasters in developed and developing countries is growing up of possible environmental problems. Hard structures have a negative impact on aesthetical aspect of coastlines, prevent interaction between human, nature and sea and have significantly high construction costs.

Seawalls and Groins

Seawalls are structures built parallel to shoreline in order to protect land side against wave actions. Seawalls can be built up of concrete and stone fill with sloping front. Concrete seawalls are constructed with a curved slope or with steps. Groins are built in order to prevent coastal erosion or to support beach nourishment works. Together with seawalls, they are helpful to reduce tsunami effect.

Sometimes, these seawalls cannot provide enough protection against tsunami. The main reason of this situation is that dimensions of seawalls are inefficient when compared to run-up height of tsunami. For example, a seawall with a height of 4.5 m has been built on Okushiri Island in order to protect Aonae Peninsula; however, the tsunami took place in 1993 has climbed over this seawall and caused more than 185 people death (Dalrymple and Kriebel, 2005). Besides, it has been mentioned above that run-up distance is as important as run-up height in with a gentle slope regions. In such regions, seawalls, groins and plantation can be used together in order to reduce tsunami effect. In the study carried out by Sonkar and Kumar (2008) for Karaikalda which is located on a smooth region on eastern coastline of India and where Indian Tsunami caused a lot of people death in 2004, 10 groins with an average length of 150 m have been proposed for construction in Chnnai Region while two rows of masonry blocks have been suggested for Pudukuppam Region 200 m behind coastline and plantation between these blocks and coastline. Kottilpadu Colachel located in Kanyakumari Region is one of the most adversely affected villages by tsunami. One of these suggestions is a seawall of 4 m width and 6 m height above MSL. This seawall which has a very strong toe with a width of 3 m will be supported with plantation and a crown wall approximately 7 m above MSL. Another example of using seawalls and plantation has also been indicated in Edward et al., (2006). A seawall with a height of 4 m has been built immediately after sandy beach in

Nami-ita Beach, which is 300 m length and 50 m width, in Funakoshi Bay. Behind this wall, an area with a width of 30 m has been covered with dense pine trees.

As it can be clearly understood from these examples, seawalls can eliminate tsunami risk totally only if its crest point is higher than run-up height of tsunami. It is also clear that such a seawall would destroy all coastal properties and would make it impossible for people to benefit from sea. This is why a couple of measures are proposed together.

Breakwaters

Breakwaters are the most important engineering applications being used as coastal structures since antique ages. As known, the most common type is rubble mound breakwaters and they can provide complete protection against wave forces by based on resistance of stone weights. For dimensioning such breakwaters based on storm surges, there is adequate amount of knowledge in literature (SPM 1984, CEM 2002). Stability of breakwaters' toe, protective layer and crest mainly depends on the diameter of characteristic stone blocks. Diameter of characteristic stone blocks can be calculated under given conditions for normal waves, for example Van der Meer (1988, 1997). However, these formulas cannot be applied while determining stone stability under tsunami conditions (Van der Plas, 2007). On the other hand, currently, there is a limited amount of information available about dimensioning of breakwaters under long wave conditions such as tsunamis.

Stability of rubble mound breakwaters under tsunami conditions is a lot higher than their stability under storm surge conditions if tsunami is not overtopping these breakwaters. Tsunami causes almost no damage to rubble mound breakwaters. However, if tsunami overtopping this type of breakwaters, exactly opposite situation takes place and it may even be destroyed completely based on overtopping quantity. On the contrary with storm surges, even if tsunami overtopping this breakwater with a limited quantity, the rear inclined surface of breakwaters should be protected with stone blocks of adequate weight (Cuypers, 2004).

Tsunami load on an rubble mound breakwater on harborside consists of two components as flow on the breakwater and flow between breakwater stones. In order to prevent a possible collapse in a breakwater, it should be built on a strong and impermeable core to minimize or eliminate internal flow.

There are only a few literature available focusing on the behavior of rubble mound breakwaters against tsunami. In these documents, such as Cuypers (2004), Bas (2007) and Yuce (2007), there is no formula given in this context; however, information about destruction models and weak points of rubble mound breakwaters is given. As it can be seen in these studies, sea side of an rubble mound breakwater for protection against tsunami should be designed according to short-period. The protective layer on a rubble mound breakwater only on its harbor side should be designed in according to tsunami load because tsunami waves damage mostly the rear inclined surface of stone breakwaters. In the literature, there are breakwater project designs being carried out to get protected against tsunamis. Under tsunami conditions, the advantage of caisson breakwaters is their stabile characteristic. In rubble mound breakwaters, the structure may collapse if the protective layer or the core material is washed away. The biggest advantage of rubble mound breakwaters is that they can easily be built. There is almost no information about dimensioning criteria of caisson type breakwaters under tsunami conditions. It has been shown that pressure distribution created by the tsunami especially on vertical walls is valid for tsunami. The maximum value of a dynamic pressure by waves on a vertical face structure under tsunami effect is generally visible on still-water level and the relation between offshore wave heights and maximum dynamic pressure applied on the structure can be explained as a linear. The dynamic pressure distribution by tsunami on a coastal structure with a vertical surface becomes uniform as waves getting higher when compared to the structure and maximum pressure values under still water level reaches to its maximum pressure value on still-water level. If height of a berm, which will be built in front of a coastal structure with a vertical surface especially under tsunami effect capable of overtopping this structure, increases, the dynamic wave force that will be applied on this structure will increase significantly. In other words, berm application has a negative effect on dynamic forces that will be applied on a coastal structure under tsunami condition (Yuce, 2007).

It has been found out that overtopping has a very significant negative effect on stability, permeability of a breakwater section do not have any contribution to the stability and, if available, crown wall will effect stability negatively based on reflection (Bas, 2007).

Especially, tsunami barriers in Japan are caisson, combination of caisson with rubble mound and high seawalls. The effect of offshore breakwaters depends on the height of these structures and the correlation between the gap in the detached breakwaters and tsunami height. Another way to reduce wave effect is submerged breakwaters having crests lower than SWL. However, submerged breakwaters have a limited effect on tsunamis which its wave length is kilometers (Van der Plus, 2007).

Natural Barriers

Natural barriers refer to coastal forests, mangroves and plantation. These forests block the water flow despite of their porous structure and break the energy by means of the turbulence. Coastal forests act as a natural barrier protecting human life, settlements and agricultural lands from tsunami effect. Absorption of tsunami energy depends on forest density, height, diameter of tree trunk and roots, tree age, width of the forest, beach slope, bathymetry and characteristics of incident waves. Just as hard structures, these barriers reduce life losses and economical loses caused by tsunami in coastal regions by creating resistance against water body movement and increasing drag force. Studies by Kathiresan and Rajendran (2005) in 18 coastal villages of India immediately after the tsunami disaster on 26 December 2004 confirm this statement. In Latief and Hadi (2007) are shown that open lands have been completely destroyed by 4-5 m high tsunami while less damage has occurred in forestry areas Pangandaran coasts after West Java Tsunami in 2006.

Natural barriers are mostly preferred by developing countries because they are economical and have positive effects on environment. Moreover, coastal forests prevent coastal erosion caused by waves, have no negative effect on ecosystem, enhance biodiversity and sand dune and prevent sea water intrusion to ground water resources supplying fresh water for people living on coastal regions. In Japan, coastal forests on different coast were planted around 3-4 centuries ago in order to protect agricultural lands from tsunami, abnormal tides and storm surges (Edward et al., 2006).

When tsunami waves proceed landward, their wave current, wave velocity and height reduce by coastal forest effect. Furthermore, an increase in forest width can reduce not only inundation depth, but also the hydraulic forces behind the coastal forest. Harada and Imamura (2005) quantitatively evaluate the hydrodynamic effects and damage-prevention functions of coastal forest against tsunamis with a view to using them as tsunami counter-measures. They also performed numerical simulations, including an evaluation of the quantitative effects of coastal forest in controlling tsunami reduction and damage. Hiraishi and Harada have shown by means of numerical studies in 2003 that if 30 trees are planted in an area of 100 m^2 100 m width, maximum tsunami flow pressure is reduced by more than 90%. However, if the tsunami is extremely high (H_{in} >6m), then, coastal forests are not adequate in reducing damage caused by tsunami. H_{in} is wave height above ground and called as inundation height. Although they reduce damage created by tsunami, these coastal forests have some disadvantages, too. When serious tsunami disasters proceed to landward, they carry away tree roots and branches and therefore, may cause increased damage to the coastal region. Breaking up or uprooting of trees depends on the elasticity of tree itself, its root, trunk and branches, its resistance against breaking up, root depth, root dimensions and mass, resistance of soil, drag and impact force of the wave and wave height. Another disadvantage of coastal forests is that there are open areas, namely gaps, created for farm houses and roads in the forests. When these gaps are not large enough, they increase flow velocity and pressure. For this reason, such gaps should be avoided in coastal planning or should be designed very well (Forbes and Broadhead, 2007).

Mangrove

In Equatorial Climate and in subtropical regions, the forests consisting of various tree types creating a dense forest on muddy coasts, river mouth and wetlands are referred as "Mangrove". They are very common in India, Pakistan, Sri Lanka, Thailand, Malaysia, Singapore and Indonesia. There are different mangrove types on Pacific and Atlantic coasts of America continent and on West African coasts. Mangrove trees grow up to 6-18 m and very resistant against salty sea water. Most of mangroves, such as *Rhizophora Mangle*, grow supporting roots in order to hold on loose mud of wetlands. When these roots reach to the mud underground, they stool. Therefore, mangroves become very dense forests impossible to walk through (Mammadov, 2004).

This type of forest covers more than half of Equatorial coastlines and reduces the effect of tide storms and hurricanes creating a natural barrier between sea and coastal land. They prevent coastal erosion and ensure stability of coastline. Moreover, nutritious waters of mangrove forests provide a good habitat and shelter for most of fishes and shrimps. Mangrove forests are highly important in terms of protection and development of coastline. Mangroves, growing with an adequate width, length and density alongshore, may eliminate destructive effect of tsunami disaster. For example, when a wave with an inundation depth of 3 m reaches to an area which is 400 m wide and covered with 0.2 trees having a diameter of 15 cm in each square meter, it has been seen that inundation depth of tsunami is reduced by 30% (Yanagisawa et al., 2009).

Mangroves act as buffer zones protecting natural sea habitat from land effects and protecting coastal lands from sea effects. However, mangrove forests are destroyed in some regions for aquaculture and tourism. Dahbou Guebas (2006) has reported that 30 of 418 villages on Andaman coasts have been seriously damaged by the tsunami disaster took place in December 2004. The special land use plan in coastal villages is highly important for protecting these villages from natural disasters and mangroves and shelterbelt plantation are a part of this land use plan. This plan also ensures that sand dunes, which are also neglected as an ecosystem element on coastal regions, are stabilized and protected (Sonkar and Kumar, 2008).

Mangrove forest should be wide enough to absorb tsunami wave energy in order to reduce tsunami effect. For example, it has been seen in Thailand that mangroves with an adequate width can even reduce effects of a tsunami wave which is 10 m high. Only first 50 m of *Rhizophora* mangrove in Phang Nga was destroyed by the tsunami which was 8 meters high. In Sri Lanka, first 2-3 meters of mangroves (*Rhizophora ssp and ceriops spp*) were seriously damaged by 6 m high tsunami while less damage has been observed on remaining 3-4 meters. However, only the width of a mangrove forest may not be adequate in reducing tsunami effect. Various factors, such as diameters of tree trunks, height and resistance, should be taken into consideration. Additionally, it has been reported that *Sonneratia spp* mangroves with a width of 200 m were uprooted and destroyed under tsunami effect in Sri Lanka (Forbes and Broadhead, 2007). Mangroves survival rate depends on diameter of tree trunk. Under same tsunami conditions, only 19% of mangroves having a tree trunk diameter of 15-20 cm can survive while 72% of mangroves with a tree trunk diameter of 25-30 cm can remain standing (Yanagisawa et al., 2009).

Other Natural Barriers

Trees, such as *Casuarina equisetifolia, Waru, Pandalis Odoratissimus and Cocos Nucifera*, can also act as a natural barrier against tsunami. Contrary with mangroves, *Casuarina Equisetifolia* grows fast on sandy coastlines, needs less care and used for coastal stabilization. This tree is always green, around 15-20 m high with a diameter of 40-50 cm and used as windbreak on coastal areas. The only area which was not damaged just after Indian tsunami in 2004 in Hambantota City was Casuarinas region and local people state that casuarinas has increased dimensions of sand dune around 1 m (Zoysa, 2008). Casuarinas are an alternative protection method against small scale tsunamis (H_{in} <6m). Wolanski (2007) reported that Casuarinas absorb wave energy during 2 or 3 m high tsunamis. During bigger tsunamis, trees break up or uprooted. For example, *Casuarina equisetifolia* plantation was not able to protect most densely populated areas from tsunami effect in Sri Lanka. This shelterbelt with a width of 10-15 meters has been terribly damaged under 6-9 m high tsunami waves (Forbes and Broadhead, 2007).

In tropical regions, coastal forests consisting of Waru trees against tsunami attack are suggested. Waru tree is a kind of Zelkova tree and mostly planted on Java islands. It grows rapidly and reaches to a few meters in several years. For this reason, waru trees can be used for protecting residences on coastal regions from tsunami (Hiraishi and Harada, 2003). A forest simulation created with Sisano, Papua New Guinea Waru (*Hibiscus Tiliaceus*) trees has shown a significant decrease in hydraulic force and inundation depth of tsunami. 4 big waru trees in an area of 100 m^2 reduce hydraulic force from 275000 Newton to 90000 Newton (by 67%) (Forbes and Broadhead, 2007).

Pandanus Odosratissimus and Cocos Nucifera, dominant coastal vegetation types in Sri Lanka, can also be used for reducing tsunami effect (Nandasena and Tanaka, 2007). After

Showa Sanriku Tsunami took place in 1933, coniferous trees have been planted on northern regions of Japan, affected by this tsunami disaster, as a protective coastal structure (Edward, 2006).

Effect of Coastal Forests on Tsunami Run-Up Height

There are studies being carried out in order to determine performance of artificially planted coastal forests in reducing tsunami effects. In this context, a detailed experimental study has been published by Irtem et al. (2009). In this study, the effect of coastal forest on tsunami run-up height has been experimentally analyzed and empirical equations have been proposed. The experimental setup used in this experiment has been given in Section 2.

The coastal forest model consisting of artificial pine trees, with a height of approximately 9 cm and a diameter of 4.6 cm, is called as FM-I (Figure 4). Artificial trees are placed on inclined beach in three different layouts: rectilinear (Case I), staggered (Case II) and dense rectilinear (Case III). The plan and the geometrical characteristic of these cases are shown on Figure 5.

Figure 4. Artificial trees used in the model (Irtem et al., 2005).

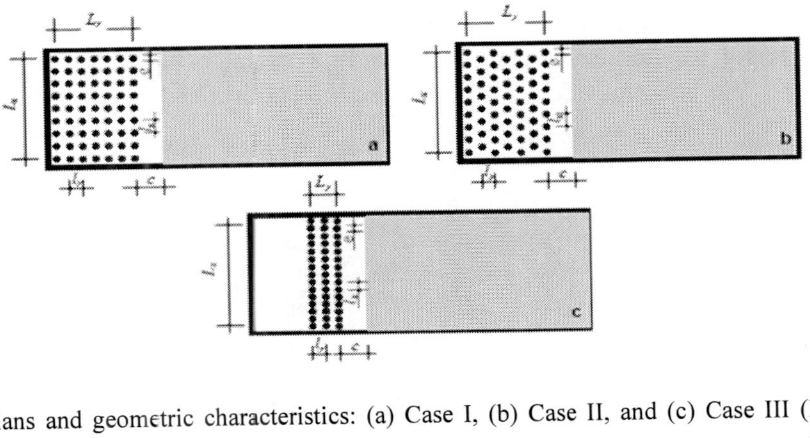

Figure 5. Plans and geometric characteristics: (a) Case I, (b) Case II, and (c) Case III (Irtem et al., 2005).

In Figure 5, c is the distance of the coastal forest model from SWL, l_x is the horizontal distance between trees, l_y as the perpendicular distance between trees, e is the diameter of artificial trees, L_x is the width of the coastal forest model and L_y is the length of the forest perpendicular to the coastline of the model.

The number of trees per unit area (ø) is obtained from Eq. 5.

$$\phi = \frac{N}{L_x L_y} \tag{5}$$

where, N is total number of trees.

Empirical equations obtained from regression analyze for Case I, II and III are given below (Eqs. 6-8).

$$\frac{R}{d} = 0.0048\left(\frac{HG_{sb}(\cot\beta)^{a}\, c(l_x - e)(l_y - e)}{H_{veg}\, DV_{veg}\,\phi\, d}\right)^{0.4158} \quad \text{for Case I} \tag{6}$$

$$\frac{R}{d} = 0.004\left(\frac{HG_{sb}(\cot\beta)^{a}\, c(l_x - e)(l_y - e)}{H_{veg}\, DV_{veg}\,\phi\, d}\right)^{0.4273} \quad \text{for Case II} \tag{7}$$

$$\frac{R}{d} = 0.0376\left(\frac{HG_{sb}(\cot\beta)^{a}\, c(l_x - e)(l_y - e)}{H_{veg}\, DV_{veg}\,\phi\, d}\right)^{0.2984} \quad \text{for Case III} \tag{8}$$

where, H_{veg} is height of a tree and V_{veg} is volume of a tree while a = 5/ Tan β.

In order to analyze effect of tree leaves on run-up height, the coastal forest model consisting of solid cylindrical timber sticks, with a height of approximately 9 cm and a diameter of 1 cm and equivalent to the volume of the artificial trees, is called as FM-II (Figure 6).

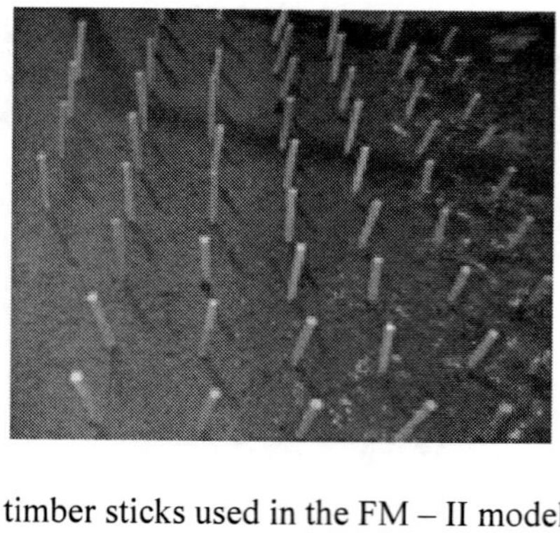

Figure 6. Cylindrical timber sticks used in the FM – II model (Irtem et al., 2005).

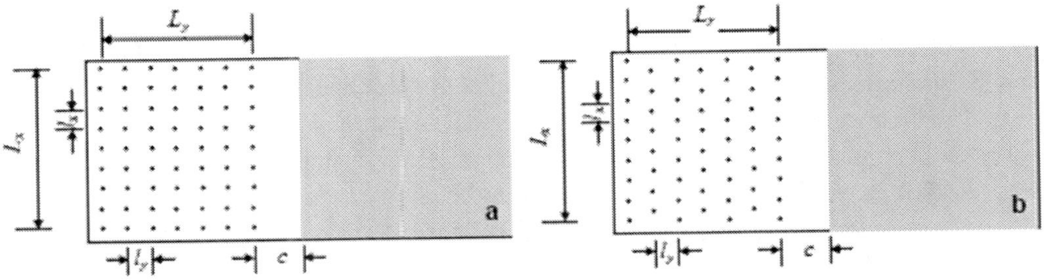

Figure 7. Plans and geometric characteristics: (a) Case IV and (b) Case V (Irtem et al., 2005).

Two patterns of sticks were used rectilinear (Case IV) and staggered (Case V). The plan and the geometrical characteristic of these cases are shown on Figure 7.

Empirical equations obtained from regression analyze for Case IV and V are given below (Eqs. 9-10).

$$\frac{R}{d}=0.0001\left(\frac{HG_{sb}(\cot\beta)^{a}\,c^{0.1}(l_x-e)(l_y-e)}{H_{veg}DV_{veg}\phi\,d^{0.1}}\right)^{0.6701} \quad \text{for Case IV} \qquad (9)$$

$$\frac{R}{d}=0.0011\left(\frac{HG_{sb}(\cot\beta)^{a}\,c^{0.1}(l_x-e)(l_y-e)}{H_{veg}DV_{veg}\phi\,d^{0.1}}\right)^{0.483} \quad \text{for Case V} \qquad (10)$$

Reduction rates on run-up heights in above mentioned five cases are given in Table 1. Accordingly, below mentioned results can be obtained:

Table 1. Extents of reduction of run-up height for different cases (Irtem et al., 2005)

CASES	c (cm)	Without Tree	CASE I c (cm)		CASE II c (cm)		CASE III c (cm)		CASE IV c (cm)		CASE V c (cm)	
			10	20	10	20	10	20	10	20	10	20
CASE I	10	0.37	-	-	-	0.11	-	-	0.15	0.21	0.16	0.18
	20	0.21	-	-	-	-	-	-	0.01	0.06	0.05	~0
CASE II	10	0.42	0.08	0.26	-	-	-	0.11	0.23	0.27	0.22	0.25
	20	0.30	-	0.11	-	-	-	-	0.10	0.12	0.06	0.10
CASE III	10	0.45	0.10	0.28	0.02	0.19	-	-	0.23	0.29	0.24	0.27
	20	0.35	0.03	0.17	-	0.07	-	-	0.15	0.18	0.13	0.16
CASE IV	10	0.25	-	-	-	-	-	-	-	-	-	-
	20	0.19	-	-	-	-	-	-	-	-	-	-
CASE V	10	0.25	-	-	-	-	-	-	0.06	0.09	-	-
	20	0.22	-	-	-	-	-	-	0.03	0.06	-	-

- For trees placed with a staggered case (Case II, c=10cm), run-up height is reduced around 26% when compared with the case of rectilinear trees(Case I, c=20cm) .
- When trees are placed close to SWL and dense rectilinear layout, they act as a barrier and reduce run-up height around 45% when compared with the case of without trees.
- When trees are placed in a rectilinear layout and in a staggered layout with a distance of 10 cm from SWL, they reduce run-up height by 15% and 22% respectively when compared with results obtained from cylindrical sticks. This result emphasizes the effect of tree leaves.

Moreover, all experiments have been repeated at slope 1:3.5 in order to confirm above suggested equations. It has been found out that dimensionless run-up height obtained from Eqs. 6-8 is agree with results measured (Irtem et al, 2009).

Conclusion

Years ago, human race have developed various methods in order to get protected from tsunami, abnormal tides and storm surges. Especially after Indian Ocean Tsunami on 26 December 2004 caused hundred of thousands death, studies aiming at reducing tsunami damages became intensified. Field study, experimental and numerical studies are still being carried out and discussed in this subject.

It will be highly important to estimate run-up height of tsunamis while designing coastal structures because tsunamis are mainly effective on coastlines around their run-up zones. This is why knowledge, empirical equations about tsunami run-up height are given. Studies show that roughness also affects run-up height. In armored beaches, run-up height is reduced by 50% when compared with not armored beaches.

The protective measures against tsunami can be classified in two groups as hard structures and natural barriers. Hard structures are compositions of two or more structures such as caisson, composite, rubble mound, submerged breakwater, inclined or vertical seawalls and groins. Hard structures have a negative impact on aesthetical aspect of coastal region, prevent interaction between human, nature and sea and have significantly high construction costs. However, there is no alternative for providing protection against high tsunami waves.

When tsunami waves proceed to landward, wave velocity and height reduce by coastal forest effect. However, if the tsunami is extremely high, then, coastal forests are not adequate in reducing damage caused by tsunami. Effect of coastal forests on tsunami run-up height is investigated as experimentally and empirical equations are proposed.

Below mentioned results are obtained from studies carried out in relation with effect of coastal forests:

- It has been found out that coastal forests reduce tsunami height by approximately 40% and maximum flow pressure by 90%
- If trees close to SWL and dense, they reduce tsunami run-up height by 45% when compared to without trees.

In high tsunamis, uprooting trees and broken branches create debris and cause death and damage in coastal region. Tsunami cannot be controlled because tsunami natural hazard. However, protective measures should be taken in order to minimize uprooting or breaking of trees.

Harada and Imamura (2005) found that four main functions to reduce tsunami disaster: to stop drifts or ships carried by a tsunami, to reduce tsunami energy, to form sand dune protecting tsunamis as well as high waves, and to catch persons carried back by a tsunami to the sea.

Coastal forests should not be destroyed for urbanization, industrialization, fishing industry, agriculture and coastal protection studies. Based on destruction of coastal forests and increase in population on coastal regions, human race exposed to various risks such as tsunamis and storm surges and this case causes material and moral damages.

There is no certain or standard measure reducing damages caused by tsunami. One or more combination of measures, mentioned above, can be applied. In terms of reducing tsunami damages, various physical parameters, such as tsunami wave height, wave incident angle, coastal bathymetry, beach slope, distance between coastline and coastal forests, their width, diameter, height and resistance of trees, as well as regional socioeconomic structure is important. For the success of these measures, field should be completely analyzed, projects should be supported with physical and numerical models and take lessons from the past hazards. Moreover, nonstructural measures (education, awareness, evacuation, etc) are also important and should not be ignored.

References

Al-Banaa, K.; Liu, P. Journal of Coastal Research. 2007, Special Issue 50, 201 – 205.

Bas, B. Examining stability of rubble mound breakwaters under tsunami wave attack; MSc Thesis, ITU, Institute of Science and Thechnology , 2007 [Turkish].

Carrier, G.F.; Wu, T.T.; Yeh, H., *Journal of Fluid Mechanics*. 2003, Vol. 475, 79.

CERC, Shore Protection Manual, USA, 1984.

CERC, Coastal Engineering Manual, USA, 2003.

Cuypers, K. Breakwater stability under tsunami attack for a site in Nicaragua; MSc Thesis, Faculty of Civil Engineering, TU Delft, 2004.

Dalrymple, R.A.; Kriebel, D.L. *The Bridge*. 2005, Vol. 35 (2), 4-13.

Dahdouh-Guebas, F.McGraw-Hill *Yearbook of Science and Technology,* McGraw-Hill Professional, New York, USA, 2006, 187-191.

Edward, J.K. P.; Terazaki, M.; Yamaguchi, M. *Coastal Marine Science*. 2006, Vol. 30 (2), 414-424.

Forbes, K.; Broadhead, J. *The role of coastal forest in the mitigation of tsunami impacts*; Food and agriculture Organization of the United Nations Regional Office, for Asia and the Pacific, Bangkok, 2007.

Gedik, N.; Irtem, E.; Kabdasli, S. *Ocean Engineering*. 2005, Vol. 32 (5-6), 513–528.

Hall, J.V.; Watts, J.W. *Tech. Memo*. No.33, Beach Erosion Board, US Army Corps of Engineers, 14 pp, 1953.

Hiraishi, T.; Harada, K. *Report of the Port and Airport Research Institute*. 2003, Vol. 42 (2), 1–23.

Hsiao, S.C.; Hsu, T.W.; Lin, T.C.; Chang, Y.H. *Coastal Engineering*, 2008, Vol. 55 (12), 975-988.

Hughes, S.A. *Coastal Engineering*. 2004, Vol. *51*, 1085–1104.

Irtem, E.; Gedik, N.; Kabdasli, M. S.; Yasa, N.E. *Ocean Engineering*. 2009, *Vol. 36, (3-4)*, 213-320.

Kanoglu,U.; Synolakis, C.E. *Journal of Fluid Mechanics*. 1998, Vol. 374, 1.

Kathiresan, K.; and Rajendran, N. Estuarine, *Coastal and Shelf Science*. 2005, Vol. 65, 601-606.

Kobayashi, N.; Karjadi, E.A. Journal of Waterway, Port, *Coastal, and Ocean Engineering*. 1994, Vol. 120 (6), 645–650.

Latief, H.; Hadi, S. "The role of forests and trees in protecting coastal areas against tsunamis", Coastal protection in the aftermath of the Indian Ocean tsunami: What role for forests and trees?, *Proceedings of the Regional Technical Workshop*, Khao Lak, Thailand, 28–31 August 2006, Compiled and edited by Susan Braatz, Serena Fortuna, Jeremy Broadhead and Robin Leslie, 2007.

Li,Y.; Raichlen, F. *Journal of Waterway, Port, Coastal, and Ocean Engineering*, 2001 Vol. 127, 33.

Liu, P.L.; Cho, Y.S.; Briggs, M.J.; Kanoğlu, U.; Synolakis, C.E. *Journal of Fluid Mechanic*. 1995, Vol. 320, 259-285,.

Li, Y.; Raichlen, F. Journal of Waterway, Port, *Coastal, and Ocean Engineering*. 2003, Vol. 129 (2), 47–59.

Liu, P.L.; Synolakis, C.E.; Yeh, H. H. *Journal of Fluid Mechanics*. 1991 Vol. 229, 675.

Nandasena, K.N.A.; Tanaka, N. *Annual Research Journal of SLSAJ*. 2007, Vol. 6, 16–21,.

Madsen, P. A.; Fuhrman, D.R. *Coastal Engineering*. 2008, Vol. 55 (3), 209-223.

Mammadov, N. *Ecology Magazine*, 2004, 4.

Maiti, S.; Sen, D. *Ocean Engineering*. 1999, Vol. 26, 1063.

Pedersen, G.; Gjevik, B. *Journal of Fluid Mechanics*. 1983, Vol. 135, 283.

Pelinovsky, E.; Troshina, E.; Golinko, V.; Osipenko, N.; Petrukhin, N. *Phys. Chem. Earth (B)*. 1999, Vol. 24, 431.

Plas, T.V.D. A Study Into The Feasıbılıty Of Tsunamı Protectıon Structures For Banda Aceh and A Prelımınary Desıgn Of An Offshore Rubblemound Tsunamı Barrıer", MSc Thesis, Delft University of Technology, 2007.

Reeve, D.; Chadwick, A.; Fleming, C. *Coastal Engineering*, Taylor and Francis, Spon, London, 2004.

Shankar, N. J.; Jayaratne, M. P. R. *Ocean Engineering*. 2003, Vol. 30 (2), 221-238.

Sonkar, L. P.; Kumar, U. *Country's Action Plan on Tsunami Countermeasures*- India. Comprehensive Tsunami Disaster Prevention (CTDP) Training Course, 2008.

Synolakis, C.E.; Skjelbreia, J.E. *Journal of Waterway, Port, Coastal, and Ocean Engineering*. 1993, Vol. 119, 323.

Synolakis, C.E. *The runup of long waves*. PhD Thesis, California Institute of Technology, 1986.

Synolakis, C.E. *Journal of Fluid Mechanics*. 1987, Vol. 185, 523.

Tadepalli, S.; Synolakis, C.E. The run-up of N waves on sloping beaches. *Proc. R. Soc. Lond. A*., 1994, 445, 99-112.

Teng, M.H.; Feng, K.; Liao, T.I. Experimental study on long wave run-up on plane beaches. *Proc. of the Tenth International Offshore and Polar Engineering Conference,* Seattle, USA, 2000

Van der Meer, J.W. *Stability of Cubes, Tetrapods and Accropodes*, Design of Breakwaters, Thomas Telford, 1988.

Van der Meer, J.W.; Ligteringen, J. *Breakwater Design*, IHE, Delft, 1997.

Yanagisawa, H.; Koshimura, S.; Goto, K.; Miyagi, T.; Imamura, F.; Ruangrassamee, A.; Tanavud, C. Estuarine, *Coastal and Shelf Science.*2009, Vol. 81(1), 27-37.

Yuce, A. *Experimental analysis of the effects of long waves on vertical faced coastal structures*; MSc Thesis; ITU, Institute of Science and Thechnology , 2007 [Turkish].

Wolanski, E.,2007. "Synthesis of the protective functions of coastal forests and trees against natural hazards", Coastal protection in the aftermath of the Indian Ocean tsunami: What role for forests and trees?, *Proceedings of the Regional Technical Workshop*, Khao Lak, Thailand, 28–31 August 2006, Compiled and edited by Susan Braatz, Serena Fortuna, Jeremy Broadhead and Robin Leslie, 2007

Zoysa, M.D., *Small-scale Forestry*. 2008 Vol. 7, 17–27.

In: Tsunamis: Causes, Characteristics, Warnings and Protection ISBN: 978-1-60876-360-3
Editors: N. Veitch and G. Jaffray, pp. 131-147

Chapter 6

RESPONSE OF COASTAL VEGETATION AND THE NEED FOR GREEN BELTS ALONG THE TAMIL NADU COAST, INDIA: THE DECEMBER 2004 TSUNAMI EXPERIENCE

Antonio Mascarenhas[1]
National Institute of Oceanography,
Dona Paula 403004, Goa, India

Abstract

The Indian Ocean tsunami of December 2004 transmitted multiple lessons: whereas certain strips were destroyed in totality, others remained unaffected or practically intact. Detailed post-tsunami field surveys conducted along the coast of Tamil Nadu, south India, in April 2005 and January 2006, confirmed that coastal landforms and vegetation played a significant role in neutralizing the force of virulent waves. Impact on casuarina forests was restricted to a maximum of 25 meters from the dune line. Only frontal casuarina strips were attacked, bent and stripped of their leaves by wave up-rush. Whereas dune creepers and herbs were uprooted, coconut and palm trees remained in position. This phenomenon was verified along the entire Tamil Nadu and Pondicherry coasts. Evidence of minimal damage to casuarina plantations and coconut groves supports the view that biological buffers can serve as efficient energy dissipaters during powerful oceanographic events. Villages located behind dense plantations remained safe. In 2006, natural restoration was identified in the form of rejuvenated and healthy vegetal species. Dune vegetation had bounced back and bent casuarinas had sprouted.
The need for a protective coastal buffer zone is proposed. Its levels of effectiveness will depend on a progression of species landward from the shore. Casuarinas should not be located on dunes, but planted further backshore. Herbs – shrubs – bushes – trees form a gradation of species and a natural slope that is inferred to offer protection as natural shelter belt against any eventual extreme event. Future plantation strategies will have to consider natural bio-zonation rather than haphazard patterns that are observed at present. Green belts are beneficial for several reasons: control of erosion, stabilization of shores, alleviation of wind energy, effective buffer against the force of waves, preservation of biodiversity and advantage in terms of food, shelter and income.

---

[1] E-mail address: mascarenhas@nio.org.

It is established that physical and geological processes are intense along the open ocean. Manmade structures thus experience extreme processes as such sites become vulnerable to natural hazards. In comparison, the forested hinterland is sufficiently protected from physical forces as vegetation attenuates energy from onrushing waters. Therefore, elevated coastal stretches with protective vegetation are the only environments where risks due to extreme oceanographic events are modest.

Introduction

The alarm triggered by the tsunami of December 2004 is gradually being forgotten. Nonetheless, in the aftermath of an oceanographic episode of such a magnitude, several critical topics have been and are still being debated: sustainability of hazard-prone coasts, urbanization of vulnerable seafronts, rebuilding of devastated shores, relocation of infrastructure, need for broader buffer zones and fate of coastal ecosystems during extreme events.

Of these, the functions of coastal landforms and the protective role of coastal vegetation were the key issues. The audio visual and the print media were quick to suggest that plantations bordering the coast proved to be the saviours of seaside dwellers in India (Ganesan, 2005) and in China (Anonymous, 2005). These assumptions were based on quick observations during the few days that followed the disaster.

Several months later, the results of scientific research started pouring in. In Sri Lanka, mangrove sites played a critical role in storm protection, depending on the natural quality of the mangrove forest (Dahdouh-Guebas et al., 2005). Reports from India, Sri Lanka, Indonesia and Malaysia also suggest that communities living behind intact mangrove forests, in particular, suffered less destruction (Williams, 2005). Similarly, around the Parangipettai coast of south India, mangroves protected human life and property; death toll was the highest where the vegetal cover was the least (Kathiresan and Rajendran, 2005). The use of mangrove thickets as absorbers of wave power was further confirmed by additional data (Vermaat and Thampanya, 2006). In the Pichavaram complex, mangrove forests and woody vegetation provided protection from the tsunami (Olwig et al., 2007). It may be recalled that the utility of mangroves as a coastal defense against cyclones was already documented earlier (Clark, 1996; Nayak et al., 2001; Badola and Hussain, 2005).

Backshore, the taller inland trees played an equally defensive role. Casuarina forests remained intact, except for the narrow sea facing strip (Mascarenhas, 2006). Villages behind the dense plantations were saved, as loss of life was minimal (Danielsen et al., 2005; Mascarenhas and Jayakumar, 2007, 2008). The validity of casuarina trees as shelter belts were nevertheless questioned by others (Bhalla, 2007).

Most of the work cited above deals with mangroves, confirming that the salt tolerant frontal intertidal trees, acted as biological buffers against high energy oceanic episodes. In comparison, there are few field verification reports and empirical data related to the most conspicuous coastal forests of Tamil Nadu – casuarinas. The objective of this chapter is to document and assess the defensive role of coastal vegetation in the wake of the December 2004 tsunami, with particular reference to casuarina plantations.

Vegetation along Tamil Nadu Coast

Tamil Nadu is one of the two southernmost littoral states of India, and is bordered by the Bay of Bengal (Figure 1). The coast is characterized by linear sandy beaches mostly backed by sand dunes of varying heights and morphology. From north to south, the coast is intersected by several rivers flowing eastward. The coastal sand dunes of Tamil Nadu are colonized by a wide variety of vegetal species ranging from grasses, shrubs to tall trees. Some of the major plants are briefly summarized below:

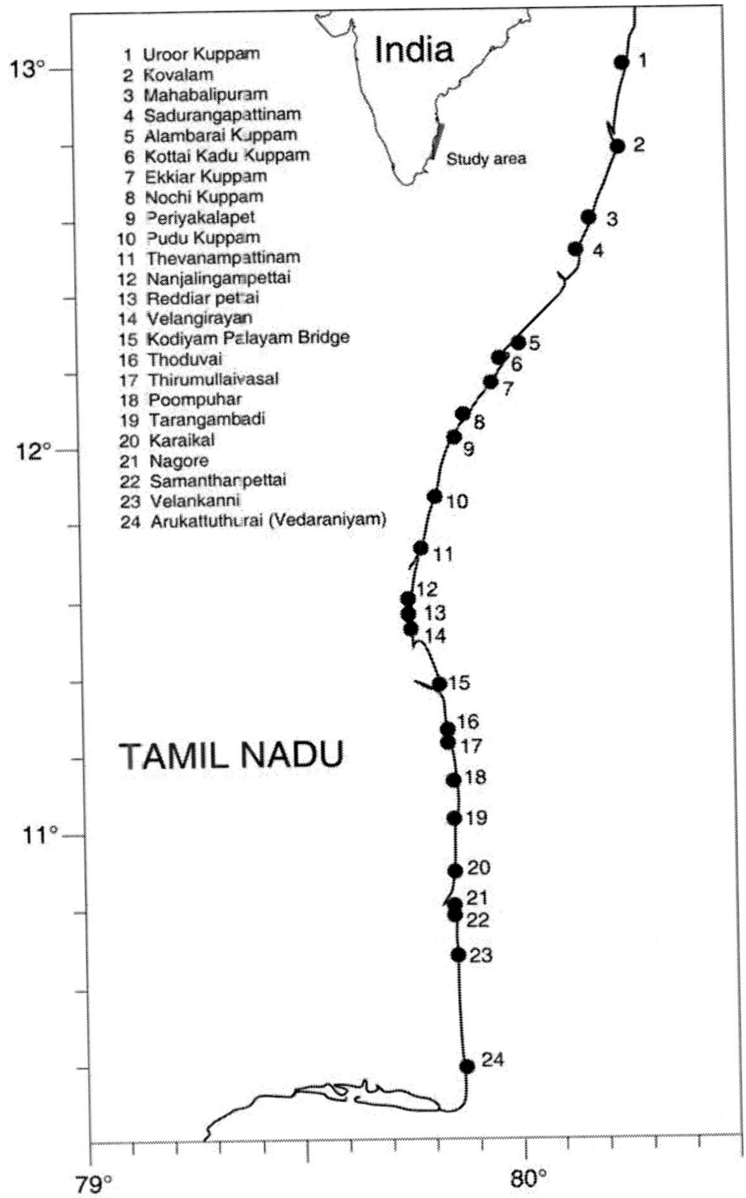

Figure 1. Location of sites along the coast of Tamil Nadu where post tsunami surveys were carried out in April 2005 and in January 2006.

a) Dune grass is often found in patches, on frontal dunes that merge into the beach,
b) Ipomoea creepers with bright purple flowers are abundant all along the coast. At places, a thick carpet covers large areas of pioneer dunes. This species does not need any special care,
c) Spinifex, the spiny variety of dune vegetation, can also be observed at many sites,
d) Spiny shrubs, about 1-2 m in height, can be noted mostly on hind dunes,
e) Casuarina trees are the most prominent all along as large stretches of Tamil Nadu coast are marked by luxuriant casuarina forests. Rising to a height of 30 to 40 m, these species form luxuriant coastal forests that can be noticed by any traveler who passes by. Some of the dense casuarina plantations are found in Mahabalipuram, Vada Nemelli, Nanjalingampettai, Karaikal, Silladi, Samanthanpettai and Puatadi,
f) Coconut groves are also found all along the coast. Dense patches of such fruit bearing trees are observed mostly amongst coastal dwellings, as in Nagore,
g) Tall palms are seen in a few places as in Mahabalipuram and Pondicherry,
h) Other types of inland trees such as acacia are occasionally seen on the coast, as in Mahabalipuram and Karaikal.

It is pertinent to note here that there is no proper zonation of vegetation as one moves in a perpendicular direction from the beach to the hinterland. Although the natural species such as dune grass and 'ipomoea' creepers are located on pioneer dunes, other man made forests such casuarinas are planted haphazardly. These trees have even been planted along the dune line, an improper habitat for casuarinas. The same random fashion is noted in the case of coconut trees. The disadvantages of such indiscriminate coastal plantations will be discussed in later sections of this chapter.

Methods

Post-tsunami surveys were carried out along the coast of Tamil Nadu from Chennai in the north to Vedaranyam in the south during April 2005 and January 2006. A total of 24 major locations were investigated (Figure 1). Several other intervening sites (not shown in Figure 1) were also studied. The survey consisted of ascertaining the inundation limits, run-up heights at different locations and measurement of cross-section profiles (Mascarenhas and Jayakumar, 2008). Damage to coastal dwellings and impact on vegetation was also assessed. In addition, various types of impacts of the tsunami on coastal landforms were noted and the width of strips of broken houses and injured vegetation were measured.

Beach profiles were recorded using a surveyor's Dumpy Level (Karl Zeiss) and a graduated staff. Beach heights along the line of sight were taken from the water line till the tsunami inundation limit. In general, profiles were measured at an interval of 5 m. Larger distances were considered where the beach is flat whereas smaller intervals were used where the beach is undulating with features like dunes, ridges or valleys. Beach profiles were taken till the inundation limit wherever this point was accessible for survey equipment.

Response of Coastal Vegetation to the Tsunami Attack

Tsunami run-up levels ranged from 0.7 to 6.5m above sea level and flooding varied from 31 to 862m from the swash zone (Chadha et al., 2005; Ramanamurthy et al., 2005). Based on our results, the average inundation was around 247 m (Mascarenhas and Jayakumar, 2008). As a consequence, the high energy tsunami waves extensively transformed the natural and the man made backdrop of Tamil Nadu coast.

Coastal vegetation reacted in various ways. The manner in which various types coastal vegetal species was affected and/or changed during and after the event is attempted here. The scenario that emerged is listed in Table 1. The post tsunami surveys carried out in April 2005 and January 2006 on damage to plantations along the coast of Tamil Nadu are briefly discussed below:

Table 1. Characteristics of some the locations surveyed along Tamil Nadu coast, from north to south, where post-tsunami damage studies were carried out. Beach width (April 2005) is measured from the water line mostly to the base of dune (wherever identified). High dunes refer to those higher than 4 m. Damage to trees refers to the strip of damaged vegetation, measured physically, mostly from the dune line. Figure 'zero' signifies that damage was nil

Coastal site	Characteristics / state of beach, dune and vegetation prior to December 2004 tsunami	Wave run-up (m)	Damage to coastal vegetation (m) (April 2005)	State of vegetation / forests (January 2006)
> Besant Nagar, Chennai > Southern part	> Inland trees on high dunes > Dunes razed; huts on beach	5.0	0 -	> Trees survived in totality; dune creepers have returned > No evidence of vegetation
Vada Nemelli	High dunes capped by vegetation; extensive casuarina forests	-	13 - 25	Only frontal trees perished; casuarinas have revitalized; uprooted dune plants are now growing normally
Mahabalipuram	A tourist spot, flattened dunes, scattered trees	5.4	0	Patches of casuarinas are still seen
Periyakalapet	Shorefront dwellings within coconut groves	5.3 - 6.5	0	All coconut trees survived wave attack
Thevanapattinam	Small dwellings within coconut groves	3.8	0	No damage to coconut trees
Nanjalingampettai	High dunes with coconut plantations; casuarinas are wide spread	3.7	10 - 17	Sea facing strip of casuarinas suffered nominal damage; bent trees are sprouting
Poompuhar	A pilgrimage site, marked by tall palm trees and coconut groves	2.2	0	Existing 19 palm trees were found unharmed

Table 1. Continued

Coastal site	Characteristics / state of beach, dune and vegetation prior to December 2004 tsunami	Wave run-up (m)	Damage to coastal vegetation (m) (April 2005)	State of vegetation / forests (January 2006)
Vanagiri	Historical shore temples; sparse vegetation	-	-	Few existing trees survived
Tarangambadi	Dense dwellings along the shore	1.9 - 3.4	0	Whereas all dwellings were flattened, existing trees remain
> Karaikal, north > South of creek	> A tourist spot, sparse vegetation > Luxuriant casuarina forests	2.6 -	0 ~15	> No damage to trees > Casuarina forest is intact except for bent trees within the frontal strip
Nagore	Houses, huts of fisher folk amidst dense coconut trees	-	0	All houses collapsed; except few, coconut trees however remained intact
Silladi	High vegetated dunes; extensive casuarinas plantations	1.7 - 3.1	7 - 19	Only frontal trees were bent and stripped of bark; rest are surviving
Samanthanpettai	Conspicuous dunes with vegetation	1.9 - 3.3	5	Casuarinas were partly affected; dune vegetation has recuperated
Nambiar Nagar	Prominent, high sand dunes	-	0	Backshore trees behind dunes are intact
Nagapattinam lighthouse	Flattened dunes, huts,	-	0	Nominal damage to a few trees
Akkaraipettai - Keechankuppam	A thickly populated village on the coast; scattered trees	-	-	Severe damage to dwellings; a patch of casuarinas still standing; scattered coconut trees survived
Kallar	A low, flat, flood prone area	-	0	Few shrubs were uprooted
Velankanni	A pilgrimage site; flat beach with thatched huts and make-shift restaurants	2.1 - 4.7	-	Structures washed off in totality; backshore plants are still standing
Puatadi	A pristine coast with well formed, high vegetated sand dunes; thriving casuarinas forests	-	0	Casuarina trees remained unaffected; presently growing normally

Dune grass: almost all sites visited revealed a complete removal of dune grass. Being frontal species often found on the upper beach, the wave up rush left no trace of frontal dune vegetation anywhere on the shorefront.

Dune creepers: The most affected dune vegetation comprises 'ipomoea' creepers and 'spinifex' that carpet frontal sand dunes. At Uroor Kuppam near Chennai, the wide beach stretch was found with shredded 'ipomoea' creepers. Large beach areas were rendered bare. At Nanjalingampettai, 'spinifex' was uprooted, whereas at Nagore and Silladi, there were no signs of any living green material on damaged dunes (Mascarenhas and Jayakumar, 2008).

Casuarina forests: Most remarkably, coastal casuarina plantations survived the tsunami invasion (Mascarenhas, 2006; Figures 2A and 2B). Only frontal casuarina strips were attacked, bent, and stripped of its leaves by wave up-rush (Mascarenhas and Jayakumar, 2008; Table 1). Some of the notable places where such a phenomenon was confirmed along the Tamil Nadu coast are Vada Nemelli, Mahabalipuram, Nanjalingampettai (Figure 2A) and Silladi (Figure 2B) in particular, Samanthanpettai near Nagapattinam and at Puatadi, south of Velankanni. Similarly, southern Karaikal in Pondicherry that is marked by large stretches of casuarina plantations, revealed that the few trees that perished were those found within a narrow seaward frontal strip.

Coconut trees: Similarly, coconut trees also showed appreciable resistance against the tsunami wave attack. Our observations reveal that coconut plantations remained intact, although, at places, these trees were located dangerously close to be beach. Nagore can be cited as an example where flood waters penetrated 860 m inland (Jayakumar et al., 2005), and dwellings up to 132 m from the dune line were destroyed. Here, all coconut groves survived. Thevanapattinam is another site where coconut trees were unharmed.

Other plants: Palm trees are a common feature along Tamil Nadu coast. These species also displayed an analogous reaction. At Poompuhar, all the original 19 palm trees are standing. A similar situation was noted in Periyakalapet in Pondicherry. At Cuddalore, within a children's' park located backshore, all play facilities were severely damaged whereas all inland adult trees within the complex survived.

Figure 2. Continued on next page.

Figure 2. The response of casuarina plantations in the wake of the tsunami was noteworthy. 2A: At Nanjalingampettai, only a few trees from the sea facing narrow strip were uprooted; 2B: At Silladi, the tsunami up rush came to a halt at the sea side boundary of the dense plantation (Photos: A. Mascarenhas and S. Jayakumar, NIO).

Role of Sheltered Coasts Compared to Built-Up Coasts

It is well known that the east coast of India has been annually by struck recurring cyclones of varying intensity over centuries (Shreshta, 1998; see Mascarenhas, 2004). The tsunami of December 2004 served as an eye opener. The sea shore is a site of extraordinary release of energy (Pilkey et al., 2000). Although our coastal managers may not admit, field evidence unequivocally confirms that there is no way of protecting a coastal structure from a direct hit by virulent oceanic hydrometeorological events.

However, our observational analyses of the extensive surveys supported by measurements of tsunami damage reveals that natural coasts indeed possess inherent protective properties. As such, coastal vegetated landforms are capable of significantly attenuating incoming wave energy. Coastal sand dunes and associated vegetation is of prime importance here.

To prove this point, two sites along Tamil Nadu coast have been selected. Beach profiles from the water line to the uppermost accessible point are presented. Both places are in fact coastal villages with contrasting geomorphology, the first being a village with an intervening forest, and the second a cluster of dense dwellings along the open sea front.

1) *Nanjalingampettai:* The coast is characterized by prominent sand dunes more than 5 m high and very steep seaward gradients as seen in the profile (Figure 3). Coastal vegetation is marked by luxuriant casuarina forests interspersed with dense coconut groves (Table 1). The wave up-rush was as high as 3.7 m. It is evident that the wave stopped at the dune, the dune crest being higher. Thick vegetation further aided a complete attenuation of the wave onslaught (Mascarenhas and Jayakumar, 2008).

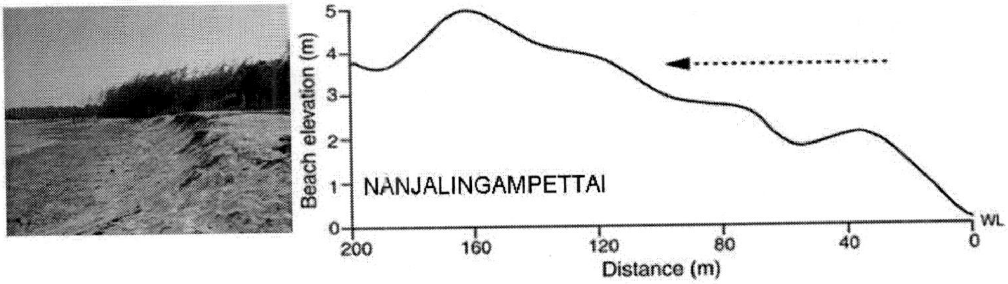

Figure 3. An example where a village remained unscathed, being located behind dense casuarina plantations (and high sand dunes) at Nanjalingampettai. The arrow indicates the tsunami wave run up (Photo: A. Mascarenhas and S. Jayakumar, NIO).

It is pertinent to note here that an entire village is located behind the dense plantation. Natural forested landforms such as well preserved dunes and casuarina forests acted as natural biological shields by saving life and property. There was no damage to any habitation behind the forested dune complexes as houses behind densely vegetated high dunes are all intact, and loss of life was nil.

Several similar situations can be cited. Silladi, Samanthanpettai and Puatadi are some examples where villages behind casuarina forests remained secure.

2) *Tarangambadi:* Before the tsunami, this sea front presented a dismal setting. The entire coastal strip, right up to the beach, was occupied by dense dwellings occupied by a large fishing community. Coastal geomorphology was obliterated, as sand dunes were flattened in favour of houses (Figure 4). There was no evidence any noteworthy vegetation. Here, the wave run-up of 2.4 m bypassed the flat beach thus razing whatever came its way. As they lacked natural protection, all the beach front houses disappeared (Mascarenhas and Jayakumar, 2008).

The above is a case where the interference with geomorphic features (dunes) and lack of natural protection (trees) is the cause of human disaster. Loss of life was heavy. Similarly, several coastal hamlets that lacked natural protection were smashed by the powerful incoming waves. Our surveys showed complete destruction along the kilometre long Akkaraipettai – Keechankuppam sector, Nagore and Velankanni in particular. Having been hit directly, these structures which were dangerously close to the sea were eliminated in totality.

Published work also shows that areas with coastal tree vegetation were markedly less damaged than areas without (Danielsen et al., 2005; Olwig et al., 2007). The villages on the coast were completely destroyed, whereas those behind the mangrove suffered no destruction even though the waves damaged areas unshielded by vegetation north south of these villages. Similarly, five villages located within casuarina plantations experienced only partial damage, as these plantations were undamaged, except for rows of 5 to 10 trees nearest to the shore. This observation is in conformity with our results. Mangroves and casuarina plantations attenuated tsunami-induced waves and protected shorelines against damage, as tree vegetation shielded coastlines from tsunami damage by reducing wave amplitude and energy (Danielsen et al., 2005).

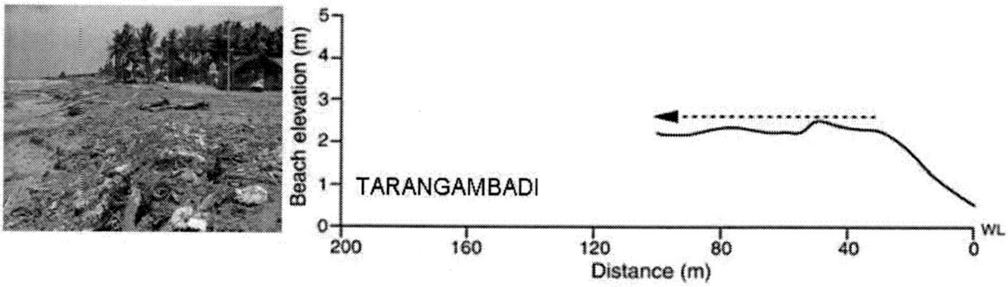

Figure 4. An example where lack of natural protection resulted in a complete destruction of sea front dwellings at Tarangambadi. The arrow indicates the tsunami wave run up (Photo: A. Mascarenhas and S. Jayakumar, NIO).

Earlier research had also contended that Sunderban mangrove of Bengal suffer less from wind and surges (Clark, 1996). The defensive role of mangroves during cyclones as natural barriers protecting the life and property of coastal communities has been demonstrated at Bhitarkanika on the northern Indian east coast (Badola and Hussein, 2005), and by post-cyclone imageries of Orissa coast (Nayak et al., 2001).

A question that is being raised is the levels of natural protection that a dense woody coastal plantation can offer. Empirical evidence for casuarina is definitely lacking. Experimental studies however show that mangrove forests attenuate wave energy. The rate of dissipation of waves depends on the density of forests and the diameter of tree roots and trunks (Massel et al., 1999). Currents thus flow around rather than through the forest (Furukawa et al., 1997). In similar experiments in Indonesia (Hiraishi, 2004), the tsunami height and velocity was estimated in the original topography and in the coastal line with greenbelt 100m wide. In the numerical simulation, the pressure value significantly decreases as the number of N increases (N is the density, defined as the number of trees in a 10m ×10m plot). Maximum flow pressure is high for N=0; pressure decreases as 'N' increases; for N=30, flow pressure value is negligible. The reduction of water flow pressure demonstrates the role and validity of a greenbelt to reduce the tsunami inundation and damage along Indonesian coastlines (Hiraishi, 2004).

Nevertheless, the profusely vegetated stretches of Tamil Nadu coast displayed an exceptional resilience by dissipating high waves. Our field measurements confirmed that only the frontal strip of casuarina woodland ranging from 0 to 25 m were twisted, bent and stripped of leaf cover (Figures 2A and 2B). Detailed observations on the damage inflicted on sea side forests are given in Table 1. These comments reiterate that coastal vegetation in general and casuarinas in particular played an essential role as efficient biological buffers against powerful oceanographic episodes (Mascarenhas, 2004, 2006; Danielsen et al., 2005; Mascarenhas and Jayakumar, 2007, 2008).

Post Tsunami Restoration Efforts

In the aftermath of the tsunami, fear and grief was written large. Majority of people who perished unfortunately belonged to the lower strata of society, mostly the fisher folk. As such, these communities were neither aware of the impending dangers from natural hazards nor had a proper access to any management or mitigation techniques.

Having realized that coastal plantations were indeed beneficial as bio shields, massive plantation programs were initiated, both at the government level as well as by local communities. One year later, significant rebuilding activity was noted along the entire coast of Tamil Nadu, as noted in January 2006.

Vast sandy stretches from Chennai to Mahabalipuram were covered by plantation programs. Thousands of casuarina trees have been planted on the bare dunes impacted by the tsunami. Unfortunately, at several sites as in Silladi, casuarinas were planted too close to the beach, sometimes by leveling dunes which the tsunami had only partially damaged. Similarly, at some sites, the typical dune plants 'spinifex' was noted. Planting these species on damaged dunes is however noteworthy.

Haphazard methods adopted for greening the coast leaves a lot to be desired. In the past, wrong trees were fixed at wrong places. Hinterland vegetal species were planted along the beach. Being too close to the coast, it is doubtful if these saplings will survive. Cases of inappropriate methods of afforestation are plenty. Obviously, the selection of species for coastal afforestation lacks prudent planning.

The worst scenario was noticed at Akkaraipettai, where an entire coastal village was wiped out, inflicting a heavy loss of life and property. Apparently, a panic reaction by the frightened people lead to a selection of wrong sites for plantation without a suitable plan (Figure 5A). Planting coconut trees on the beach is a most unscientific instance that one can encounter. An existing dune was razed, and a "sand barrier" was created along the water line by the villagers, apparently for "protection" against future killer events. During our surveys in April 2005, we identified several rows of coconut saplings planted on the beach very close to the waterline (Figure 5B). In January 2006, the "barrier" was gone, an erosive scarp took its place, and all the coconut saplings had perished (Figure 5C). A lack appropriate administrative, management and plantation guidelines is evident.

Figure 5. Continued on next page.

Figure 5. An inappropriate planting strategy adopted by the local community at Akkaraipettai. 5A: Panic in the aftermath of the tsunami induced some villagers to resort to afforestation for 'protection' against future tsunamis (Photo: Rajendran, Press Information Bureau, 12 January 2005); 5B: In April 2005, small coconut saplings were identified along the beach, close to the waterline; 5C: In January 2006, all the saplings had perished. Note that the long wall was built subsequent to the tsunami, apparently as a 'protection' to infrastructure of Nagapattinam fishing harbour (Photos 5B and 5C: A. Mascarenhas and S. Jayakumar, NIO).

Need for Functional Green Belts

Although this chapter makes a case for casuarina forests as a potential natural buffer, certain doubts about the utility of these trees in the wake of extreme oceanic events have been raised. Extensive casuarina plantations established (in the 1990's) as a cyclone protection measure along the Orissa coast were ineffective in preventing damage; rather, they caused destruction to Olive Ridley sea turtle (see Badola and Hussein, 2005). Casuarina trees have been found to erode during storms even though they survive wind speeds greater than 100 km per hour (Dash, 2002). Similarly, the idea for large-scale plantation of casuarina as bioshields is questioned as vegetation did not have a significant impact on the inundation distance (Bhalla, 2007). Therefore, although casuarina trees did perform as an amazing natural barrier against the tsunami, the same may not hold in the case of a cyclone.

India does not have a science oriented coastal hazards policy. A national legislation on the conservation of coastal vegetation does not exist either (Mascarenhas, 2004). One can only find scattered suggestions, rarely followed by the littoral states of the country. A question thus arises whether the plantation programs should be undertaken and, if so, the manner in which these should be accomplished, by taking the coastal morphology into account.

The Government of Tamil Nadu has launched an ambitious project under the environment and forest department (Vaithilingam, 2005). One of the objectives is to raise shelterbelts in coastal areas as protective shield against the tsunami. An amount of Rupees 648 lakhs has been allocated in 2006. It has also been proposed to cover the remaining parts of the coast with a bio-shield in phased manner, for which a proposal costing about Rupees 10 crores for repair and reconstruction of coastal shelterbelt has been forwarded to the World Bank for funding.

One year after the tsunami, natural restoration processes were identified as follows: (a) Dune vegetation that was torn to shreds by the tsunami was actively growing in January 2006. At places, the entire backshore that was devoid of creepers one year ago, is presently characterized by a rich carpet of 'ipomoea' creepers, (b) Remarkably, frontal casuarina trees that were bent and stripped by the tsunami impact have bounced back to life. In January 2006, casuarinas have revitalized (Figures 6A and 6B). Sufficient evidence can be forwarded in favor of natural vegetal rejuvenation along Tamil Nadu coast.

Figure 6. Continued on next page.

Figure 6. Examples of natural rejuvenation of casuarina plantations. 6A: One year after the tsunami, bent and damaged frontal trees have revitalized, and 'ipomoea' creepers have carpeted the dune at Nanjalingampettai; 6B: Stripping of the bark of trees denotes the height of the wave. Over a year later, in January 2006, a majority of such trees had sprouted back to life (see the encircled portions) (Photos: A. Mascarenhas and S. Jayakumar, NIO).

If a transect across a normal, undisturbed dune coast is considered, a natural disposition of vegetal species is noticed. Grass and creepers cover the dune whereas taller bushes are found further backshore. Such an inherent zonation of floral species is rarely observed, particularly where human interference such trampling, sand mining or dune leveling is rampant. That is the reason why plantation programs tend to follow a haphazard path. As such, even 10-15 m tall casuarinas are sometimes found right along the dune line. Such inappropriate methods are to be avoided.

Green shelter belts can be created by strip planting of shorelines and this concept should be based on social forestry, eco-development and participatory planning (Clark, 1996). Such buffer zones serve several purposes: (a) Shrubs control erosion and stabilize the shore; (b) Green belts significantly alleviate wind energy thus protecting the hinterland from oceanic forces; (c) A greenbelt of trees effectively reduces the force of devastating storm surges and waves; (d) Trees are beneficial for biodiversity and can induce habitats for wild life (Clark, 1996); (e) People inhabiting hazard prone coasts would benefit from green belts in terms of security, access to food, materials for shelter; and even income (Valdiya, 2001); (f) Strips behind the green belts serve as areas of peace and tranquility. As such, barren strips have to be reserved for afforestation; sea side strips can be acquired and converted into recreation parks, forests, gardens, agriculture farms or wild life sanctuaries (Valdiya, 2001).

For functional green belts, a gradation of species from the edge of open sea coastline towards the hinterland is needed: a pioneer zone of shallow rooted herbs such as grass, 'ipomoea' and 'spinifex'; a midshore zone of medium rooted shrubs, bushes and dune plants; and a backshore zone of deep rooted hydric species of taller trees as casuarinas, coconut, and eventually fruit bearing trees on higher land. This gradation forms a natural slope that: (a) forces winds (as those associated with cyclones) to deflect upwards, and (b) induces onrushing waters (such as those from powerful hydrometeorological events) to attenuate energy.

Therefore, preservation of natural landforms and encouragement of afforestation are crucial. A suitable policy for hazard management by considering forested natural landforms may well be the only option to save inhabitants in the Indian coastal zone from the ravages of hydrometeorological events.

Conclusion

(1) The ecosystem services provided by coastal vegetation have rarely been considered by the polity that governs littoral states and the officials who deal with management issues. (2) Field surveys and measurements confirm the role of coastal forests as efficient energy dissipaters during the December 2004 powerful tsunami event. (3) Coastal hamlets shielded by dense forests remained unharmed by the tsunami wave attack as compared to those subjected to a direct hit. (4) Future plantation programs have to necessarily follow a natural zonation of coastal vegetal species; casuarinas should be located backshore. (5) Scientific designing of an effective shelter belt along hazard prone coasts is imperative for an efficient management of Indian coasts.

Acknowledgement

The author is grateful to the Director, NIO, Goa, for permission to publish this chapter. Post tsunami field work was carried out along with colleagues S. Jayakumar and R. Gowthaman, NIO, Goa. The author is thankful to Dr. Frank Columbus, Nova Science Publishers, New York, for the invitation to contribute this chapter. This is NIO contribution 4572.

References

Anonymous, 2005. Forest defenses help shelter coast from disaster. *China Daily,* **21** September.

Badola, R. and Hussain, S.A., 2005. Valuing ecosystem functions: an empirical study on the storm protection function of Bhitarkanika mangrove ecosystem, India. *Environmental Conservation,* **32**:85-92.

Bhalla, R.S., 2007. Do bio-shields affect tsunami inundation? *Current Science,* **93**:831-833.

Chadha, R.K., Latha, G., Yeh, H., Peterson, C. and Katada, T., 2005. The Tsunami of the Great Sumatra Earthquake ofM9.0 on 26 December 2004 – Impact on the East Coast of India. *Current Science,* **88**:1297–1301.

Clark, J.R., 1996. *Coastal zone management handbook.* Lewis Publication, USA, p. 694.

Dahdouh-Guebas, F., Jayatissa, L.P., Di Nitto, D., Bosire, J.O., Lo Seen, D. and Koedam, N., 2005. How effective were mangroves as a defence against the recent tsunami? *Current Biology,* Volume 15, Issue 12, 21 June 2005, p. R443-R447

Danielsen, F., Sørensen, M.K., Olwig, M.F., Selvam, V., Parish, F., Burgess, N.D., Hiraishi, T., Karunagaran, V.M., Rasmussen, M.S., Hansen, L.B., Quarto, A., Suryadiputra, N., 2005. The Asian tsunami: a protective role for coastal vegetation. *Science,* **310** (5748), p.643.

Dash, J., Exotic tree species backfiring for Orissa beaches: greens. *Indo-Asian News Service,* 21 April, 2002.

Furukawa, K., Wolanski, E. and Muller, H., 1997. Currents and sediment transport in mangrove forests. *Estuarine Coastal and Shelf Science,* **44**:301-310.

Ganesan, S., 2005. Shelter belt plantations take on killer waves. *The Hindu,* January 2005.

Hiraishi, T., 2004. Tsunami risk and counter measures in Asia and Pacific areas: applicability of greenbelt tsunami prevention in Asia and Pacific regions. *Report, Port and Airport Research Institute, Japan,* vol. 43, 6 pp.

Kathiresan, K. and Rajendran, N., 2005. Coastal mangrove forests mitigated tsunami. *Estuarine Coastal and Shelf Science,* **65**:601-606.

Mascarenhas, A., 2004. Oceanographic validity of buffer zones for the east coast of India: a hydrometeorological perspective. *Current Science,* **86**, 399-406.

Mascarenhas, A., 2006. Extreme events, intrinsic landforms and humankind: post-tsunami scenario along Nagore – Velankanni coast, Tamil Nadu. *Current Science,* **90**, 1195-1201.

Mascarenhas, A. and Jaykumar, S., 2007. Protective role of coastal ecosystems in the context of the tsunami in Tamil Nadu coast, India: implications for hazard preparedness. In: Tad S. Murty, U. Aswathanarayana and N. Nirupama, Editors, *The Indian Ocean Tsunami,* pp. 423-436.

Mascarenhas, A. and Jaykumar, S., 2008. An environmental perspective of the post-tsunami scenario along the coast of Tamil Nadu, India: Role of sand dunes and forests. *Journal of Environmental Management,* **89**:24–34.

Massel, S.R., Furukawa, K. and Brinkman, R.M., 1999. Surface wave propagation in mangrove forests. *Fluid Dynamics Research,* 219-249.

Nayak, S.R., Sarangi, R.K. and Rajawat, A.S., 2001. Application of IRS-P4 OCM data to study the impact of cyclone on coastal environment of Orissa. *Current Science,* **80**:1208-1213.

Olwig, M.F., Sørensen, M.K., Rasmussen, M.S., Danielsen, F., Selvam, V., Hansen, L.B., Nyborg, L., Vestergaard, Parish, F., and Karunagaran, V.M., 2007. Using remote sensing to assess the protective role of coastal woody vegetation against tsunami waves. *International Journal of Remote Sensing,* **28**:3153–3169.

Ramanamurthy, M.V., Sundaramoorthy, S., Pari,Y., Rao,V.R., Mishra, P., Bhat, M., Usha,T.,Venkatesan, R. and Subramanian, B.R., 2005. Inundation of seawater in Andaman and Nicobar Islands and parts of Tamil Nadu coast during 2004 Sumatra Tsunami. *Current Science,* **88**:1736-1740.

Pilkey, O.H., Bush, D.M. and Neal, W.J., 2000. Storms and the coast. In: Pielke, R. and Pielke, R., Editors, *Storms,* Routledge Hazards and Disaster Series, New York, pp. 427-448.

Shrestha, M. L., Editor, 1998. *The Impact of Tropical Cyclones on the Coastal Regions of SAARC Countries and their Influence in the Region.* SAARC Meteorological Research Centre, Agargaon, Bangladesh, pp. 329.

Valdiya, K.S., 2001. Public policy for natural hazard management. *Current Science,* **80**: 486-487.

Vaithilingam, R., 2005. *Policy note 2005-2006.* Environment and Forest Department, Tamil Nadu, 34 pp.

Vermaat, J.E. and Thampanya, U., 2006. Mangroves mitigate tsunami damage: A further response. *Estuarine Coastal and Shelf Science,* **69**:1-3.

Williams, N., 2005. Tsunami insight to mangrove value. *Current Biology,* Volume 15, Issue 3, p. R73.

In: Tsunamis: Causes, Characteristics, Warnings and Protection ISBN: 978-1-60876-360-3
Editors: N. Veitch and G. Jaffray, pp. 149-167

Chapter 7

2004 TSUNAMI INUNDATION AND EVIDENCE FOR EARLIER EVENTS –A CASE STUDY FROM SRI LANKA

Nayomi Kulasena[1] and Kapila Dahanayake[2]
[1]Roughton International (UK), Southern Transport Development Project (ADB funded section), Baddegama , Sri Lanka
[2]Department of Geology, University of Peradeniya , Peradeniya, Sri Lanka

Abstract

The tsunamigenic Sumatra-Andaman earthquake (moment magnitude $M_W = 9.3$) of December 26/2004 caused destruction and human casualties in many coastal Indian Ocean countries with more than 35,000 deaths in the island of Sri Lanka. Tsunami waters started inundating many coastal regions of the island at different times after the event. The initial water movement was characterized by a rapid drawdown or lowering of the sea surface at the coast as waters moved into the area of seabed displacement. In different parts of Sri Lanka, the drawdown resulted in recession of the sea about 500 m from the present coastline. According to eyewitness accounts, this phenomenon had lasted for about 20 minutes before the return of massive turbulent tsunami waves inland with speeds of about 30 to 40 km per hour bringing behind them long trains of water from deep sea environments. The waters carrying deep sea materials such as dark suspensions of fine sediments encroached inland via embayments such as harbours, bays, estuaries, lagoons and upstream along rivers before depositing them. Run up heights of about 3 meters or more were reported from different coastal locations. The tsunami event has left significant geological signatures with changes in coastal geomorphology and deposition of sediments along and across the coast.

The sediments consisting of discontinuous sheets of differing thickness are found at coastal depressions and similar low lying areas. Such sediments are poorly sorted and contain heterogeneous mixtures of debris of buildings and vehicles, tree trunks, shells of organisms etc., derived from coastal and deep marine environments. However, at other locations such as depressions where water had been stagnant for a few hours before retreating back to the sea , thin films of brownish clayey silt materials from suspensions had been deposited. Sediments with comparable textures and compositions were located in previously empty bottles, stacked on a bench of about 50 cm from the floor of a house close to the beach, which had been subsequently filled with tsunami waters. The recent tsunami sediments when studied under the

Scanning Electron Microscope (SEM) showed the occurrence of foraminiferal facies and other deep sea fauna. Drilling carried out deeper into coastal depressions in different locations revealed sediments with textures, structures and microfossils characteristic of 2004 deposits in at least two lower horizons signaling the occurrence of paleo-tsunami events. Although historical evidence also indicates at least two tsunami events, radiocarbon dating yielded anomalous results.

Introduction

The undersea tsunamigenic Sumatra-Andaman earthquake (moment magnitude (M_w) = 9.3) of December 26/2004 (Kruger and Ohrnberger, 2005; Ishii et al., 2005; Subarya et al., 2006) caused destruction to property and human casualties in many coastal Indian Ocean countries with more than 35,000 deaths reported in Sri Lanka. Tsunami waters started inundating many coastal districts at different times after the event (Figure 1). The initial water movement was characterized by a rapid recession of the sea or lowering of the sea surface at the coast as waters moved towards lower reaches of the ocean. In some parts of Sri Lanka the drawdown resulted in recession of the sea about 500 m from the present coastline (Figure 2).

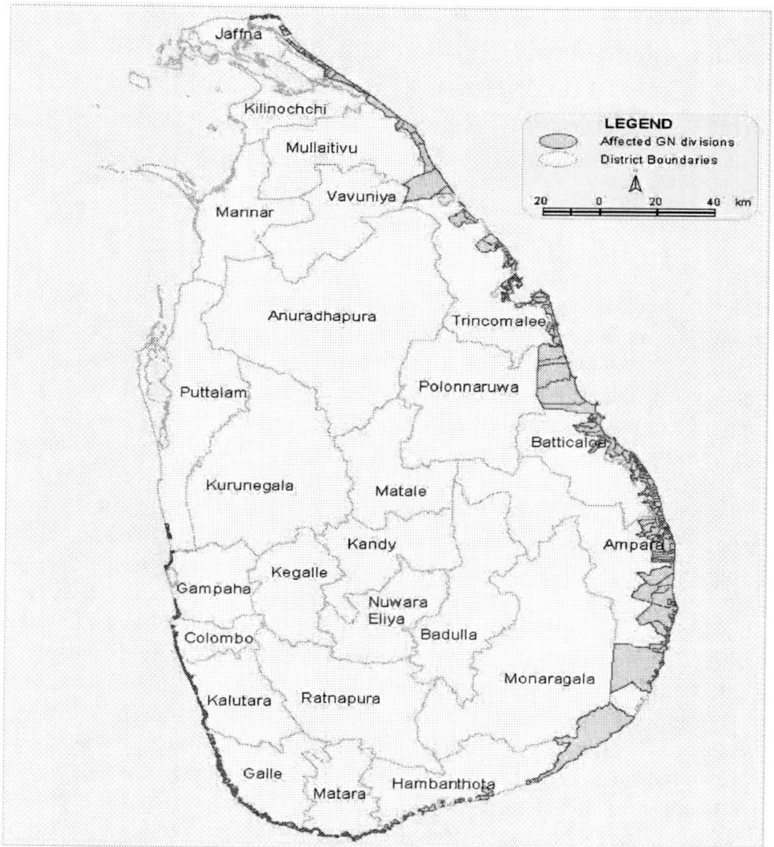

Figure 1. Map of Sri Lanka indicating the coastal districts of Sri Lanka inundated (in blue) by the 2004 tsunami event in different Grama Niladari (GN) or Village Headmen's divisions. The present study was in Galle District mainly at Peraliya and Denuwala.

According to accounts of affected villagers, this phenomenon had lasted for about 20 minutes before the return of massive turbulent tsunami waves inland with speeds of about 30 to 40 km per hour with run up heights of about 3 meters or more being reported from different coastal locations (Tanioka et al., 2004; Liu et al., 2005; Wijetunga, 2006). These waves had brought behind them long trains of water carrying dark colored suspensions of fine grained materials from deep sea environments. The waters carrying these materials encroached inland sometimes as much as 8 km via embayments such as harbours, bays, estuaries, lagoons and upstream along rivers and had deposited sediments with varying compositions (Figures 1, 2). The tsunami waters commenced to recede from inland areas soon after inundation and the process continued for several hours depending on the topography of the coastal locations concerned. The tsunami event has left significant geological signatures with changes in coastal geomorphology and deposition of sediments along and across the coast.

Methods of Study

For the present study, a generalized tsunami inundation map was prepared using the data available in the Department of Census and Statistics, Sri Lanka (Figure 1). The detailed in inundation map of Peraliya study area (Figure 2) in Sinigama was based on field observations and interviews with the villagers of the area.

The recent tsunami event was studied with emphasis on recent sediments collected at different sites. Reliable data on the grain size distribution and microfossil content of recent tsunami sediments was obtained from the study of tsunami water accidentally preserved in glass bottles at a house near the beach in the coastal study area of Galle (Dahanayake, 2006). Using these observations as a pilot study, several sites of occurrence of recent tsunami sediments were located and studied. In addition near shore sand deposits as well as those from storm surge were studied for comparative purposes.

Sediment samples collected from sites described above were air dried and grain size analysis was done using 1 mm, 0.5 mm, 0.25 mm, 0.212 mm, 0.125 mm, 0.063 mm sieves. Cumulative curves (as in Tickell, 1965) were constructed for typical samples from each 2004 tsunami sediment location. Each size fraction of the sediments was studied for the microfossil content using a reflected-light microscope. The 0.125 fraction of the tsunami sediments contained the highest concentration of foraminiferal assemblages. Therefore only this size fraction was used for more detailed identification of microfossils using LEO 1420 VP Scanning Electron Microscope (SEM) on gold-sputtered mounts. This was done with typical samples collected from each 2004 tsunami sedimentation site.

Augur drilling of depressions overlain by recent tsunami sediments showed the occurrence of potentially older tsunami sediments at depths of 30 to 75 cm from the surface. These sediments were also studied for their grain size distribution and microfossil contents. This paper reports how the findings from a detailed sedimentological and micropaleontological study of recent tsunami sediments were extrapolated to determine the existence of paleo-tsunami sediments in some locations of Sri Lanka. In the light of historical evidence of at least two ancient tsunami events (Geiger, 1934; Suraweera, 2000), radiocarbon age studies were undertaken.

Identification of Recent Tsunami Sediments

In highly populated coastal areas, the tsunami sediments had mixed with not only beach sands but also with materials such as debris from broken buildings, boats, ships as well as land vehicles, trunks of trees etc., Such sediments showed a highly heterogeneous nature. However, even in unpopulated areas, the sediments were not homogeneous, due to mixing occurring during landward and seaward movements of tsunami waters. On the other hand, at other locations such as depressions where water had been stagnant for a few hours before retreating to the sea, there exists thin films of brownish clayey silt materials deposited on grasslands. Such sediments deposited from suspensions can be located on careful field observations in specific locations. Thus, our observations had to be concentrated more on sediments found in coastal depressions as well as in other potential sites where sedimentation had occurred from suspensions.

In our search for uncontaminated recent tsunami sediments, field studies were centered on potential locations of sedimentation from suspensions in the known inundated locations such as (i) houses or partially destroyed dwellings (ii) containers (iii) kitchen sinks (iv) concrete table tops (v) interiors of cabinets/cupboards (vi) Unaffected houses that had remained closed during the tsunami inundation (Dahanayake and Kulasena, 2008).

Figure 2. Dec. 26, 2004- Satellite imagery showing the drawdown of the ocean ~500 m on a portion of the southwestern coast of Sri Lanka, near Kalutara. (Image source: Digital Globe).

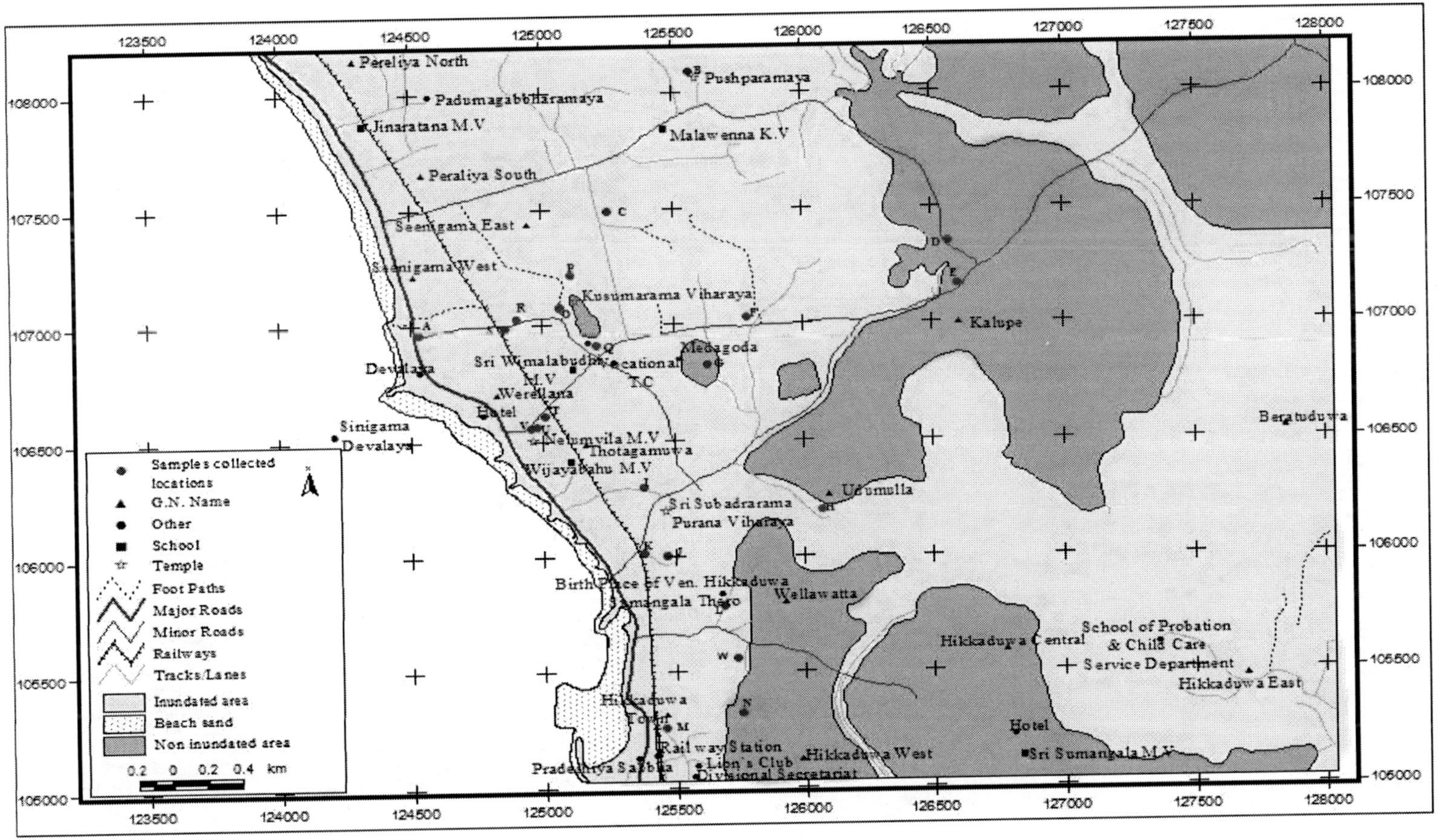

Figure 3. Inundation map of the coastal study area of Sinigama (Peraliya) in Galle District, Sri Lanka (See Figure 1).

Figure 4. The bottles filled with tsunami water are displayed by the owner of the house where they were located. The plastic cases containing more such bottles lie by his side.

Figure 5. An abandoned house is located at Akurala beach bordering the Colombo-Galle Road (A2). Note that the roof is completely damaged (suggesting the tsunami height) and the sediments settled in the kitchen sink are indicated by the person in the photograph.

Figure 6. The person is separating fine tsunami sediments from other domestic debris. According to the villagers, the house situated near the coast had remained closed during tsunami inundation and the sediments had settled from the suspensions in tsunami water.

Figure 7. Man (in trousers) stands on a sheet of tsunami sediments deposited from suspensions in retreating tsunami waters. Note the highly heterogeneous character of the tsunami sediment (debris of buildings, parts of vehicles etc.) at this coastal location on Colombo-Galle highway (Figure 1).

Figure 8. A coastal depression newly developed after the tsunami inundation. In such locations tsunami sediments have been deposited from suspensions.

Figure 9. An excavated depression reveals the recent fine grained white tsunami sediment (1) underlain by a heterogeneous coarse sediment lying on a humic soil horizon (See also Smoot et al., 2000). Note the series of depressions (D) in this coastal area underlain by coral formation. Depressions could be attributed to underlying sink holes/ subsided locations due to solution of coral limestones over a period of time and characteristic of karst topography.

During our field studies to locate "genuine" tsunami sediments we visited coastal households or the destroyed houses to determine the tsunami run-up heights with emphasis on the occurrence of tsunami sediments. During the field visit to a house in the tsunami-inundated coastal village of Mahamodera near the city of Galle, several previously empty *arrack* bottles in a container case now filled with water carried by the tsunami flood were found (Dahanayake, 2006). The bottles contained thin layers of fine grained sediments and organic debris at their bottoms. The plastic container case was found on a flat bench at a height of about 1m from the ground floor of the house. The house is situated at a location (about 5 m above MSL) and above the normal highest tide level and about 50 m from the southern beach. We were told by the residents their house was not flooded during subsequent storm tides. Thus we satisfied ourselves that the water has not been contaminated subsequent to the tsunami event. The sediments in the bottles were later separated out from water which had a composition significantly different from the sea water collected from the vicinity. The different size fractions of the sediment were studied for their contents. In the size fraction, a large number of microfossils such as foraminifera, diatoms and radiolarians characteristic of a deep ocean provenance were observed. The observations were confirmed by later SEM studies (Dahanayake, 2006; Dahanayake and Kulasena, 2008; Kulasena and Dahanayake, 2008). The results of the study of the sediments in bottles served as the pilot study for later observations in other potential sites listed below:

a) Houses, other buildings and shipping containers located above the highest tide level which had remained closed during the tsunami event. Floors of such buildings had been flooded by tsunami waters. They had remained closed/abandoned until the time of collection of sediment samples.
b) Concrete slabs (located about 50 cm to 1 m above ground surface) of kitchen sinks and the pantry cupboards abandoned/destroyed during the tsunami. A brownish yellow silty fine sand had been deposited on these slabs.
c) Tsunami waters had subsided several hours after flooding. Topographic depressions had been potential sites for gravity settling/deposition of finer tsunami sediments carried by tsunami waters.

As described earlier, recent tsunami sediments consisting of discontinuous sheets/layers of differing thickness are found at coastal depressions as well as the abandoned houses etc., as detailed above.

Paleo-Tsunami Sediments from Sri Lanka

In coastal depressions where water had been stagnant for a few hours before retreating back to the sea , thin films of brownish clayey silt materials from suspensions had been deposited on grasslands. The 2004 tsunami sediments when studied under the Scanning Electron Microscope (SEM) showed the occurrence of foraminifera and other deep sea fauna (Uchida et al., 2005; Hawkes et al., 2006; Nagendra et al., 2006; Dahanayake and Kulasena, 2008). Random drilling carried out and deeper into the coastal depressions in different locations indicated sediments with textures, structures and microfossils characteristic of 2004 deposits in at least two lower horizons of these soil profiles signaling the occurrence of paleo-tsunami events.

Several randomly selected soil profiles located at about 100 m inland from the southern coast were studied.The locations were at the coastal villages of Peraliya and Denuwala situated 50 km apart. The profiles showed comparable horizons as follows from top to bottom: relatively thin yellowish brown horizon of Recent Tsunami sediments; black clayey calcareous humic soil; thin yellowish brown Paleo-tsunami horizon; black clayey calcareous humic soil; yellowish brown clayey fine sand paleo-tsunami horizon; mollusk rich coral reef; lateritic bedrock (Figure 10).

Detailed microscopic observations of yellowish brown soil horizons (paleo-tsunami sediments) indicated the presence of some of the well preserved microfossil species which we observed in recent tsunami sediments (Figures 11, 12, 13 a, b; Tables 1, 2).

Figure 10. A dug pit from the coastal study area of Denuwala showing the recent and ancient tsunami horizons. See also Figure 12.

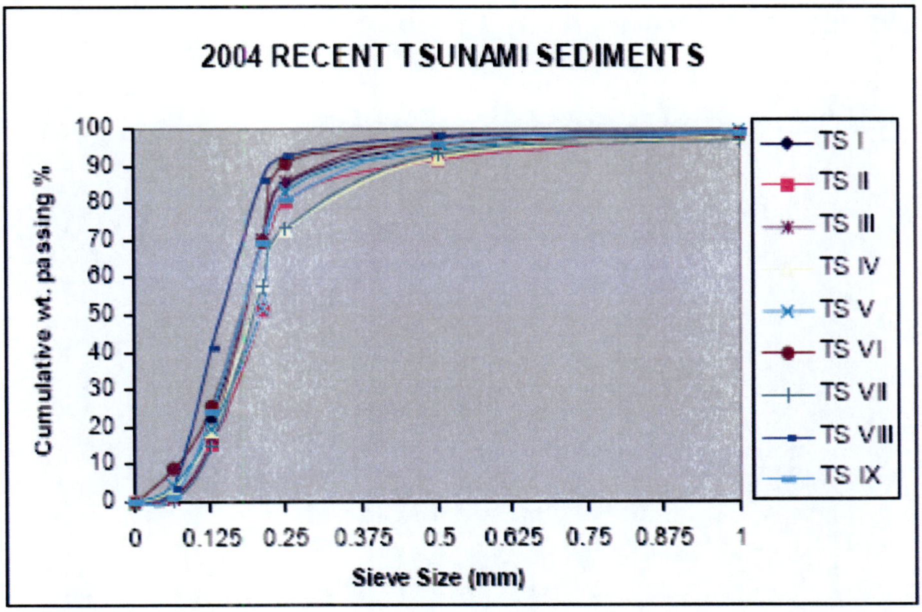

Figure 11. Grain size distribution curves for recent tsunami sediments collected from following different locations of the study areas of Peraliya, Galle and Denuwala: TS I-floor of an abandoned house at Denuwala, TS II- an abandoned house at Denuwala, TS III-kitchen sink, Nelumvila, Peraliya, TS IV-kitchen slab, Nelumvila, Peraliya, TS V-at Denuwala, on site, TS VI-at Denuwala, on site, TS VII-arrack case, Mahamodara, Galle, TS VIII-floor of a kitchen, Akurala, TS IX-at Akurala, on site.

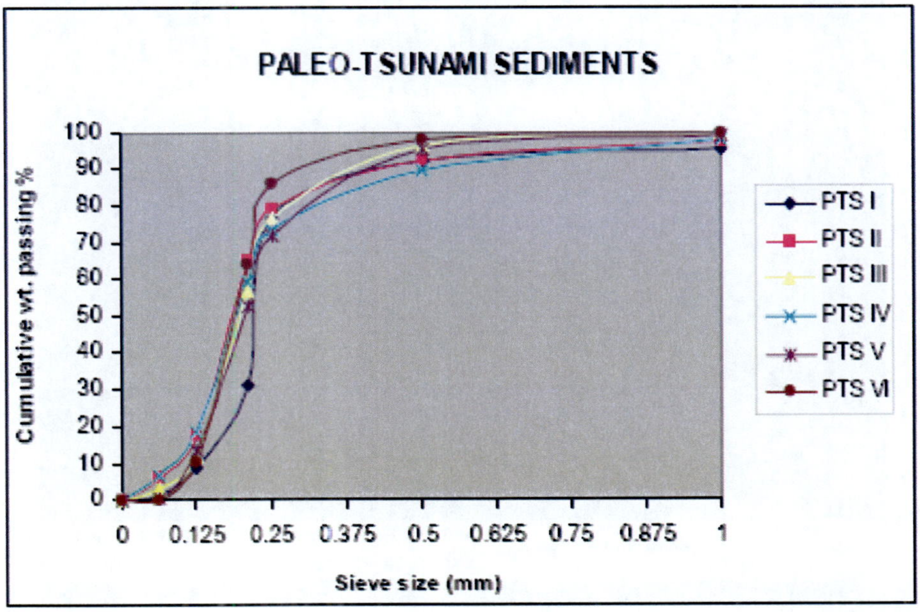

Figure 12. Grain size distribution curves for paleo- tsunami sediments collected from dug pits of Peraliya, and Denuwala: PTS I, II, V, VI-at Peraliya and PTS III, IV-at Denuwala. Note that PTS I, III, V are located at about 30 cm below the present ground level, and the others PTS II, IV, VI are at about 75 cm below.

Assemblages of Microfossils in Tsunami Sediments

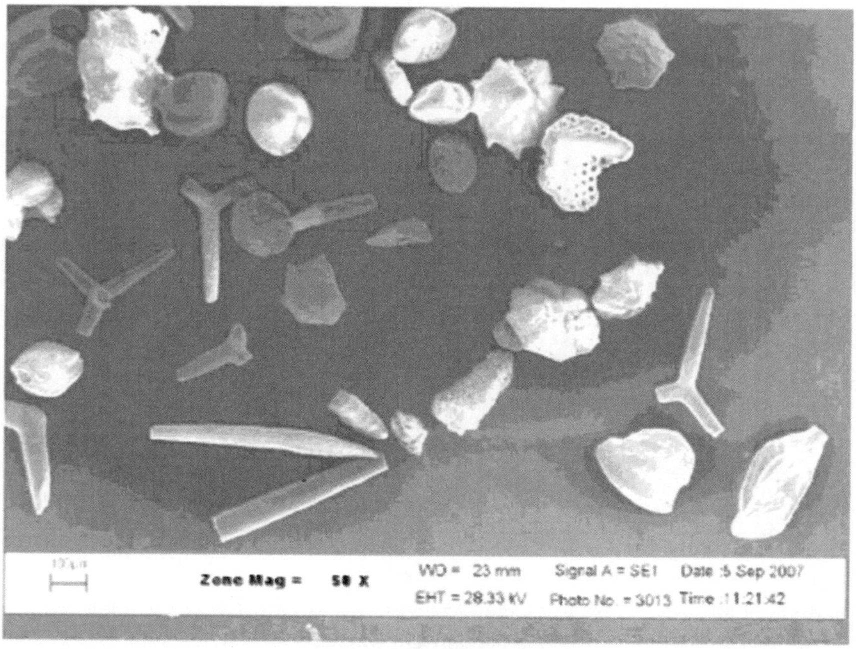

a) Recent Tsunami Sediments

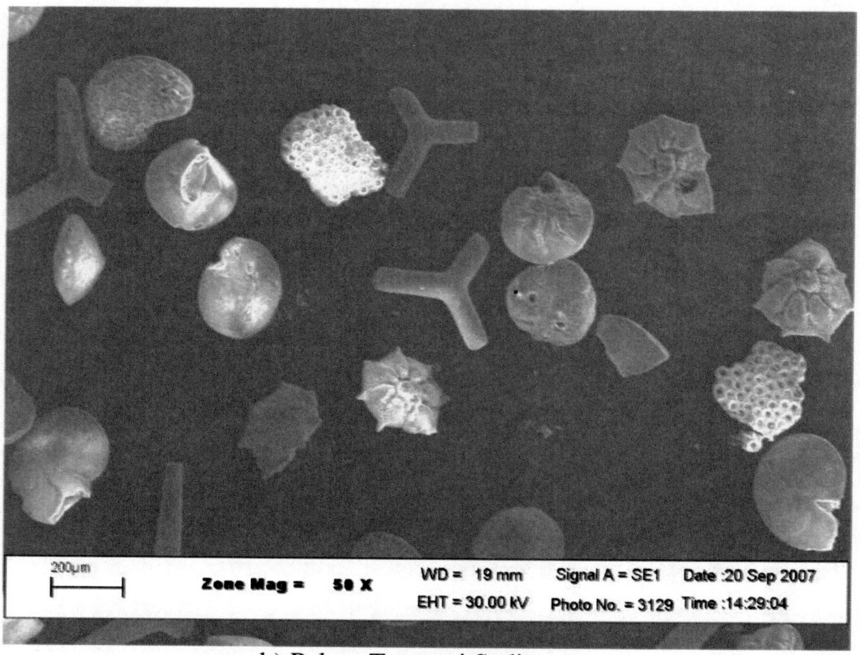

b) Paleo- Tsunami Sediments

Figure 13 a, b. SEM photographs of deep ocean microfossil assemblages (f-foraminifera, r-radiolaria, s-spicules of diatoms).

Distribution Of Microfossils In Tsunami Sediments

a. Recent Tsunami Sediment

Table 1. Distribution of microfossils in 2004 Recent Tsunami Sediments (RTS).

Microfossil Species	2004 Tsunami Sediments
Foraminifera	
Ammonia sp.-b	✓
Ammonia beccarii (Linne')-b	✓✓
Amphistegina radiate-b	✓
Amphistegina lessonii-b	✓
Cibicides lobatulus-b	✓
Elphidium margaritaceum-b	✓✓
Elphidium excavatum-b	✓✓
Elphidium delicatulum-b	✓✓
Elphidium advenum (Cushman)-b	✓✓
Elphidium discoidale-b	✓
Elphidium crispum-b	✓
Elphidium macellum (Fichtel and Moll)-b	✓
Elphidium galvestonense-b	✓
Globigerinita glutinata-p	✓✓✓
Globigerinella calida-p	✓
Globorotalia menardii (Parker, Jones and Brady)-p	✓
Hantkenina sp.-p	✓✓✓
Miliolinella subrotunda-b	✓✓
Nonion grateloupi-b	✓
Pararotalia nipponica-b	✓✓✓
Quinqueloculina lamarckiana d'Orbigny-	✓✓
Quinqueloculina seminula (Linné)-b	✓✓
Quinqueloculina sp.-b	✓✓✓
Radiolarians	
Acantharia-p	✓✓

* b-benthic, p-planktonic, ✓✓✓- abundant, ✓✓- moderately present, ✓- rare.

b. Paleo-Tsunami Sediment – 30 cm below the Present Ground Level

Table 2. Distribution of microfossils in paleo-tsunami sediments located at about 30 cm below the surface of the present ground level (PTS 1).

Microfossil Species	Paleo-tsunami Sediments
Foraminifera	
Ammonia sp.-b	✓
Ammonia beccarii (Linne')-b	✓
Amphistegina lessonii-b	✓
Elphidium margaritaceum-b	✓
Elphidium excavatum-b	✓
Elphidium delicatulum-b	✓✓
Elphidium advenum (Cushman)-b	✓✓
Elphidium macellum (Fichtel and Moll)-b	✓
Elphidium galvestonense-b	✓
Eponides repandus-b	✓
Globigerinita glutinata-p	✓✓
Globigerinella calida-p	✓
Hantkenina sp.-p	✓✓
Miliolinella subrotunda-b	✓
Pararotalia nipponica-b	✓✓
Quinqueloculina lamarckiana d'Orbigny-	✓
Quinqueloculina seminula (Linné)-b	✓
Quinqueloculina sp.-b	✓✓
Radiolarians	
Acantharia-p	✓✓

* b-benthic, p-planktonic, ✓✓✓- abundant, ✓✓- moderately present, ✓- rare.

c. Paleo-Tsunami Sediment – 75 cm below the Present Ground Level

Table 3. Distribution of microfossils in paleo-tsunami sediments located at about 75 cm below the surface of the present ground level (PTS 2).

Microfossil Species	Paleo-tsunami Sediments
Foraminifera	
Ammonia sp.-b	✓
Ammonia beccarii (Linne')-b	✓
Cibicides lobatulus-b	✓
Discorbis williamsoni-b	✓
Elphidium margaritaceum-b	✓
Elphidium advenum (Cushman)-b	✓✓
Elphidium galvestonense-b	✓
Globigerinella calida-p	✓
Hantkenina sp.-p	✓✓
Miliolinella subrotunda-b	✓
Nonion grateloupi-b	✓
Pararotalia nipponica-b	✓✓✓
Quinqueloculina lamarckiana d'Orbigny-	✓
Quinqueloculina seminula (Linné)-b	✓
Quinqueloculina sp.-b	✓✓
Radiolarians	
Acantharia-p	✓✓

* b-benthic, p-planktonic, ✓✓✓- abundant, ✓✓- moderately present, ✓- rare.

Ages of Tsunami Sediments

The buried sand sheets in the dug pit soil profiles are comparable in texture and composition to the sands washed up by the 2004 tsunami. Our findings based on such sheets imply the occurrence of at least two paleo-tsunami events of different ages in Sri Lanka originating apparently from a common source. Historical texts of Sri Lanka refer to at least two past tsunami events that had occurred between 2000 to 3000 B.C. (Geiger 1934; Suraweera 2000; Stoddart, 2005; Dahanayake, 2006). To authenticate such historical events by present observations and findings, the radiocarbon dating of the identified two paleo horizons were attempted. Also it should facilitate to place Sri Lankan tsunamis in a historical and global perspective and will stimulate to do further researches on paleo-tsunami sediment deposits in other coastal belts of Sri Lanka since detailed research studies on Sri Lankan tsunami deposits are rare. However, dating of tsunami sediments can be difficult and further work needs to be done to obtain more accurate and precise ages (Bondevik, 2008). Our preliminary attempts at finding radiocarbon ages for paleo-tsunami sediments were not very promising as the ages shown for recent as well as old sediments were found to be anomalous. Some of our observations for these anomalies are listed below:

- Tsunami events may deposit sediments with components belonging to different periods of time and formed in different environments-deep marine, shallow marine or continental. Hence the heterogeneous character of sediments- texturally as well as compositionally. Further, waves can cause extensive mixing of sediments adding to their heterogeneity. Therefore anomalous ages could well be recorded for the same sample.
- Since sediments can have different proportions of modern and old (detrital) carbonates, or if the organic matter is of mixed marine (older) and terrestrial (younger) origin, then it becomes again difficult to interpret the ages. The organic C could have come from both silt and clay fractions presumably from land and ocean sources during different periods of time. Organic fraction dates are considered more reliable, if the terrestrial and marine organic fractions are properly recognized (Bondevik, 2008). In this context, it should also be noted that the carbonate fraction C could have come from coral fragments as well as foraminifera which lived and died during different periods of time.
- When tsunami waves pass across the land areas during inundation with velocities ranging from 30 – 40 kmph, there is the likelihood of strong currents stripping away years of organic deposition, just before the currents lose velocity and deposit their suspended fine sand load during a calm period. Therefore if the samples of organic/humic soil samples were collected below the tsunami sediment sheets, the ages of the organic/humic layer would then be older than the true age of the tsunami concerned. Jankaew et al. (2008) found leaf fragments in the tsunami sediment which were several thousand years older than the age obtained from bark fragments in the soil just below the tsunami sediment layer. The age of the leaf fragments means that older material must have become mixed into the tsunami deposits, and these observations highlight the importance of selecting the right material for radiocarbon dating of such deposits.

Conclusions

After a tsunami event, the sediment came with the train of oceanic waters had deposited and spread totally throughout the flooded region, they had been mixed with different sediments such as terrestrial pre-tsunami sediments, beach sands, building debris, etc. since the chaotic nature of the tsunami waves. So care should be taken to select the unmixed tsunami sediments. In this case, sediments settled with suspension on the elevated areas during the inundation event were selected for the collection of samples.

Our emphasis in this study was on sedimentological, stratigraphic and paleontological aspects of sediments carried and deposited during the 2004 tsunami inundation. Authentic recent sediment samples were found in particular environments such as empty bottles filled with water during the inundation, shipping containers, abandoned closed houses, kitchen sinks, table tops, offices etc., Grain size distributions of tsunami sediments were compared with beach sands, dune sands and storm surge deposits. Such comparisons were supported by observations with reflection microscopy as well as SEM studies. Thus it was possible to conclusively identify recent as well as paleo-tsunami sediments due to the marked presence of deep marine benthic and planktonic microfossils such as foraminifera and radiolarians e.g. *Acantharia* (Hickman et al., 2001; Nagendra et al., 2005; Prendergast, 2006; Satyanarayana et al., 2007).

To receive a zero or very low age for the 2004 tsunami sediments from the absolute age determination techniques requires the presence of organic material (foraminifera, radiolarians, diatoms, etc.) transported onshore by the tsunami from the ocean. Given that the sediments settled by suspension contain a mixture of deep ocean microfossils that had been killed at different times, the result would be the average age of the sediment or the age of the particular microfossil that was subjected to radiocarbon dating. To overcome this issue it is required to choose the microfossils that had been killed after the tsunami event. But due to the microscopic size and considerable mixing of tsunami sediments with materials of different ages and the time elapsed from the tsunami event to the commencement of research studies, it is nearly impossible to select the microfossils with certainty about their deaths.

In dating tsunami sediments using carbonate C also, it is very important to choose the right components. This is a complicated exercise given the different processes such sediments have undergone .e.g the amount of detrital carbonate material of different ages derived from different sources viz. microfossil tests, coral reefs etc.

Acknowledgements

We thank Prof. P. Wickramagamage and Ms Kumudu Kumarihamy of Center for Environmental Studies, University of Peradeniya for their assistance in the preparation of maps. Mr G.G.Chandrabhanu, Materials Engineer at Southern Transport Development Project (STDP), Baddegama kindly provided some photographs. This work was funded by the National Science Foundation (NSF), Sri Lanka by way of a generous research grant (RG/2005/DMM/05) to the senior author (K.D).

References

[1] Bondevik, S. (2008). Earth Science: The sands of tsunami time. *Nature* 455, 1183-1184.

[2] Dahanayake, K., (2006). Science at the Solstice: A day in the life of a scientific planet, Abstract on Tsunami sediments. *Nature,* v.441, 1040-1045.

[3] Dahanayake, K. and Kulasena, N., (2008). Recognition of diagnostic criteria for recent- and paleo- tsunami sediments from Sri Lanka, *J. Marine Geology (Elsevier)* v.254, 180-186.

[4] Dahanayake, K. and Kulasena, N., (2008). Geological Evidence for paleo-tsunamis in Sri Lanka, *Science of Tsunami Hazards* v.27, 54-61.

[5] Geiger, W. (Translator), (1934). *The Mahawamsa* (The Great Chronicle of Ceylon-Sri Lanka) Oxford University Press London, (in Sinhala).

[6] Hawkes, A.D., Horton, B.P., Robinson, R., Bird, M., Nott, J., (2006). The sediments deposited by the Indian Ocean tsunami along the Malay-Thai Peninsula. *Marine/Coastal Science* v.38, 376.

[7] Hickman, C. P., Roberts, L. S., Larson, A., (2001). *Integrated Principles of Zoology*. McGraw-Hill New York.

[8] Jankaew, K., Atwater, B.F., Sawai, Y., Choowong, M., Charoentitirat, T., Martin, M.E., Prendergast, A., (2008). Medieval forewarning of the 2004 Indian Ocean tsunami in Thailand, *Nature* v.455, 1228-1231.

[9] Kulasena, N. and Dahanayake, K., (2008). Identification of Recent- and Paleo- Tsunami Sediments in Sri Lanka. Abstr. Proceedings, 24th Annual Sessions, Geological Society of Sri Lanka (GSSL) p.3.

[10] Kruger, F., Ohrnberger, M., (2005). Tracking the rupture of the M_w = 9.3 Sumatra earthquake over 1150km at teleseismic distance, *Nature* v.435, 937-939.

[11] Liu, P.L.F., Lynett, P., Fernando, H., Jaffe, B.E., Fritz, H., Higman, B., Morton, R., Goff, J., Synolakis, C., (2005). Observations by the International Tsunami survey Team in Sri Lanka. *Science,* v.308, 1595.

[12] Nagendra, R., Kamalak Kannan, B.V., Sajith, C., Sen, G., Reddy, A.N., Srinivasalu, S., (2005). A record of foraminiferal assemblage in tsunami sediments along Nagappattinamcoast, Tamil Nadu. *Current Science* v.89, 1947-1952.

[13] Prendergast, A., (2006). Echoes of ancient tsunamis. Aus.Geo News #83.

[14] Satyanarayana, K., Nallapa Reddy, A., Jaiprakash, B.C., Chidambaram, L., (2007). A note on foraminifera, grain size and clay mineralogy of tsunami sediments from Karaikal-Nagore-Nagapattinam beaches, Southeast Coast of India. *Journal Geological Society of India,* v.69, 70-74.

[15] Seyfert, C. K. and Sirkin, L. A., (1979). *Earth History and Plate Tectonics –An introduction to Historical Geology* 7-8 (Harper and Row Publishers, New York).

[16] Shrock, R.R. and Twenhofel, W.H., (1953). *Principles of Invertebrate Paleontology*. McGraw- Hill Book Company New York, p.816.

[17] Smoot, J. P., Litwin, R.J., Bischoff, J.L., Lund, S.J., (2000). Sedimentary record of the 1872 earthquake and "Tsunami" at Owens Lake, southeast California. *Sedimentary Geology* v.135, 241-254.

[18] Stoddart, J., (2005). Tidal waves in Ceylon resulting from the Eruptions in the Straits of Sunda, August 1883. Surveyor General's Report, Ceylon (Sri Lanka). Island Newspaper, Sri Lanka Feb. 15.

[19] Subarya, C., Chileh, M., Prawirodirdjo, L., Avouac, J.P., Bock, Y., Sieh, K., Meltzner, A.J., Natawidjaya, D.H., McCaffrey, R., (2006). Plate-boundary deformation associated with the great Sumatra-Andaman earthquake. *Nature,* v.440, 46-51.

[20] Suraweera, A. V., (2000). *Rajavaliya*- A Comprehensive account of the kings of Sri Lanka. Vishvalekha Publication, Sri Lanka.

[21] Tickell, F. G., (1965). *The Techniques of Sedimentary mineralogy*. Elsevier, Amsterdam.

[22] Uchida, J., Abe, K., Hasegawa, S., Fujiwara, O., Kamataki, T., Irizuki, T., Hirakawa, K., (2005). Characteristics of Faunal Succession of Foraminifera in Tsunami-deposits and Recognition of Sauce Area of Particles- A Case Study of the Holocene Tsunami Deposits at Tateyama, Southern Part of the Boso Peninsula, Central Japan. Abstract American Geophysical union-#T11A-0361.

[23] Vigny, C., Simons, W.J.F., Abu, S., Bamphenyu, R., Satirapod, C., Choosakul, N., Subarya, C., Socquet, A., Omar, K., Abidin, H.Z., Ambrosisu, B.A.C., (2005). Insight into the 2004 Sumatra-Andaman earthquake from GPS measurements in Southeast Asia. *Nature,* v.436, 201-206.

[24] Wijetunge, J.J., (2006). Tsunami on 26 December 2004: Spatial distribution of tsunami height and the extent o inundation in Sri Lanka. *Science of tsunami hazards,* v.24, 225-239.

Web References

[25] Thomas, E., 2005. Deep-sea benthic foraminifera http://ethomas.web.wesleyan.edu/BFhandout.htm

[26] MIRACLE (Microfossil Image Recovery and Circulation for Learning and Education) website, university college of London
http://www.ucl.ac.uk/GeolSci/micropal/foram.html
http://www.ucl.ac.uk/GeolSci/micropal/radiolaria.html

In: Tsunamis: Causes, Characteristics, Warnings and Protection ISBN: 978-1-60876-360-3
Editors: N. Veitch and G. Jaffray, pp. 169-213

Chapter 8

INDIAN OCEAN EARTHQUAKE AND TSUNAMI: HUMANITARIAN ASSISTANCE AND RELIEF OPERATIONS*

Rhoda Margesson

Abstract

On December 26, 2004, a magnitude 9.0 undersea earthquake off the west coast of northern Sumatra, Indonesia, unleashed a tsunami that affected more than 12 countries throughout south and southeast Asia and stretched as far as the northeastern African coast. Current official estimates indicate that more than 160,000 people are dead and millions of others are affected, including those injured, missing, or displaced, making this the deadliest tsunami on record. News reports suggest that the death toll may be well above 200,000. Sections of Indonesia, Sri Lanka, India, and Thailand have suffered the worst devastation. Eighteen Americans are confirmed dead, with another sixteen presumed dead, and 153 remain unaccounted for.

In response, the United Nations, the United States, and other donor nations have organized what some have called the world's largest relief and recovery operation to date. President Bush pledged $350 million in aid and mobilized the U.S. military to provide logistical and other assistance.

Funding the Indian Ocean tsunami relief and reconstruction effort is likely to be a challenge faced by the 109th Congress. Even before the disaster struck, Congress was expected to struggle to find the resources to sustain U.S. aid pledges amid efforts to tackle rising budget deficits by, among other measures, slowing or reducing discretionary spending. Congress also may wish to consider debt relief as a means of helping those nations hit by the tsunami to recover economically. Additionally, there have been calls to institute a tsunami detection and warning system in the Atlantic and/or Indian Oceans, both of which would require allocations of funds. The large-scale U.S. response to the tsunami is unlikely to reverse the decline in the U.S. image abroad since the September 11 attacks, because this decline primarily is due to American policies in the Middle East. However, the scale and scope of U.S. assistance could provide a positive example of U.S. leadership and military capabilities. Additionally, the disaster relief cooperation between the U.S. and Indonesian militaries is likely to be

* Excerpted from CRS Report RL32715

mentioned during the annual congressional deliberations over renewing restrictions on U.S.-Indonesian military-to-military relations, which the Bush Administration has sought to restore since the September 11, 2001 attacks.

This chapter summarizes the extent of the disaster and relief effort and includes descriptions of the U.S. and international assistance efforts. It also examines protection mechanisms for children and separated orphans. A section is devoted to the situation in each of the affected countries followed by an analysis of selected issues for Congress. A *Most Recent Developments* section at the beginning of the chapter is updated through February 10, 2005. The rest of the chapter is updated through January 21, 2005.

Contributing Authors and Subject Areas

Name	**Subject**
Alan Kronstadt	Asian Affairs
Bruce Vaughn	Indonesia and Sri Lanka
Emma Chanlett-Avery	Thailand
Larry Niksch	Burma and Indonesia
Larry Nowels	Budget and Policy Issues
Mark Manyin	Asian Affairs
Nicolas Cook	African Affairs
Rhoda Margesson	Humanitarian Assistance and Child Protection
Wayne Morrissey	Early Warning Systems

Most Recent Developments

This section provides a brief summary of the most recent developments on the status of humanitarian assistance and relief operations for the tsunami disaster. It builds on the information provided in the body of this report, beginning on page 8, which is updated only through January 21, 2005.

The Indian Ocean earthquake and tsunami created a natural disaster of historic proportion. It is estimated that more than 160,000 lives were lost and possibly 140,000 remain missing. The massive relief and reconstruction effort underway also departs from previous emergency operations in its scope and scale. The initial objectives of the relief operation involving search and rescue, treatment and survival are thought to have been met: in the immediate post-tsunami period, basic needs were addressed and further deaths were prevented. Although it is early to determine "lessons learned," the assessment of the response to the tsunami disaster so far has been positive on many levels — from meeting basic humanitarian needs, to civilmilitary coordination, information sharing, and working with national governments and indigenous organizations. The operation has not been without its challenges, such as bottlenecks in aid delivery, but all things considered, it is currently viewed by many as largely successful.

In addition to working closely with the national governments of the countries affected by the disaster, The United Nations Office for the Coordination of Humanitarian Affairs (UNOCHA) has been the lead agency working with actors on the ground, coordinating with the military, and enlisting donor support. As the immediate humanitarian requirements of the operation have been fulfilled, a transition to recovery and reconstruction is now taking place and the operation is shifting from using military to civilian capacity for delivery of assistance. For the foreseeable future, UNOCHA will continue as the lead agency.

The transition phase of the post-tsunami period will be challenging. While emergency assistance and the need to guard against the outbreak of disease will continue for some time, there is a new emphasis on conducting assessments and planning for long-term reconstruction, and with that, priorities and funding are beginning to shift. [1] Initial assessments focused mainly on basic assistance needs; now long-term challenges, such as creating jobs and housing, are become more pressing. Host governments are also taking more of a lead in determining the outcome of this next phase.

And with this transition, there are other issues to consider such as security and political tensions, access for aid workers, and the return of displaced populations. Within the relief operation, transparency and accountability at the United Nations, but also with any organization receiving funds, remains a point of focus. Coordinating the assessments, projects, and capabilities of numerous actors with host governments will become more difficult as the complicated task of reconstruction takes hold.

It is well known that in previous disasters, pledges made by governments have not always resulted in actual contributions. Billions of dollars have been pledged to help the victims of the tsunami disaster. Reconstruction will be costly and take time. Maintaining enough pressure on donors to honor their pledges while securing funds needed for other disaster areas requires a delicate balance, particularly if donor fatigue is to be avoided. [2]

The table below reflects the most recent data available on relief and reconstruction pledges and contributions.

Legislation [3]

Several bills pertaining to the Indian Ocean tsunamis and their after-effects have been introduced in the 109th Congress. [4] One of these bills, H.R. 241 (Thomas), entitled *To Accelerate the Income Tax Benefits for Charitable Cash Contributions for the Relief of Victims of the Indian Ocean Tsunami*, was the first legislative measure passed by the 109th Congress to be signed into law; it became P.L. 109-1. As of February 8, 2005, other pending bills included the following:

- **H.Res. 12 (Hyde).** Introduced and passed by the House on January 4, 2005; entitled *Expressing condolences and support for assistance to the victims of the earthquake and tsunamis that occurred on December 26, 2004, in South and Southeast Asia.*
- **H.R. 60 (Jackson-Lee).** Introduced and referred to the House Committee on the Judiciary on January 4, 2005; entitled *To designate Sri Lanka, India, Indonesia, Thailand, Somalia, Myanmar, Malaysia, Maldives, Tanzania, Seychelles, Bangladesh, and Kenya under section 244 of the Immigration and Nationality Act in order to render nationals of such foreign states eligible for temporary protected status under such section.*

Table 1. International Governmental, Inter-Governmental, and Private Tsunami Relief and Reconstruction Pledges and Contributions (millions of U.S. dollars)

Country/Agency Donor	Governments*	Private**
Australia	815.50	177
Germany	664.47	619.8
Japan	500.55 -	NA
France	442.77	90
United States	362.09	700
Canada	350.68	122
Netherlands	266.45	150
Norway	170.48	61
Saudi Arabia (Kingdom of)	163.50	101
Italy	113.27	20
Kuwait	100.00	NA
United Kingdom	95.69	375
Denmark 77.10	35	
Sweden	74.68	75
Spain	71.65	–NA
Finland	69.20	28
Austria	69.17	26
China	64.25	1.8
Korea (Republic of)	50.60	13
Switzerland	50.26	110
Taiwan	50.00	–NA
New Zealand	45.50	7
Greece	25.96	22.5
Qatar	25.00	–NA
India	23.00	–NA
Russian Federation	22.00	–NA
Portugal	16.22	5
Belgium	16.13	40
Singapore	13.66	–NA
Ireland	13.32	67
Malta	10.85	–NA
Czech Republic	10.53	8
Other Countries; National contributions and pledges of less than $10 million each.†	48.65	4.05
National Totals	**4,893.16**	**2858.15**
International Financial Institutions (IFIs)	**IFI Pledges and Contributions***	
European Investment Bank (prospective pledge)	1,275.94	-
International Monetary Fund (prospective pledge)	1,000	-
Asian Development Bank (initial support)	675	-
World Bank (first phase support)	672	-
IFI Totals	**3,622.94**	-

Table 1. Continued

Country/Agency Donor	Governments*	Private**
International Intergovernmental Pledges and Contributions	**Intergovernmental Organization Pledges and Contributions***	
U.N. and U.N.-Affiliated Organization Allocations	205.65	-
European Commission	645.34	-
Arab Gulf Fund	0.1	-
Intergovernmental Organization Totals	851.09	-
Total Pledges and Commitments by Category	9,367.19	2858.15
Total Pledges and Commitments	12,225.34	

Note: Some pledges are conditional or prospective, and data on both pledges and commitments is currently subject to change on a daily basis. In addition to the pledges noted above, numerous countries, including the United States, have made in-kind and other contributions for which no value is specified in available reporting data. The value of the resources that affected countries are devoting to their own tsunami relief and reconstruction are not included above.

Compiled by Nicolas Cook, African Affairs Specialist, Foreign Affairs, Defense and Trade Division.

*UN OCHA, "Table VII: Total Humanitarian Assistance for Indian Ocean Earthquake-Tsunami 2005," Indian Ocean Earthquake-Tsunami 2005 [financial tracking tables], February 7, 2005, [http://www.reliefweb.int/fts]; international organization data; and supplementary national government information. Totals shown may differ from the sum of individual entries, due to rounding. A Reuters news report (Reuters, "Nations Pledge Aid after Tsunami Disaster," Jan. 28, 2005) indicates that some countries may have pledged different amounts than those reported by UN OCHA, the source of the national data reported above. If all higher pledge figures reported by Reuters are taken into account, the aggregate country governmental pledge level would be $174.18 million higher than that noted above.

** Except as noted, the source for all private donation figures is Reuters, "Nations Pledge Aid...," Jan. 28, 2005. Source for U.S. entry is InterAction, "Disaster Response Relief Barometer," Feb. 4, 2005, [http://www.interaction.org/disaster/relief_barometer.html], which is currently being updated weekly. Source of entries for Italy, Sweden, France, China, and South Korea is BBC News, "Tsunami aid: Who's giving what," Jan. 27, 2005; source for Germany is Reuters, "German Private Tsunami Aid Exceeds 475 Mln Euros," Jan. 25, 2005.

†Countries contributing $10 million or less as of Feb. 7, 2005, in rank order are: Luxembourg; Turkey; Iran; Brunei Darussalam; Iceland; Poland; Macedonia, Republic of; Algeria; Libya; Trinidad and Tobago; Hungary; Israel; Brazil; Malaysia; Thailand; Slovakia; Azerbaijan; Nigeria; Papua New Guinea; Romania; Liechtenstein; Estonia; Lithuania; Monaco; Niger; Jamaica; Equatorial Guinea; Bulgaria; Senegal; Latvia; Democratic People's Republic of Korea; Belarus; LAO PDR; Madagascar; Mauritania; Mexico; Mozambique; Nepal; Slovenia; South Africa; Georgia; Guyana; Palau; Timor Leste; and Kazakhstan.

- **H.R. 397 (Menendez).** Introduced and referred to the House Committee on International Relations on January 26, 2005; entitled *To amend the Foreign Assistance Act of 1961 to provide assistance to children who are orphaned or unaccompanied as a result of the tsunamis that occurred on December 26, 2004, in the Indian Ocean.*
- **H.R. 465 (Faleomavaega).** Introduced and referred to the House Committee on Resources on February 1, 2005; entitled *To provide for the establishment of a tsunami hazard mitigation program for all United States insular areas.*
- **H.R. 499 (Shays).** Introduced and referred to the Committee on International Relations, and in addition to the Committee on Resources on February 1, 2005;

entitled *To provide for the development of a global tsunami detection and warning system, to improve existing communication of tsunami warnings to all potentially affected nations, and for other purposes.*

- **S.Res. 4 (Frist).** Introduced and passed in the Senate on January 4, 2005; entitled *A resolution expressing the sympathy and pledging the support of the United States Senate and the people of the United States for the victims of the powerful earthquake and devastating tsunami that struck Bangladesh, Burma, India, Indonesia, Kenya, Malaysia, the Maldives, the Seychelles, Somalia, Sri Lanka, Tanzania, Thailand, and other areas of South Asia, Southeast Asia, and Africa, on December 26, 2004.*
- **S. 34 (Lieberman).** Introduced and referred to the Committee on Commerce, Science, and Transportation on January 24, 2005; entitled *A bill to provide for the development of a global tsunami detection and warning system, to improve existing communication of tsunami warnings to all potentially affected nations, and for other purposes.*
- **S. 50 (Inouye).** Introduced on January 24, 2005; ordered to be reported an original measure by the Committee on Commerce, Science, and Transportation on February 2, 2005; entitled *A bill to authorize and strengthen the National Oceanic and Atmospheric Administration's tsunami detection, forecast, warning, and mitigation program, and for other purposes.*

In addition, on February 9, 2005, President Bush announced plans to request $950 million as part of a supplemental request to support the countries affected by the tsunami. This request includes the commitment of $350 million and adds $600 million as follows [5]:

- $339 million: large-scale reconstruction projects (rebuilding infrastructure such as roads and bridges)
- $168 million: shelter and food aid, rebuilding housing, schools and clinics, developing livelihood programs
- $35 million: early warning systems
- $62 million: capacity building and reconstruction planning assistance
- $346 million: replenish costs incurred by USAID ($120 million) and by DOD ($226 million)

Early Warning

International science ministers will finalize plans for a global observing system in Brussels, Belgium February 14-16, 2004. That system would be the backbone on which a regional tsunami early warning system for the Indian Ocean would be built. The United States is not expected to provide details of its commitment to the internationally sponsored global tsunami early warning network prior to the convening of the G-8 summit in July 2005.

Recent Developments on Selected Countries

Indonesia

The U.S. military is lowering its role in humanitarian aid and a U.S. aircraft carrier has withdrawn, replaced by a Navy hospital ship. The Singapore and Australian militaries also are also withdrawing. None have objected to Indonesia's March 26 deadline for ending foreign military activities in Aceh. The latest toll: 115,000 dead; over 100,000 missing.

Foreign donor countries have pledged billions for tsunami reconstruction to the region in 2005. This is on top of $3.4 billion in development aid to Indonesia. Indonesia's plans to establish relocation centers to initially house 30,000 Acehnese tsunami refugees, then an additional 60,000, remains controversial. The military will have a role in operating the centers. In the past, the military has practiced forced relocation of Acehnese as a counter-insurgency tool. Foreign NGOs are reluctant to be involved in this program. There are an estimated 380,000 refugees in Aceh. The military has admitted that it has continued to carry out operations against Free Aceh insurgents, despite the military's self-proclaimed cease-fire after December 26. The Indonesian government and the Free Aceh (GAM) political organization have begun peace talks in Helsinki, Finland.

Secretary of State Condoleeza Rice reportedly plans to certify that Indonesia is cooperating in investigating the killings and woundings of an American teacher in Papua in August 2002. This will end the congressional restriction on Indonesian participation in the IMET program. The Bush Administration has viewed militaryto- military cooperation in tsunami relief as an opportunity to restore full military-tomilitary relations with Indonesia.

Sri Lanka

The Sri Lankan government has issued guidelines for construction near the coast. Residential and commercial construction must be at least 100 meters from the coast in the western and southern coastal zones. In the north and east it must be 200 meters from the coast. The LTTE have established a buffer of between 300 and 500 meters in areas under their control.

The U.S. Government has provided $62 million in "emergency food assistance, relief supplies, shelter, water and sanitation, health, livelihoods recovery, psychological and social support, protection and anti-trafficking, logistics and coordination, and cleanup and rehabilitation activities" in Sri Lanka. [6] An estimated 30,974 are dead, 4,698 are missing and a further 553,287 persons have been displaced in Sri Lanka. [7] Of the 33 Americans thought killed by the tsunami, 9 are thought to have died in Sri Lanka. [8]

Maldives

An estimated 82 are dead, 26 are missing and a further 12,558 persons have been displaced in the Maldives. It is thought that reconstruction will take two to three years. [9] The USAID/DART Field Officer closed operation in the Maldives on January 28. The U.S. government has provided $1.36 million in assistance to the Maldives. [10]

Malaysia

In Malaysia 68 are dead, 6 missing and 8,000 have been displaced due to the tsunami.

India

Indian administrators continue to receive harsh criticism for perceived interference with relief efforts in the Andaman and Nicobar Islands, interference that allegedly has caused considerable and unnecessary suffering for tsunami victims there. [11]

Thailand

The United States military has begun to wind down its relief efforts in Thailand and Sri Lanka and is likely to wrap up its operations in Indonesia by the end of February.

On February 7, 2005, Thai Prime Minister Thaksin Shinawatra was elected to a second term. Despite pubic criticism in 2004 over his handling of sectarian violence in the country's Muslim-dominated south, Thaksin popularity surged after he rushed to Tsunami disaster areas and provided hands-on leadership, delegating authority and consoling survivors.

Background [12]

Introduction

On December 26, 2004, a magnitude 9.0 undersea earthquake off the west coast of northern Sumatra, Indonesia, unleashed a tsunami that affected more than 12 countries throughout south and southeast Asia and stretched as far as the northeastern African coast. Within six hours the deadly waves traveled more than 3,000 miles and carved a trail of death and destruction as they arrived on land. Current official estimates indicate that more than 160,000 people are dead, and millions of others are affected, including the injured, missing, or displaced. [13] Recent news reports suggest that the death toll may be well above 200,000. The World Health Organization (WHO) indicates that an estimated three to five million people lack the basic necessities for survival; between one and two million people may be displaced. In many places the physical environment is badly damaged or destroyed, including entire communities, homes, businesses, tourist areas, and infrastructure (roads, bridges, power and telephone systems, and public buildings). For many their means of livelihood and way of life has been wiped out. In the hardest hit areas, social services are severely compromised or nonexistent. Experts have said this is the most powerful earthquake in 40 years and the fourth (and perhaps the second) most deadly in the last century. Estimates of the dead make it the worst tsunami disaster on record.

A massive, global relief and recovery operation is underway. According to the United Nations, the relief operation is the largest ever undertaken. Indonesia, Sri Lanka, India, and Thailand have suffered some of the worst devastation. Within a day, all were declared a disaster by their respective U.S. ambassador, which allowed U.S. aid to be immediately released through the Office of Foreign Disaster Assistance (OFDA). For information on current conditions and latest developments, view the reports of governments, private voluntary agencies, and U.N. agencies on the web at [http://www.reliefweb.int.]

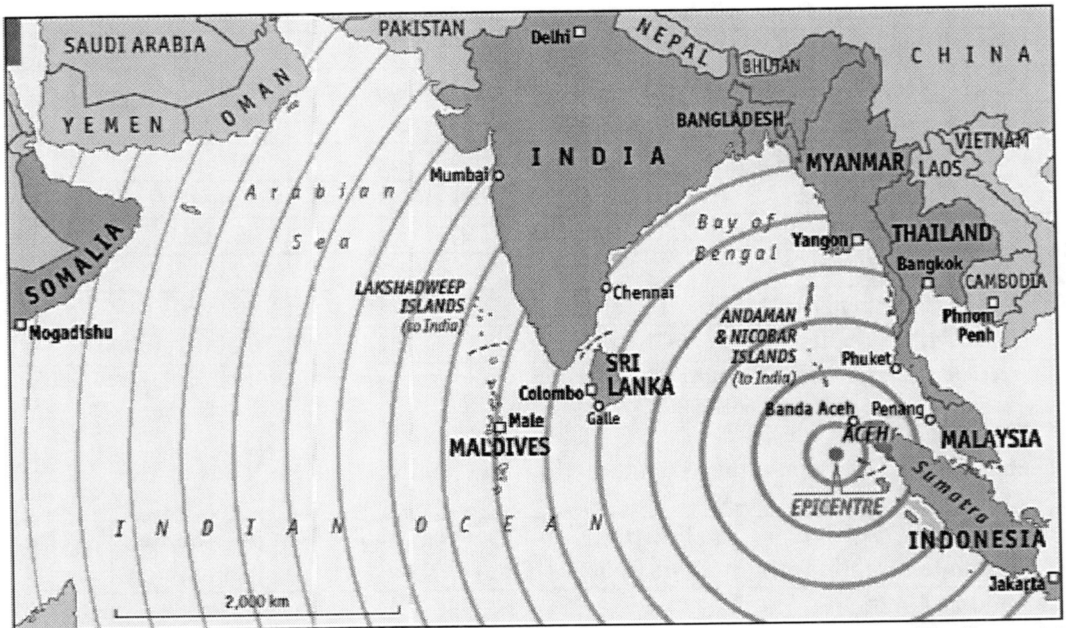

Source: Based on a map provided by the *Struggling to Bring Relief to the Survivors* report, *The Economist*, January 2005. Adapted by CRS. (K. Yancey 1/4/05).

Figure 1. Map of the 2004 Indian Ocean Earthquake and Tsunami.

Comparisons to Past Disasters [14]

In terms of estimated fatalities, the Indian Ocean tsunami ranks among the world's worst natural disasters, though it falls below other events. (See Table 2) The unique feature of this tsunami is the extent of the damage and the number of countries affected. Unlike the damage caused by other disasters, which tended to be highly localized, the Indian Ocean tsunami struck thousands of miles of populous coastline in nearly a dozen countries, affecting millions of people. The devastation was particularly acute in several island areas, where at times, entire land masses were flooded. The very nature of the tidal waves, combined with the lack of warning, made women, children, the elderly and others unable to swim particularly vulnerable. Also, the potential deaths of thousands of tourists from the industrialized world vacationing in southern Thailand and Sri Lanka — mostly Europeans but also many Americans and Japanese — has given the Indian Ocean tsunami a higher profile than previous disasters.

No natural disasters in recent memory compare with the magnitude and scope of this earthquake and tsunami. Table 3 provides context, detailing the large-scale U.S. assistance that followed after a previous natural disaster, the October 1998 Hurricane Mitch, which inflicted severe destruction upon several countries in Central America.

Table 2. Deadliest Natural Disasters

Year	Location	Event	Estimated Death Toll
1931	Huang He River, China	flood	3.7 million
1970	Bangladesh	cyclone	300,000
1976	Tangshan, China	earthquake (magnitude 7.5)	255,000
1920	Ningxia-Kansu, China	earthquake (8.6)	200,000
1927	Tsinghai, China	earthquake (7.9)	200,000
2004	Indian Ocean	earthquake (9.0) and tsunami	150,000+
1923	Kanto region, Japan	earthquake (7.9)	143,000
1991	Bangladesh	cyclone	139,000
1948	Turkmenistan, USSR	earthquake (7.3)	110,000
1908	Messina, Italy	earthquake (7.2)	70,000-100,000

Sources: *Washington Post*, December 30, 2004; U.S. Geological Survey. * Official death toll. Unofficial estimates range as high as 655,000.

Table 3. U.S. Governmental Assistance after Hurricane Mitch (millions of U.S. dollars)

Country Assisted (Estimated Death Toll)	Existing U.S. Resources and Debt Relief at Time of Disaster	Supplemental Appropriation	Total
Honduras (14,000)	238.3	324.9	563.2
Nicaragua (3,500)	57.4	113.0	170.4
Guatemala (440)	42.5	35.9	78.4
El Salvador (370)	19.4	35.1	54.5
Costa Rica (6)	-	9.0	9.0
Central America Regional	-	27.3	27.3
Total	357.6	545.2	902.8

Even as the emergency response gains momentum, discussion of the medium and long-term reconstruction of the area has begun and will likely continue at international meetings and within the U.S. government. Preliminary damage assessments are underway in the affected countries. Experts had already estimated the total damage to the region in the billions of dollars. In Indonesia, a joint report issued by the government of Indonesia and the international donor community estimates that the total cost of damages and losses is $4.45 billion. [15] Secretary- General Kofi Annan said it could take ten years to bring parts of the region back to full capacity.

The reconstruction effort will likely attempt to reduce the vulnerability of these countries to similar disasters in the future. Although countries in the Pacific region have a warning system for tsunamis (which are a relatively frequent occurrence), the countries in the Indian Ocean lack such a coordinated response. In an effort to improve disaster preparedness a review of the response to the earthquake and tsunami may include an examination of the dissemination of information by national governments to other governments and to their populace, communication between regional governments about the course and damage of the storm, and local governmental disaster response plans and procedures. See the section on early warning systems later in this report.

Current Situation [16]

As the disaster unfolded, the dilemma involved in prioritizing resource allocations began to take shape: on the one hand, to try to save as many lives as possible and on the other, to identify and dispose of bodies as death tolls continued to rise. Multiple challenges have arisen because of the large number of countries affected across a wide geographic area. Moreover, Indonesia, Sri Lanka, and Somalia have been in conflicts that are as yet unresolved and present potential difficulties in the distribution of aid. And there are millions of people displaced, separated from their families and left with nothing. Critical problems vary by country, including the condition of the infrastructure and response system, the scope of destruction, and degree of access. The situation in each country is discussed later in the report. Initial assessments indicated that the most urgent priorities in the affected areas were for potable water, sanitation (and waste disposal), food, and shelter.

Table 4. Estimated Number of Persons Affected by the Earthquake and Tsunamis

	Country Death toll (estimated)	Missing (estimated)	Displaced (estimated)
Indonesia	114,978	12,070	555,156
Sri Lanka	30,922	5,565	437,482
India	10,749	5,640	112,558
Thailand	5,318	3,199	
Burma (Myanmar)	60-80		
The Maldives	81	26	21,633 300,000 affected
Malaysia	68	6	8,000
Tanzania	10		
Bangladesh	2		
Somalia	150-298		5,000 displaced 102,000 affected
Kenya	1		
Seychelles	3		40 households displaced

Sources: Statistical data provided by USAID Indian Ocean Earthquake and Tsunamis Report, and BBC News online, January 19, 2005.

Health

The World Health Organization (WHO), which is the lead agency for the coordination of international public health response to disasters such as the tsunami, and the United Nations Children's Fund (UNICEF), along with international organizations and nongovernmental organizations (NGOs), are all working to meet the public health needs of the affected region. In the first week after the disaster, WHO warned that the death toll could double if clean water, sanitation, medical treatment, and relief supplies were not provided to the affected areas. [17]

WHO continues to stress the need to guard against the risk of disease and further deaths through surveillance and early warning systems. WHO remains particularly concerned about disease outbreaks among the many vulnerable populations from contaminated water sources and crowded, unsanitary living conditions, including cholera, dysentery, malaria, and dengue fever. The numbers of injured are estimated to be twice or three times the death toll. So far there do not appear to be any signs of epidemics. WHO has also identified the need to address mental health issues and rebuild the capacity of health systems as critical to recovery. [18]

Relief Operations and Aid Delivery

Experts break relief operations into several phases: search and rescue; treatment and survival; relocation and rehabilitation; and long-term reconstruction. [19] As with any massive undertaking that has many moving parts, it can take days to get a relief effort underway. Delays in transportation and congestion, lack of transportation infrastructure, bureaucratic problems, lack of access, all can cause bottlenecks at key points in the system. While timing is critical to save lives, to enable a network of this size to function efficiently requires the coordination of assessments and appropriate responses with local governments, communities, and the international community.

In general, the relief effort has been viewed positively and the convening authority of the United Nations has been well received. The sheer scale of this relief effort has brought together tremendous capacity and willingness to help, but experts generally caution that an ongoing effort and strategic planning is required at each phase to work out coordination and logistics issues. The relief effort is now beginning to focus primarily on recovery and rehabilitation. [20]

Three weeks after the tsunami hit the region, more detailed interagency assessments are underway, the information from which will be critical for planning recovery and reconstruction initiatives, developing strategies for the use of funding, and determining whether personnel are in place with adequate resources. In certain areas, particularly in Indonesia, access and logistics problems continue. There are logistical bottlenecks, and the lack of transportation and adequate infrastructure remain a challenge. Concerns about disease and the need for sanitation and medical capacity are still critical.

Impediments to aid in Indonesia appear to be particularly challenging for several reasons. There are the obvious logistical difficulties. The destruction of transportation infrastructure has made it difficult to extend assistance to all of the affected areas. [21] The coordination of national and local level government with the military and over 50 relief groups presents problems. The conflict between secessionists and the government has also complicated the relief effort. The Indonesian military feels it has to look to both relief and counter-insurgency operations. There is also the issue of national pride. Indonesia was, like India, a leading member of the non-aligned movement. This may be, in part, a reason for Indonesia's decision to ask providers of foreign military assistance to leave the country by March.

Protection for Children and Separated Orphans [22]

Background

Trafficking in children goes on worldwide and may even be increasing. Statistics on child trafficking, however, are very unreliable and official estimates may reflect only a part of its actual extent. The Department of State's 2004 *Trafficking in Persons Report* says that of the 600,000 - 800,000 persons trafficked across international borders each year, 70% are female and 50% are children. In addition, according to that report, many more people (probably millions) are trafficked within countries. The International Labor Organization (ILO) puts the number of children trafficked both internally and across borders annually at 1.2 million. All these numbers are estimates and no country is immune from trafficking, including the United States. [23]

According to the United Nations, human trafficking is a highly lucrative global industry controlled by powerful criminal organizations from which they derive many billions in revenues annually. This places human trafficking just behind drug and arms trafficking in terms of illicit revenues. Global experience in addressing child trafficking, and distinct focus on the problem separate from the overall human trafficking issue, is relatively new. The problem is huge in scope, multifaceted and sensitive, both culturally and politically.

Both boys and girls are trafficked, as are children of all ages — some very young children and some nearly adults. Trafficking in children is directly linked to their subsequent exploitation. The forms of exploitation vary including commercial sexual exploitation (for prostitution or pornography), use as domestic servants, as bonded laborers, as beggars, in other illicit activities from drug running to burglaries, as well as child soldiers. In addition, babies may be trafficked for adoption, and older teens for marriage. In all cases constraints are put on the movement of the children involved who are virtually enslaved. Girls are the chief victims of trafficking for sexual exploitation, domestic work and marriage. Boys and girls, however, are subjected to trafficking and most forms of exploitation. [24]

The root causes of sale and trafficking of children are complex, and include conditions of conflict and population movements, poverty, lack of employment opportunities, low social status of the girl child, impunity from prosecution, and a general lack of education and awareness. Children from minority groups, or those who are undocumented, are particularly vulnerable to being trafficked. [25] Situations of massive dislocation due to natural disasters, like the recent tsunami in the Indian Ocean, provide opportunities for syndicates to take advantage of the chaos and breakdown of protection mechanisms that leave orphans and children separated from their parents particularly vulnerable.

Tsunami Orphans: The Tsunami Generation

UNICEF, and other organizations focused on the fate of children orphaned or separated from their families amid the chaos of a disaster, acknowledge it is a multifaceted problem that will take time to resolve. The scope of the problem in the tsunami-affected countries is not fully known, although some believe the reports of child exploitation have been exaggerated. There are only estimates of the number of children orphaned or separated from their parents. UNICEF refers to these children as the Tsunami Generation.

The United Nations, international organizations and NGOs have issued warnings of the risks to children left unprotected in the aftermath of the tsunami. They are working on high-alert prevention mechanisms, including raising awareness at camps, providing guidelines to officials and volunteers, urging governments in affected countries to act, and identifying police and community officials to be of assistance. International adoptions are considered very premature and are not considered the best option for the child. Governments of affected countries are working with UNICEF to prevent illegal adoptions and trafficking.

UNICEF has developed five key steps to protect children from exploitation, including identification and registration; provision of immediate, safe care; tracing and reunification with extended family members; alerting police and community authorities; and working with governments of the affected countries to monitor the problem. UNICEF is also encouraging children to go back to school as soon as possible as a way of creating a more normal environment and beginning to deal with the mental trauma of the disaster.

Humanitarian Response: U.S. and International Assistance [26]

U.S. Emergency Assistance to the Region

Offers of assistance have greatly increased since December 26, 2004, as the international community has come to realize the growing scale of the disaster. In the case of the United States, American Ambassadors responsible for Sri Lanka, the Maldives, India and Indonesia provided $400,000 in immediate assistance in the wake of the Indian Ocean tsunami. The United States Government then provided $4 million in additional assistance to the Red Cross. The United States Agency for International Development (USAID)'s Office of Foreign Disaster Assistance (OFDA) immediately sent Disaster Assistance Response Teams (DARTs) to the region to assess needs in the areas of sanitation, health, and other kinds of relief supplies.

On December 28, $10 million was allocated for the relief effort for a total estimated initial contribution by the United States of around $15 million. [27] As reports of the growing scale of the disaster came in, the United States raised its pledge to $35 million. [28] By December 31, this number had increased to $350 million. Of this amount, as of January 19, 2005, USAID reports that close to $100 million has been committed. [29] For the latest breakdown of U.S. government assistance to the region, see [http://www.usaid.gov].

Military assistance to the region, in coordination with international organizations and NGOs, includes flights with relief aid, medical supplies, personnel, and equipment to affected areas. [30] Initially, the U.S. Navy dispatched P-3 patrol aircraft and an aircraft carrier to assist with relief operations. Since then, helicopters have been used to deliver relief supplies and evacuate the injured. In addition, surface ships, landing crafts and inflatable boats were positioned to provide relief supplies, including the capacity to produce potable water, transport vehicles, generators and other equipment. Military forensic teams are in Thailand and preventive medicine units are conducting assessments in Indonesia. As of January 19, 2005, more than 11,600 military personnel are involved in the relief operation with 17 ships and 75 aircraft. The cost of total military spending to date is not yet available, although it is estimated that the U.S. military contribution is more than $5 million a day with a cost to date of $165 million. It is expected that an additional $175 million may be required to cover costs through the end of February. [31] It remains to be seen what military costs are included in the $350 million pledge.

On December 29, 1004, President Bush announced the formation of a donor group consisting of the United States, Australia, India and Japan to coordinate relief and military capabilities in the region in the first weeks of the crisis. On January 6, the Core Group joined the efforts of the United Nations Office for the Coordination of Humanitarian Affairs (OCHA) as the lead agency on the relief effort.

Two interagency task forces have been established — one to coordinate U.S. government relief efforts and the other to assist in tracking missing Americans. Eighteen Americans are confirmed dead, with another sixteen presumed dead, and 153 remain unaccounted for.

Private sector assistance has already been substantial and is expected to continue to grow. [32] On January 3, President Bush announced that former Presidents George H.W. Bush, and Bill Clinton would lead a fundraising effort in the U.S. private sector in support of the tsunami crisis. Cash donations are being encouraged. It is too soon to estimate the value of private relief supplies, which will be transported by DOD under the Denton program. [33]

U.S. Secretary of State Colin Powell, Florida Governor Jeb Bush, and USAID Administrator Andrew Natsios visited the affected region in early January 2005 to assess the situation and whether the response is sufficient to meet the needs on the ground. Several U.S. Congressional delegations also traveled to the region over the past few weeks.

The U.S. Emergency Response Mechanism

The United States is generally a leader and major contributor to relief efforts in humanitarian disasters. [34] In 2004 the United States contributed more than 2.4 billion to disaster relief worldwide. In the case of the Indian Ocean earthquake and tsunami, it is clear that the response will require a major long-term effort beyond the relief and recovery operation currently underway. [35]

The President has broad authority to provide emergency assistance for foreign disasters and the United States government provides disaster assistance through several U.S. agencies. The very nature of humanitarian disasters — the need to respond quickly in order to save lives and provide relief — has resulted in an unrestricted definition of what this type of assistance consists of on both a policy and operational level. While humanitarian assistance is assumed to provide for urgent food, shelter, and medical needs, the agencies within the U.S. Government providing this support expand or contract the definition in response to circumstances. Funds may be used for U.S. agencies to deliver the services required or to provide grants to international organizations (IOs), international governmental and non-governmental organizations (NGOs), and private or religious voluntary organizations (PVOs.) USAID is the U.S. agency charged with coordinating U.S. government and private sector assistance. [36] It also coordinates with international organizations, the governments of countries suffering disasters, and other governments.

OFDA in USAID's Bureau of Humanitarian Response can respond immediately with relief materials and personnel including personnel and materiel already located in various countries around the world. [37] It is responsible for the provision of nonfood humanitarian assistance and has disaster response teams (DARTS) which can be assembled quickly to conduct assessments of the situation. OFDA has wide authority to borrow funds, equipment and personnel from other parts of USAID and other federal agencies. USAID has two other offices that administer U.S. humanitarian aid: Food For Peace (FFP) and the Office of

Transition Initiatives (OTI). USAID administers Title II of the FFP under P.L. 480 and provides relief and development food aid that does not have to be repaid. OTI provides post-disaster transition assistance, which includes mainly short-term peace and democratization projects with some attention to humanitarian elements but not emergency relief.

The Department of Defense (DOD) Overseas Humanitarian, Disaster and Civic Aid (OHDACA) appropriation funds three DOD humanitarian programs: the Humanitarian Assistance Program (HAP), the Humanitarian Mine Action (HMA) Program, and Foreign Disaster Relief and Emergency Response (FDR/ER). The office provides humanitarian support to stabilize emergency situations and deals with a range of tasks including the provision of food, shelter and supplies, and medical evacuations. In addition the President has the authority to draw down defense equipment and direct military personnel to respond to disasters. The President may also use the Denton program to provide space available transportation on military aircraft and ships to private donors who wish to transport humanitarian goods and equipment in response to a disaster.

Generally, OFDA provides emergency aid which lasts 30-90 days. The same is true for Department of Defense humanitarian assistance. Aft the initial emergency is over, assistance is provided through other channels, such as the regular country development programs of USAID.

The State Department also administers programs for humanitarian relief with a focus on refugees and the displaced. Emergency Refugee and Migration Account (ERMA) is a fund available until spent [38] and provides wide latitude to the President in responding to refugee emergencies. Emergencies lasting more than a year come out of the regular Migration and Refugee Account (MRA) through the Population, Migration and Refugees (PRM) bureau. PRM [39] covers refugees worldwide, conflict victims, and populations of concern to the United Nations High Commissioner for Refugees (UNHCR), often extended to Internally Displaced Persons (IDPs). Humanitarian assistance includes a range of services from basic needs to community services.

Legislation [40]

Several bills pertaining to the Indian Ocean tsunamis and their after-effects have been introduced in the 109th Congress. One of these bills, H.R. 241 (Thomas), entitled *To Accelerate the Income Tax Benefits for Charitable Cash Contributions for the Relief of Victims of the Indian Ocean Tsunami*, was the first legislative measure passed by the 109th Congress to be signed into law; it became P.L. 109-1. As of January 13, 2005, other pending bills included the following:

- **H.Res. 12 (Hyde).** Introduced and passed by the House on January 4, 2005; entitled *Expressing condolences and support for assistance to the victims of the earthquake and tsunamis that occurred on December 26, 2004, in South and Southeast Asia.*
- **H.R. 60 (Jackson-Lee).** Introduced and referred to the House Committee on the Judiciary on January 4, 2005; entitled *To designate Sri Lanka, India, Indonesia, Thailand, Somalia, Myanmar, Malaysia, Maldives, Tanzania, Seychelles, Bangladesh, and Kenya under section 244 of the Immigration and Nationality Act in*

order to render nationals of such foreign states eligible for temporary protected status under such section.

- **S.Res. 4 (Frist).** Introduced and passed in the Senate on January 4, 2005; entitled *A resolution expressing the sympathy and pledging the support of the United States Senate and the people of the United States for the victims of the powerful earthquake and devastating tsunami that struck Bangladesh, Burma, India, Indonesia, Kenya, Malaysia, the Maldives, the Seychelles, Somalia, Sri Lanka, Tanzania, Thailand, and other areas of South Asia, Southeast Asia, and Africa, on December 26, 2004.*

International Emergency Assistance to the Region

International recovery efforts are typically complex because they require coordination among numerous different actors. Those responding to humanitarian crises include U.N. agencies, international organizations, NGOs, PVOs, and bilateral and multilateral donors. A great deal of assistance is provided by other governments and international agencies. The U.N. OCHA tracks worldwide contributions to disasters. [41] According to the U.N., as of January 19, 2005, pledges from the international community for the Indian Ocean tsunami stand at over $7 billion.

Initially, the European Union pledged $40.5 million dollars. Australia pledged $7.6 million dollars while France, Germany, Russia, Britain, Pakistan, and Italy initially reacted by sending plane loads of assistance supplies. The International Red Cross and the Red Crescent Societies were focused on an initial appeal of $6.6 million. [42] Since then, donations have increased enormously (see Table 1). Australia and Japan have stated that they will help build a tidal wave warning system which is thought will cost tens of millions of dollars to establish. [43]

The U.N. agencies are also conducting damage assessments and reconstruction estimates which will likely be used at donor conferences and planning for the future. The United Nations Under-Secretary-General for Humanitarian Affairs and Emergency Relief Coordinator, Jan Egeland, has stated that "the cost of the devastation will be in the billions of dollars. It would probably be in the many billions of dollars," making it one of the largest humanitarian relief efforts in history. [44] On January 6, the United Nations and its partners launched a flash appeal for $977 million.

International Donor Conferences

On January 6, 2005, the Association of Southeast Asian Nations (ASEAN) held an emergency meeting to discuss coordination of international relief efforts and managing logistical obstacles that have delayed the delivery of aid in certain areas. [45] A meeting of summit leaders took place in Jakarta on January 6 and focused on increasing donor contributions and coordination of the relief effort. [46] A large international donors conference took place on January 11 in Geneva.

Situation Report on Countries Affected by the Tsunami

The current situation, as of January 18, 2005, in each affected country is described below with brief background descriptions, reports of the damage, and highlights of the emergency response.

Indonesia [47]

The northern part of the Indonesian island of Sumatra, especially the northernmost province of Aceh, was closest to the epicenter of the Indian Ocean earthquake. Successive tidal waves of 30 to 50 feet high slammed into Aceh's west coast of nearly 200 miles. As of January 2, 2005, the Indonesian government estimated the death toll at nearly 100,000, mostly in Aceh. Aerial surveys of Aceh's west coast from Banda Aceh, the provincial capital, southward for about 150 miles revealed near total destruction of towns and villages with many of them underwater. The coastal area was isolated with no aid getting through until January 1, 2005. The same is true of a number of small islands off Sumatra's west coast. Initial international aid is coming through the re-opened Banda Aceh airport and the airport at Medan, a major city south of Aceh. The United States, Australia, and Singapore were supplying the bulk of aid, and non-government humanitarian groups were also active.

By January 2-3, there were signs of recovery in Banda Aceh: the reopening of markets, the restoration of power and water to 40 percent of the city, and shipments of fuel supplies into the city. Indonesian government efforts to remove massive debris and bury thousands of dead people were making progress, although much remained to be done. Beginning on January 1, U.S. SH-60 Bravo helicopters flying off the U.S. aircraft carrier, Abraham Lincoln, were delivering food and water to the isolated towns and villages down Aceh's west coast from Banda Aceh. On January 2, U.S. navy helicopters, numbering about 25, flew 27 missions and delivered 80,000 pounds of supplies. Indonesian navy helicopters also were delivering supplies to these towns and villages, but the Indonesian military only has few helicopters in Sumatra. Providing adequate water to the thousands of Acehnese stranded along the west coast remains difficult. Medical treatment of numerous injuries also has been difficult. Many of the injured have to be transported by helicopter to medical facilities at Banda Aceh, which strains the helicopter fleet available.

A main problem in relief efforts was the backup of relief supplies at the airports at Banda Aceh and Medan. Hundreds of tons of food, water, medicines, and tents were at the airports and reaching destitute people, including approximately 150,000 homeless people in 20 refugee camps, very slowly or not at all. Unloading equipment at the airports was described as inadequate. There reportedly was a severe lack of trucks for distribution. The Indonesian central government and the Aceh provincial governments had little infrastructure to facilitate distribution of aid.

International private aid officials in Aceh also accused the Indonesian military of delaying distribution of relief supplies. The military (TNI) controls the relief supplies at the Banda Aceh and Medan airports. Until January 1, the TNI initially refused to allow foreign relief airplanes to land at Banda Aceh. Indonesian President Susilio Yudhoyono apparently overrode military opposition to foreign relief deliveries. Since then, several TNI commanders have cooperated with American, Australian, and Singaporean military units, and they have praised the U.S. military relief effort.

Nevertheless, on January 11 and 12, the Indonesian military and government officials announced restrictions on future foreign relief operations in Aceh. The main restrictions are termination of all foreign military relief operations by March 26; restrictions on plans by U.S. Marines to move significant quantities of aid and manpower into the west coast of Aceh; the establishment of TNI operational control over all foreign relief operations; a requirement that a TNI officer be on board any foreign aircraft engaged in relief; confinement of foreign aid workers to the towns of Banda Aceh and Meulaboh unless they receive TNI permission to operate elsewhere; and a requirement that aid workers operating outside Banda Aceh and Meulaboh must be accompanied by TNI personnel.

TNI attitudes are governed by an insurgency in the province that has gone on since 1976. Anti-Indonesia forces (the Free Aceh Movement — GAM) seek independence for the province and cite decades of repressive Indonesian rule as justification for their uprising. The Indonesian military (TNI) long has been accused of committing atrocities and other human rights abuses in Aceh and being involved in corrupt practices there. In May 2003, the Indonesian government, under pressure from the TNI, ended a six-month long cease-fire with the insurgents and declared martial law. The TNI suppressed separatist political activity and reported resumed severe human rights violations. The TNI also banned foreigners from Aceh, including aid workers. The government lifted the ban on foreign aid workers on December 27, 2004; but the restrictions announced on January 11 and 12, 2005, appear motivated, at least in part, by a desire of the TNI to restore Aceh as much as possible to the pre-tsunami situation of closure to foreigners. TNI commanders justify the restrictions as needed to protect aid workers from the GAM and prevent relief supplies from falling into the hands of GAM. However, GAM has declared a cease-fire and asserts that it welcomes the foreign presence. Experts on Indonesia predict that the next round of TNI restrictions will be aimed at the foreign press, which entered Aceh after the tsunami.

The TNI also has facilitated the entrance into Aceh of Islamic militant groups, allegedly for relief operations. The TNI provided air transport, provisions, and housing to these groups. One of these groups, the Mujahideen Council of Indonesia (MMI) is viewed by U.S. terrorism experts, such as Zachary Abuza (currently with the U.S. Institute of Peace) as a political front for Jemaah Islamiya, Al Qaeda's regional terrorist arm in Southeast Asia. [48] The TNI's support of MMI's entrance into Aceh raises questions regarding the TNI's relations with and policies toward Islamic terrorist groups.

Sri Lanka [49]

The Indian Ocean tsunami hit Sri Lanka particularly hard, killing 30,899. An estimated 40% of those killed in Sri Lanka were children. Some 6,034 people were still missing while between 441,410 and 504,440 were homeless as of mid-January 2005. Of these, an estimated 186,000 are thought to have been taken in by friends and family while some 250,000 have been placed in welfare centers and makeshift camps. Tsunami related damages have been estimated at $1.8 billion. Sri Lanka has requested some debt forgiveness and a two-year hold on its $8.82 billion debt. [50]

In the immediate wake of the disaster, President Bush expressed his condolences to the victims over the "terrible loss of life and suffering." As of January 6 it was estimated that one quarter of the disaster victims had yet to be reached. [51] The State Department issued a travel advisory warning Americans to avoid Sri Lanka. [52] The Sri Lankan Ambassador to the United States, Devinda Subasinghe, stated that up to 70% of the Sri Lankan coast was damaged. The inundation led to looting and a prison break of some 200 inmates from a coastal prison. By January 18 the situation in Sri Lanka had improved significantly and widespread disease had not emerged. Deputy Secretary of Defense Paul Wolfowitz reportedly observed during his visit to Sri Lanka that the nation was moving from the relief and rescue stage of operations to the reconstruction and rehabilitation stage. [53]

The United States *Bonhomme Richard* Expeditionary Strike Group, which had been in Guam, was ordered to the Bay of Bengal to provide assistance to affected countries. The seven ships in the strike group have 25 helicopters, 2,100 marines and 1,400 sailors which has provided assistance. The head of the Pacific Command, Admiral Thomas Fargo, also ordered two ships out of the squadron based in Diego Garcia to provide assistance as well as five pre-positioned ships located in Guam. Each pre-positioned ship can produce 90,000 gallons of fresh water per day. [54]

Sri Lanka has apparently mobilized its resources to deal with the disaster in a more effective way than was originally thought likely. In the initial post disaster phase Andrew Natsios, Administrator, U.S. Agency for International Development, stated that "I think the Sri Lankans basically are telling us this is so massive, they are being overwhelmed by it." [55] It was estimated on December 30 that some 10,000 to 12,000 Sri Lankans were injured. Sri Lanka's transportation links to the affected areas has reportedly collapsed. Rail connections to the south had closed. Truckers refuse to travel south for fear of another tsunami. Some of the estimated one million land mines set during ongoing Sri Lanka's civil war — between the government and ethnic Tamil rebels in the north and east — were reportedly unearthed and shifted during the flood. The Tamil rebel group, the Liberation Tigers of Tamil Eelam (LTTE), complained that aid is not getting through to Tamil areas. [56] The Sri Lankan army has a fleet of only 12 helicopters. [57]

By January 18, visiting Deputy Secretary of Defense Paul Wolfowitz traveled to Sri Lanka and observed Sri Lanka's recovery efforts and reportedly stated that Sri Lanka may now be at the point where it no longer needs U.S. military assistance. U.S. helicopters have run 1,500 disaster relief missions across the region. In connection with secessionist strife in Sri Lanka and Indonesia, Wolfowitz also remarked that "... hopefully they realize the stakes for which they're fighting are trivial in comparison." [58] U.S. military assistance has reportedly stayed away from Tamil areas of Sri Lanka in an effort to avoid the Liberation Tigers of Tamil Eelam. India has reportedly been providing assistance to Tamil areas of Sri Lanka. [59] As the relief effort evolves, it has moved to address issues of protection of survivors and to providing assistance for psychological social program elements.

The two sectors of the Sri Lankan economy most affected are tourism and fisheries. Hundreds of hotels are damaged or destroyed. Hotels are now estimated to be half full. Sri Lanka's fishing fleet in the affected areas has been badly damaged. Sri Lanka harvests a reported 300,000 tons of fish annually for domestic consumption. Much of this is caught by subsistence fishermen. [60] Sri Lanka announced that it is postponing the South Asian games that it had planned to host in August 2005 in order that it may focus on reconstruction efforts.

Sri Lanka is a constitutional democracy with relatively high educational and social standards. [61] The country's political, social, and economic development has been seriously constrained by two decades of ethnic conflict between the majority Sinhalese and minority Tamil ethnic groups. Since 1983, a separatist war costing some 64,000 lives has been waged against government forces by the LTTE, which has been seeking to establish a separate state in the Tamil-dominated areas of the north and east. Though Sri Lanka lost fewer people than Indonesia, it lost them out of a smaller population. Sri Lanka lost over 30,000 out of a total population of approximately 20 million while Indonesia's losses are out of a population of over 220 million. Further, Sri Lanka suffered destruction on approximately 70% of its coast while the area affected in Indonesia was much more localized. [62]

India [63]

As of January 18, 2004, India is believed to have suffered up to 16,000 deaths and $2.2 billion in financial losses as a result of the Indian Ocean tsunami. [64] Waves 12-14 feet high struck India's eastern coast approximately three hours after the first tremor. Many or most of those killed in the populous southeastern state of Tamil Nadu reportedly were women and children. [65] The city of Nagapattinam, a fishing community some 150 miles south of Madras (Chennai), was devastated by the ocean surge which advanced the shoreline up to 100 meters inland along the Tamil Nadu coast. (USAID officials reported tsunami-related destruction in Tamil Nadu more than one kilometer inland.) Nagapattinam alone eventually may account for up to 20,000 deaths, and more than 650,000 Tamil Nadu residents are said to have been displaced or otherwise affected by the tsunami. [66] The southernmost of India's Andaman and Nicobar Islands sit only 80 miles from the earthquake epicenter in the Bay of Bengal. Some 30,000 residents of the archipelago lived on the nearly flat island of Car Nicobar, where an Indian air force base was completely submerged. Car Nicobar alone may account for up to one-third of deaths in the remote archipelago; one report claims that 12 of the island's 15 villages were "obliterated" by the tsunami. [67] Severe flooding in all affected regions has contaminated water systems and, combined with the existence of many corpses floating in coastal areas, raised concerns that lethal waterborne diseases such as cholera and diarrhea may become epidemic. [68] The Tamil Nadu economy is heavily reliant on marine product exports and is expected to suffer major losses with the destruction of tens of thousands of fishing boats and nets. Shipping came to a virtual standstill at the Madras port (south India's largest), and the region's tourist industry has been devastated by physical damage and booking cancellations. Madras's 8-mile beach, said to be the world's second-longest, has been nearly deserted since December 26. [69]

India was considered by many to have had a well established disaster management system. The United States has been engaged with Indian in disaster training and technical assistance through USAID for some years. [70] However, numerous critics of the Indian relief effort have spoken out in 2005. At least one United Nations expert called the recent disaster a "wake-up call"for Indian planners who allegedly failed to learn from past experience, and Indian Red Cross officials spoke of "chaotic" relief management and the "hijacking"of aid supplies by government workers in Port Blair, the Andaman and Nicobar capital. A Hong Kongbased human rights group described India's relief efforts as "pathetic," specifying lack of interagency coordination and caste discrimination as key problems. New

York-based Human Rights Watch itself highlighted inequitable aid distribution and urged the Indian government to do more to ensure that the Dalit (so-called untouchable) community was not discriminated against in disaster-stricken areas. [71]

Following the tsunami, the Indian government immediately released $115 million for the National Contingency Relief Fund. For some days after the disaster, New Delhi did not request international assistance and turned down emergency aid offers from the United States, Russia, Japan, and Israel, saying that indigenous capabilities are sufficient. Later, the Indian government did request long-term rehabilitation aid from both the World Bank and the Asian Development Bank. As of January 15, 2005, USAID's Office of U.S. Foreign Disaster Assistance had committed just above $3.1 million for emergency relief activities in India. [72] More than 5,000 Indian navy personnel used 27 ships, 19 helicopters, and six naval aircraft to deliver many hundreds of tons of relief supplies. The Indian prime minister has promised a payment of approximately $2,300 to the next of kin of each of those killed. India also has pledged $22 million in disaster aid to Sri Lanka and $2 million for Maldives and dispatched several naval ships to Sri Lanka, Maldives, and Indonesia. [73] According to the external affairs minister, New Delhi had disbursed $250 million on relief and rehabilitation efforts in India through January 4. [74]

In early January, the Tamil Nadu government was reporting that 412 relief camps had been established and held more than 300,000 people (at least 500,000 of the state's citizens had been evacuated). That government also will provide special relief packages to families suffering loss of homes. By January 17, 41 relief camps were still hosting about 44,000 citizens. [75]

Much of the Andaman and Nicobar Islands are off-limits to foreigners due to the presence of military facilities and to protect the region's aboriginal tribes. International aid agencies have requested access to the islands, where relief efforts are hampered by the destruction of most of the islands' jetties. Emergency crews there focused on burying the dead to prevent epidemics (it is Hindu custom to cremate the dead). [76]

India is the world's second most populous country with nearly 1.1 billion residents. The U.N. Development Program's 2004 Human Development Report assigns India a ranking of 127 out of 177 world countries, a status comparable to that of Morocco or Cambodia. Despite the existence of widespread and serious poverty, many observers believe that India's long-term economic potential is tremendous, and the current growth rate of the Indian economy (8.2% for the year ending July 2004) is amongst the highest in the world. The estimated gross domestic product in 2004 was just above $3 trillion, or $2,900 per capita (both figures in purchasing power parity terms). [77] India was allocated about $177 million in U.S. assistance for FY2004 and FY2005 combined, along with another $65 million in food aid. India has recently dealt with a major disaster, an earthquake that struck the western Gujarat state in January 2001, killing some 20,000 persons, injuring another 200,000, and leaving nearly one million homeless. New Delhi reportedly intends to purchase a $29 million tsunami warning system to be functional in 2007. Some observers believe that such a purchase would be unwise, given the rarity of tsunamis in the region. [78]

Thailand [79]

Six provinces on the western coast of southern Thailand, particularly the Phang Nga province and the resort islands of Phuket and Phi Phi, were badly hit by sea surges stemming from the underwater quake. Over 5,300 dead have been identified and over 3,000 remain missing, most of whom are presumed dead. [80] Officials said that about half of the dead were foreign vacationers, many from Europe. Many oceanfront properties, particularly hotels, were destroyed in the wave. Compared to other affected nations, however, the infrastructure in Thailand was left relatively unscathed: the regional electricity grid and telecommunication network continued to function, and the transportation system and water supply in Phuket were largely unaffected.

The emergency response in Thailand has been praised by the international community: United Nations and Australian relief agency officials described effective and rapid coordination of grass roots relief teams to distribute supplies and provide first aid. Some credit Thaksin's strong political authority to command the military and police forces. Thaksin has also come out strongly in favor of establishing a tsunami alert system in cooperation with other regional governments. Scattered press reports initially accused government officials of declining to evacuate the island despite receiving a warning, based partially on fears of hurting the tourism industry. [81] Such criticism has largely subsided, however, and Thaksin's popularity ratings have increased based on his leadership in the wake of the disaster.

The diplomatic and logistical challenge of the disaster in Thailand is different from the other affected countries. Because at least 36 nationalities are represented among the victims, many consulates are directly involved in the tasks of identifying the dead. Sweden appears to be the hardest hit, with up to 1,900 missing. Other high missing national tolls include Germany (730), Austria (500), the United Kingdom (over 400) and Italy (330). [82] The Thai police have taken charge of a massive effort to identify all the victims using DNA samples, with the cooperation of several international teams of forensic specialists. DNA testing is being conducted by Chinese labs, and an American company is responsible for caring for those remains that need to be repatriated. Over 4,000 bodies were exhumed from their original burial in order to ensure that all bodies are identified using the standard set by Interpol. [83]

Thailand is the logistics hub for much of the U.S. and international relief effort. U.S. relief operations by air and sea for the entire region are being directed out of Thailand's Utapao air base and Sattahip naval base. Thailand's government immediately granted full U.S. access to the bases following the disaster. Lt. Gen. Robert R. Blackman, the overall American military commander in Okinawa, is heading the mission in Utapao, coordinating with his OFTA counterpart. Representatives from Japan, Singapore, the U.N., the World Food Program, and the World Health Organization are also working out of Utapao. A full DART team is stationed in Bangkok.

Initially, the U.S. military provided about 20 cargo planes, tanker aircraft, and search and rescue planes, flown to Thailand from Japan and Guam. P-3 surveillance aircraft conducted survey operations, including search-and-rescue efforts, and cargo planes shuttled supplies to shelter the living and dry ice to preserve the dead from Bangkok to affected areas. [84] Bangkok was the first stop by Secretary of State Colin Powell and Florida Governor Jeb Bush on their tour of countries hit by the disaster.

Beyond the immediate concern of dealing with the dead and injured, Thailand is likely to suffer economically, at least in the short term, because of the blow to its tourism industry. The industry brings in about $8 billion annually, nearly 6% of Thailand's GDP. Because the tsunami struck at the peak of tourist season in Thailand, millions of visitors are expected to cancel their plans, immediately costing operators about $750 million, analysts estimate. [85] Many analysts are optimistic, however, that the industry will rebound quickly, as only about 5-10% of Thailand's hotels were affected and rebuilding is expected to be swift. The Thai government reassured investors that it intended to spend $768 million to repair infrastructure in the area.

Thailand is a long-time military ally with ongoing relevance to U.S. logistical operations in Iraq, a key country in the war against terrorism in Southeast Asia, and a significant trade and economic partner. A proposed U.S.-Thailand Free Trade Agreement (FTA) is currently being negotiated. Despite differences on Burma policy and human rights issues, shared economic and security interests have long provided the basis for U.S.-Thai cooperation. In FY2003 and 2004, Thailand received over $20 million in economic and security assistance from the United States. For the past year, Thailand has faced an insurgency in its southern, majority-Muslim provinces; clashes between separatists and Thai security forces have left up to 560 people dead.

Burma [86]

In contrast to other governments affected by the Indian Ocean earthquake and tidal waves, the Burmese government — as of December 29 — had given out little information of the effects on Burma. An official from an international aid agency told the Agence France Presse on December 27, on condition of anonymity, that government officials were confirming 36 dead. The government subsequently issued a figure of 53 dead. On December 28, the Agence France Presse cited at least 90 killed but cited no source. The source apparently was information over the internet websites of anti-government groups. The international aid agency official speculated that the actual death toll is "far greater," given the trajectory of the tidal waves and the closeness of Burma's Indian Ocean coastline to the epicenter of the earthquake. The *London Sunday Telegraph* (reprinted in the *Washington Times*, January 2, 2005) quoted Burmese fishermen describing a major loss of life on lower Burma's coastline just north of the hard-hit Thai coast. However, U.N. officials stated on January 6 that the death toll in Burma was relatively small. The Burmese government had not issued an appeal for international aid, as of January 3, 2005. U.N. officials, Doctors Without Borders, and the International Committee for the Red Cross have sought government permission to visit the lower Burma coastline.

The issue of aid is complicated by the heavy economic sanctions imposed by the United States and the European Union on Burma because of the politically repressive policies of the military-dominated Burmese government. United Nations officials in Rangoon stated on December 27 that the United Nations was prepared to conduct relief operations. The government likely would accept humanitarian and reconstruction aid from China, Burma's main international supporter, and from regional countries like Malaysia, Singapore, and India. The government also might accept humanitarian aid from Japan, which has provided low levels of such aid despite sanctions on Japanese developmental aid and investment. However, the government likely would not allow any sizeable presence of foreign aid workers. It is also highly unlikely that the government would ask for or accept aid from the United States. A number of experts on Burma have stated that the views of Burmese military officials toward the United States have become very negative because of the absence of a positive U.S. response to the government's release from house arrest of opposition leader, Aung Sann Suu-kyi, in 1992 and the U.S. Congress' enactment of a total ban on Burmese imports to the United States in July 1993 in response to the re-arrest of Aung Sann Suu-kyi.

The Maldives [87]

The tsunami also hit the island-state of The Maldives. Initial reports put the death toll at 32. This was increased to 55 on December 29, to 80 by January 3rd, and to 86 by January 14th, 2005. A further 21 are missing and some 14,900 have been displaced by the disaster. Many outlying islands are only one meter above sea level. 10,000 persons have been evacuated off 13 low lying islands. About half of the island of Male was covered in two feet of water which closed the airport. [88] All of the Maldives is below 8 feet in elevation. Reports indicate that a 10 - 15 foot wave washed over some parts of the Maldives leaving houses smashed, wells contaminated, and power and communications infrastructure inoperable. The Maldives' outlying coral reefs reportedly protected many of the islands from the tsunami. Nevertheless the government estimates that reconstruction will cost $1 billion or the rough equivalent of two years' gross domestic product. [89] Tourism accounts for 30% of GDP in the Maldives. It is hoped that tourism can return to 80% of capacity by March. Parliamentary elections planned for December 31 were postponed. [90]

An American civil/military team was in the Maldives on the 3rd of January 2005 to make an assessment of the damage in preparation for U.S. assistance. An initial estimate called for 1,000 military personnel are to be in the Sri Lanka/Maldives area to provide disaster assistance. [91] On January 17th two military supply ships that had been providing assistance to Sri Lanka were sent to assist the Maldives. Though the Maldives managed to have a relatively low number of fatalities, its reconstruction will be particularly difficult due to its geography.

The Republic of the Maldives is a micro state of some 1,200 islands, approximately 200 of which are inhabited by a total population of roughly 310,000. The island state has less than half the land area of Washington DC and is situated in the Indian Ocean off the southwest tip of India. In 1887, the Maldives became a British protectorate. The islands became independent in 1965. The capital, Male, has approximately 70,000 residents. The overall population growth rate is about 3%. The Maldives has a 97% literacy rate. There are four main ethnic groups; Sinhalese, Dravidian, Arab and African and the main religion is Sunni Muslim.

The current president of the Maldives, Maumoon Gayoom, assumed office in 1978. [92] He was elected to a sixth five-year term in 2003 under a system where the voters vote for or against a single candidate selected by the Maldivian parliament known as the Majlis. The President appoints 8 of the 50 members of the Majlis. [93] The Republic of the Maldives is a member of the South Asian Association of Regional Cooperation (SAARC) as well as the British Commonwealth. [94]

Diego Garcia [95]

The American military base on Diego Garcia, located south of the Maldives, was one of the few places in the Indian Ocean that did receive warning of the tsunami waves. The base reportedly emerged from the event without major damage. Evidently the configuration of the ocean floor near Diego Garcia played a role in lessening the effect of the tsunami there. The base reportedly received a warning because the Navy is on the contact list of the Pacific Warning Center. [96]

Malaysia [97]

Malaysia includes the Malay peninsula in the west and to the east, and Sabah and Sarawak on the north of the island of Borneo. Malaysia has a population of some 23 million. Malaysia was spared the devastation wrecked on Indonesia as it was shielded from the tsunami by Sumatra. Despite this, some 68 were reported killed and 183 injured by the tsunami in Penang and in Kedah, Malaysia. [98] A fuel loading facility on the island of Langkawi in north western Malaysia was reportedly damaged in the tsunami. [99] Malaysia has opened its airspace and airports for international relief efforts. Malaysia also raised 4.7 million rupiah for disaster relief by December 29. [100] Prime Minister Abdullah Badawi expressed his condolences and proposed greater regional cooperation to deal with natural disasters. [101]

Bangladesh [102]

While Bangladesh has been devastated by past cyclones it was largely spared destruction from the most recent tsunami. The Bangladesh port of Chittagong was hit by large waves which caused flooding in 30 districts and left 2 dead as of December 29. [103] Bangladesh lost 300,000 in a cyclone in 1970 and a further 139,000 to another storm in 1991. [104] Bangladesh is currently working with other South Asian countries to set a new date for the South Asian Association of Regional Cooperation summit which was to be held on January 9-11 in Dhaka. Bangladesh has joined other SAARC countries to provide assistance to Sri Lanka and the Maldives. It is also hoped that the upcoming SAARC summit can provide further assistance for those affected by the disaster. [105]

Somalia [106]

Tsunami waves reached Somalia about seven hours after hitting nations in South Asia, about 4,000 miles away. Several Somali coastal towns and roads, notably in northeastern and central coastal zones, were flooded and substantially destroyed by the tsunamis. Thousands of boats and shelters were destroyed, severely damaged, and numerous persons were reported missing. U.N. and news agencies report that between 150 and 298 Somalis died as a result of the tsunamis. [107] The northern Hafun peninsula was among the worst-affected areas. The U.N.-affiliated World Food Program (WFP) sent an assessment team to the coast of the northeastern Puntland region, and OCHA led a preliminary air-based December 30 mission to assess coastal zone tsunami damage. U.N. officials estimated that about 54,000 Somalis were directly affected by the tsunamis and that about 18,000 households may require emergency aid. The WFP has sent over 277 tons of food to the affected region, where the World Health Organization deployed three emergency kits with a capacity to serve 30,000 persons' basic needs for three months. The Kenya-based Somali transitional government has reportedly made unconfirmed, possibly exaggerated claims that over 1,000 Somalis may have died as a result of the tsunamis, and announced plans to send its own assessment team to Somalia.

OCHA on January 3, reported that international tsunami-related contributions to Somalia included $50,000 from the United States, to be delivered via UNICEF, and $100,000 from Saudi Arabia, contributed through the Society of the Red Cross. Some existing U.N. drought-related and humanitarian aid was being re-prioritized to meet emerging tsunami-related needs. Somali government officials issued informal appeals for tsunami-related food and medical aid. According to a January 3 news report, a total of 24 countries had pledged to send relief aid to Somalia, but such aid had not arrived, according to a Somali presidential spokesman. [108] U.S. officials planned to respond to Somali government requests for tsunami relief aid by reviewing U.N. assessments and, if aid is warranted, to channel any U.S. aid through U.N. agencies. However, if needs prove severe and U.S. officials view the delivery of U.S. bilateral emergency aid as necessary, a U.S. emergency declaration could be made by the U.S. embassy in Nairobi. German Chancellor Gerhard Schroeder has suggested that a moratorium on debt owed by Somalia to creditor nations be discussed at a January 2005 meeting of the Paris Club of official creditors. When questioned about the proposal, U.S. officials, including President Bush, publicly did not reject it, although they did not address it in detail. [109]

Somalia, a northeastern African country of about 8.3 million, has been wracked by intermittent civil war and armed banditry since the ouster of President Siad Barre in 1991. Since then, it has lacked an effective central government, and remains politically fractious and dangerous due to the activities of diverse armed groups. It is divided into three semi-autonomous regions: Somaliland, in the northwest and Puntland in the north, both self-governed regions; and southern and central Somalia, which is divided into localities dominated by local clans, warlords, and business interests. Somalia is undergoing a process of peace making and state reconstruction. In August 2004, key warlords and politicians formed a new parliament, which appointed President Abdullahi Yusuf Ahmed in October 2004. U.S., international and Somali government access to southern Somalia is severely limited due to insecurity. U.S. interests are represented by the U.S. mission in Nairobi, Kenya.

Conventional, non-tsunami-related U.S. assistance to Somalia focuses on bolstering the capacity of civil society organizations and institutions related to local governance and adherence to the rule of law; enhancing local economic opportunities by backing a variety of projects focused on basic education, infrastructure rehabilitation, and alternative energy use; and support for healthcare delivery. U.S. Economic Support Fund monies, not shown in the aid table in the appendix, have also helped finance lengthy negotiations aimed at forming a central Somali government. The bulk of U.S. aid is delivered in the form of a various emergency, supplemental, and developmental food-related and nutrition programs. H.R. 4818, the foreign operations FY2005 appropriations bill, enacted as P.L. 108-447, did not designate a specific appropriation for Somalia, which is not mentioned in the House report (H.Rept. 108-599) or conference report (H.Rept. 108-792) associated with H.R. 4818. The Senate report (S.Rept. 108-346) that accompanied S. 2812, a Senate foreign operations FY2005 appropriations bill, later amended in relation to the passage of H.R. 4818, stated that "[t]he Committee is concerned that the budget request for assistance for Somalia under the DA account is only $986,000. The Committee requests USAID and the State Department to take a more active role to assist local efforts to promote peace and development in that country and recommends that not less than $5,000,000 in DA be provided to support secular education and strengthen civil society, particularly in Somaliland and Puntland."

Kenya

The coast of Kenya, an east African country of about 32.02 million persons, experienced tsunami waves that destroyed boats, damaged coastal properties, and reportedly killed one swimmer, a tourist. More deaths may have been averted because authorities closed coastal beaches and issued public precautions before and after the tsunami waves hit the country. Kenya has not requested tsunami-related aid. The international Committee of the Red Cross plans to ship at least 105 tons of relief supplies to Sri Lanka from Nairobi, where the organization stocks such supplies. [110]

Tanzania

In Dar es Salaam, the commercial capital of Tanzania, an east African country of about 36.59 million persons, ten young swimmers were reported killed as a result of tsunami waves. Additional persons may have died in a capsized boat. A tanker reportedly ruptured an oil pipeline as a result of the tsunamis. Tanzanian officials issued public warnings about possible further tsunami waves. Tanzania has not requested tsunami-related aid. [111]

Seychelles

Seychelles, a group of Indian Ocean islands northeast of Madagascar off the eastern African coast, sustained tsunami-related coastal floods. These destroyed two bridges, some sewer and water systems, and caused extensive damage to a port, power lines, schools, real properties, boats, and vehicles. Total damage in Seychelles is worth an estimated $23.5 million. Three tsunami-related fatalities occurred. Seychelles may formally request tsunami-related international aid, likely from the United States, according to State Department officials. [112]

Madagascar

A tsunami wave flooded a coastal village in southeastern Madagascar, a large Indian Ocean island off the coast of Mozambique, causing about 1,200 people to become homeless. Madagascar, which regularly experiences extensive typhoonrelated natural disasters, has not requested tsunami-related aid. [113]

Mauritius

Damage to property, boats, and a weather station were reported in Mauritius, where tsunami-related coastal evacuation orders were issued. Mauritius has not requested tsunami-related aid. [114]

Reunion (French Territory)

The BBC reports that tsunamis damaged about 15 fishing vessels. [115]

South Africa

South Africa reported unusually high tides, believed to be tsunami-related, in which a man perished. [116]

Issues for Congress [117]

Tsunami Aid and Reconstruction Issues

Burdensharing

A day after the south Asia crisis, U.N. Under-Secretary- General for Humanitarian Affairs and Emergency Relief Coordinator Jan Egeland, commenting on contributions by the wealthy nations to disasters in general in 2004, stated that some developed nations were being "stingy" with aid. According to the Organization for Economic Cooperation and Development, although the United States is the world's largest provider of foreign assistance, it often is one of the lowest contributors in per capita terms amongst the world's most wealthy countries. The United States has been reported as giving 0.14 percent of GNP in international development assistance as compared to Norway's 0.92 percent contribution. [118] USAID Director Andrew Natsios has refuted Egeland's statement, saying that the aforementioned data was only for development assistance and did not include disaster relief. [119]

In the first days after the tsunami, the Bush Administration was criticized by some observers for displaying a lack of urgency in its initial response. President Bush came under criticism for waiting three days before publicly speaking about the disaster during his vacation in Crawford, Texas. [120] The subsequent increase of U.S. economic and logistical assistance, along with the dispatch of Secretary of State Powell and Florida Governor Bush to the region a week after the tsunami, may help to change this perception.

In previous disasters, pledges made by governments have not always resulted in actual contributions, the Bam earthquake of December 2003 is but one example raised by the United Nations. Experts are concerned that while billions of dollars have been pledged to help the victims of the tsunami disaster, there is no guarantee that these pledges will be honored. It also cannot be assumed that the funds represent new money as it may previously have been allocated elsewhere. Some are also concerned about funding priorities and resources for other disaster areas and the very real possibility of international donor fatigue. It will take time for a more complete picture to reveal how the actual costs of the tsunami disaster will be shared among international donors. [121]

Competing Aid and Budget Priorities [122]

Funding the Indian Ocean tsunami relief and reconstruction effort is likely to be a challenge faced in the early weeks of the 109th Congress. Thus far, the Administration has been able to fund its pledge of $350 million for emergency relief by depleting most worldwide disaster contingency appropriations for FY2005. In the short term, the Administration can also shift funds from regular economic aid accounts in order to address urgent tsunami victim needs. In order to respond to future humanitarian crises, however, these resources will need to be replenished. The White House may send Congress an emergency FY2005 supplemental request in early February.

Even before the disaster struck, Congress was expected to struggle to find the resources to sustain U.S. aid pledges amid efforts to tackle rising budget deficits by, among other measures, slowing or reducing discretionary spending. During the FY2005 debate, lawmakers reduced the President's foreign assistance budget request (a subset of the larger foreign policy budget request) by $1.7 billion, or nearly 8%. This was the first time such cuts occurred during the Bush Administration. Some Members of Congress publicly have expressed concern that funding for tsunami relief and reconstruction, if not fully restored through supplemental appropriations, may jeopardize resources for subsequent international disasters or for other aid priorities from which tsunami emergency aid has been transferred. [123]

Transparency

Members of Congress have also raised concerns about transparency of donor contributions, allocation of monies, and monitoring of projects by the United Nations. The United Nations has said it will improve its financial tracking and reporting system and Price Waterhouse Coopers is reportedly assisting in that effort. Many contributions are also being made directly to international organizations and non-governmental organizations, which could raise the same questions about transparency requirements. Moreover, while earmarks and time limits may ensure greater accountability, they can also add pressure for organizations to spend contributed funds, sometimes leading to unnecessary spending, waste and duplicated efforts. Restrictions on funds also often do not allow flexibility to adapt projects to better meet the changing needs on the ground. [124]

Debt Relief

While there is an on-going need for immediate relief assistance for tsunami-affected countries, longer term aid will also be needed to assist these nations, which face substantial costs associated with rebuilding infrastructure and basic social services. Such extended aid may take the form of official debt relief or repayment moratoriums, which may free resources for reconstruction. Several creditor governments reportedly support an immediate moratorium on debt payments by affected nations while other debt-related policy options are considered. [125] While U.S. officials have not firmly committed to any large-scale program of debt cancellation or repayment term rescheduling, [126] at least one significant debt-related policy decision — the release of a communique allowing temporary credit forbearance by debtors to the consensus-based Paris Club of creditor governments, of which the United States is a member — has been made to date. [127] In addition, the World Bank, IMF, and major bilateral creditor governments, including the United States, have been considering an expansion of the Heavily Indebted Poor Country (HIPC) initiative. Although none of the tsunami-affected countries are eligible for HIPC debt relief, consideration of debt reduction proposals for these disaster-struck nations could occur during subsequent talks on HIPC enhancement. Due to the size of Indonesia's debt burden, some have argued in the past that Jakarta should be eligible for some form of HIPC debt-relief terms.

If the Paris Club decides to provide debt cancellation or the rescheduling of credit repayment terms to any of the tsunami-affected countries, or if the multilateral financial institutions recommend such relief, Congress may be called upon to consider the nature, extent, and conditions of any credit relief that may be provided by the United States. In addition, Congress may consider bilateral or multilateral debt relief as a component of U.S. policy efforts to help tsunami-affected countries to recover economically.

Implications for Other U.S. Foreign Policy Interests

The War on Terrorism

The 9/11 Commission and others have pointed out the U.S. interest in preventing regions of instability from becoming havens or recruiting grounds for Islamist terrorist groups such as Al Qaeda and Jemaah Islamiya (JI), the Southeast Asia terrorist organization that has close ties to Al Qaeda and is thought to have killed hundreds in four separate attacks since September 11, 2001. While Sumatra, in Indonesia, has not been an active base of operation for Al Qaeda or JI, the Indonesian military's support of the entrance of the Mujahideen Council of Indonesia (MMI) raises serious questions about the TNI's policy toward terrorist groups, given the MMI's relationship with Jemaah Islamiya and Al Qaeda. Moreover, any prolonged economic and political disruption, combined with potential perceptions of Jakarta's inability to deliver assistance, could open the door for a more active terrorist presence or lead the anti-Indonesian Free Aceh Movement (GAM) to establish ties to JI or Al Qaeda. Additionally, some Indonesian organizations and charities with known ties to JI have dispatched humanitarian relief teams to Aceh. In Southern Thailand, the areas most affected by the tsunami are generally considered ethnically and regionally distinct from the predominantly Muslim provinces on the western coast of peninsular Thailand, which have been the site of sectarian and antigovernment violence by Muslims over the past year.

Countering Negative Images of the United States

The large-scale U.S. response to the tsunami is unlikely to reverse the decline in the U.S. image abroad since the September 11 attacks, because this decline primarily is due to American policies in the Middle East. However, the scale and scope of U.S. assistance could provide a positive example of U.S. leadership and military capabilities. The decline in the U.S. image abroad has been particularly acute in the Muslim world, especially in Indonesia, where according to one series of polls, only 15% of those polled in 2003 said they had a favorable opinion of the United States, down from 61% in 2002. [128] Additionally, the U.S. tsunami relief effort could help counter the perception among some Southeast Asians that the United States not only has placed too much emphasis on terrorism in its Southeast Asia policy, but also has relied too heavily on "hard" (military) power to combat terrorism. The 9/11 Commission and others have recommended expanding U.S. public diplomacy programs as a way to help win the global battle for "hearts and minds" especially in the Islamic world from which the Muslim terrorists seek to draw recruits and support. The restrictions on foreign relief activities announced by the Indonesian military and government on January 11 and 12, 2005, potentially raise the reverse issue of negative U.S. reactions to Indonesia. Commentary in the U.S. press and on radio talks shows has been very negative toward Indonesia because of the restrictions.

Early Warning Systems: International Scientific, Technological and Other Challenges [129]

Nations affected by the December 26, 2004 tsunami, assisted by others, are pursuing a multilateral effort through the U.N. Environmental Programs to develop a tsunami detection and early warning network for the Indian Ocean.

Decisions about whether and how to proceed with establishing an international tsunami early warning system for the Indian Ocean (and elsewhere) will likely be complicated for a number of reasons. One reason is because of the number of different potential international parties that would be involved with the need to coordinate data collection and warning dissemination, and a second is the funding needed to establish a tsunami warning system in that region. A third is that nations, including some in the Indian Ocean, might charge for access to critical satellite data that may help in warning potential victims. Some in Congress assert that the costs of acquiring those data could be well worth it, in terms of lives saved; while others counter that access to those proprietary data should be provided free of charge, especially when the United States and other nations provide disaster relief and propose funding tsunami detection and warning activities for the region. [130]

The greatest challenge is likely be to establishing local or regional emergency management infrastructures for inhabitants in coastal regions bounding the Indian Ocean to receive tsunami alerts in sufficient time to evacuate, and to be notified when to return after the dangers have subsided. Many question who would be responsible for building and maintaining such systems. Other challenges could include standardizing instrumentation for detecting tsunamis and other related technology globally. As a possible threat for U.S. homeland security, routine, unrestricted access to international telecommunications networks that would relay sensitive environmental data as well as issue tsunami warnings could compromise domestic intelligence-gathering operations.

After the Indian Ocean tsunami disaster, some Members of Congress became concerned about the possible vulnerability of U.S. coastal areas to tsunamis, and the adequacy of early warning for costal areas of the western Atlantic Ocean. A few of them, and now the Bush Administration, proposed expanding tsunami warning networks in Pacific coastal areas, and adding coverage for the Atlantic seaboard. Others question the risks of a tsunami hitting the U.S. Atlantic coast. [131] Assessing the probability as low, they assert that risk factor should be important when conceptualizing a cooperative early tsunami warning system for the U.S. Eastern Seaboard. Although additional tsunami detection and warning instrumentation for the

United States (and elsewhere) could run into the millions of dollars, experts suggest that existing weather buoy and regional coastal and ocean observation networks and their telecommunications capacity might be shared. Others have proposed that the European Union, Canada, and the United States engage jointly establish coverage for the North Atlantic.

The Indian Ocean tsunami has led some to call for instituting a tsunami detection and warning system in the Atlantic and/or Indian Oceans. Affected nations assisted by others may consider a multilateral effort to develop a detection and warning network for the future. Australia and Japan have stated that they will help build a tidal wave warning system for the Indian Ocean. By some estimates, this will cost tens of millions of dollars to establish. [132] Some Members of Congress also have proposed such a network for the U.S. Atlantic seaboard. Although instrumentation costs could run into the millions of dollars, existing weather buoys and developing state and local coastal and ocean observation networks might serve as possible platforms for instrumentation. Accordingly, the European Union, Canada, and the United States might consider bi-lateral efforts to establish tsunami coverage of the North Atlantic.

Aid to Indonesia and the Leahy Amendment

U.S. economic aid to Indonesia for fiscal years 2002, 2003, and 2004 totaled $412 million. The Bush Administration budgeted $158 million for FY2005. Much of this aid has gone to programs supporting the development of democratic political institutions in Indonesia with a recent emphasis on Indonesia's education system. The Administration reportedly will tap this existing bilateral aid program to help fund the U.S. relief effort in Indonesia. [133] Congress can be expected to receive new aid requests from the Administration focusing on humanitarian and reconstruction aid, especially directed at Aceh. Such requests undoubtedly would turn the attention of the Administration and Congress to the political situation in Aceh, especially the insurgency and the role of the Indonesian military (TNI).

Additionally, the disaster relief cooperation between the U.S. and Indonesian militaries is likely to be mentioned during the annual congressional deliberations over renewing restrictions on U.S.-Indonesian military-to-military relations, which the Bush Administration has sought to restore since the September 11, 2001 attacks. For more than a decade, Congress has restricted the provision of military assistance to Indonesia due to concern about serious human rights violations by the TNI, most notably the massacre of hundreds of people participating in a pro-independence rally in Dili, East Timor, in November 1991.

In a press briefing on January 6, 2005, Secretary Powell said that the U.S. is trying to provide the Indonesian government with enough spare parts to repair five Indonesian C-130 Hercules transport aircraft that currently are not operational. This would raise Indonesia's number of operational C-130s to twelve. As discussed below, current U.S. legislation places strict controls on the provision of military equipment to Indonesia. When pressed on the issue of whether Jakarta in the future might use repaired planes in its conflict with the GAM rebels in Aceh, Secretary Powell said that "the humanitarian need ... trumps, right now, the reservations we have." He added his "hope" that the Indonesian government's desire to receive additional military parts in the future would serve as a disincentive for using aircraft against the GAM. [134]

Although the language has varied from year to year, in general, the Leahy amendment bans arms sales to Indonesia, U.S. military training with the TNI, and TNI participation in the U.S. International Military Education Training (IMET) program unless the President certifies that the Indonesian government and the TNI are taking actions against the TNI's reported human rights abuses, including prosecution of abusers. The Leahy amendments for fiscal years 2002 and 2003 specifically mentioned Aceh in this context. About a week after the tsunami hit, the head of the Indonesian military's relief operations, Major General. Adam Damiri, was replaced, apparently because of concerns that his indictment for war crimes by a U.N.-backed tribunal in East Timor would complicate U.S.-Indonesian military relief cooperation. [135]

Appendix 1. U.S. Assistance to Selected Countries Affected by the Indian Ocean Tsunami

(Note: Totals may not add due to rounding)

Table 5. U.S. Assistance to Indonesia, 2001-2005 (millions of U.S. dollars)

Account	FY2001	FY2002	FY2002 S.A.a	FY2003	FY2004 estimate	FY2005 estimate
CSH	19.6	35.6	—	32.0	34.0	32.3
DA	51.5	38.7	—	39.0	31.3	32.7
ESF	49.9	50.0	—	59.6	49.7	65.0
IMET	0.0	0.4	—	0.0	0.0	0.6
NADR	0.0	0.0	8.0	1.0	5.8	6.0
INCLE	0.0	0.0	4.0	0.0	0.0	10.0
Totals	121.0	124.7	12.0	131.6	120.8	146.6
Food Aid (not including freight costs)						
P.L. 480 Title I USDA Loan	15.0	19.0	—	0.0	0.0	n/a
P.L. 480 Title II Grant	12.2	10.4	—	29.5	2.2	23.0
FFP	5.1	10.9	—	0.0	5.6	n/a
Section 416(b)	0.0	11.2	—	7.9	17.7	n/a

Sources: U.S. Department of State, USAID, U.S. Department of Agriculture
a. Supplemental Appropriations (P.L. 107-206)

Table 6. U.S. Assistance to Sri Lanka, 2001-2005 (millions of U.S. dollars)

Account	FY2001	FY2002	FY2003	FY2004 estimate	FY2005 estimate
CSH	0.3	0.3	0.3	0.3	0.3
DA	3.4	5.2	6.2	4.8	6.6
ESF	0.0	3.0	4.0	11.9	10.0
FMF	0.0	0.0	0.0	1.0	0.5
IMET	0.3	0.3	0.3	0.5	0.5
NADR	0.0	0.0	2.4	1.9	1.9
PKO	0.0	0.0	0.0	1.0	1.0
Totals	4.0	8.7	13.1	21.3	20.8

Table 6. Continued

Account	FY2001	FY2002	FY2003	FY2004 estimate	FY2005 estimate
Food Aid (not including freight costs)					
P.L. 480 Title I USDA Loan	7.9	8.0	0.0	0.0	n/a
P.L. 480 Title II Grant	0.0	1.3	0.6	2.7	0.0
FFP		0.0	2.8	0.0	n/a
Section 416(b)	6.0	0.0	0.0	0.9	n/a

Sources: U.S. Department of State, USAID, U.S. Department of Agriculture.

Table 7. U.S. Assistance to India, FY2001-FY2005 (millions of U.S. dollars)

Program or Account	FY2001 Actual	FY2002 Actual	FY2003 Actual	FY2004 Estimate	FY2005 Estimate
CSH	24.6	41.7	47.4	48.3	43.4
DA	28.8	29.2	34.5	25.7	25.4
ESF	5.0	7.0	10.5	14.9	15.0
IMET	0.5	1.0	1.0	1.3	1.4
NADR-EXBS	0.9	0.9	1.0	0.7	0.7
Totals	$59.8	$79.8	$94.4	$90.9	$85.9
Food Aid (Not including freight costs)					
P.L.480 Title II*	78.3	93.7	44.8	20.2	44.8
Section 416(b)*	-.-	12.0	-.-	-.-	-.-

Sources: U.S. Departments of State and Agriculture; U.S. Agency for International Development.

Table 8. U.S. Assistance to Thailand, FY2002-FY2005(millions of U.S. dollars)

Account	FY2001	FY2002	FY2003	FY2004 estimate	FY2005 estimate
CSH	0.0	1.0	1.5	0.0	0.0
DA	0.0	0.8	1.3	0.0	0.0
ESF	0.0	0.0	0.0	0.0	1.0
FMF	0.0	1.3	2.0	1.0	1.5
IMET	1.9	1.7	1.8	2.5	2.5
INCLE	4.1	4.0	3.7	2.0	2.0
NADR	1.3	0.7	0.2	0.4	0.8
Peace Corps	1.1	1.3	1.8	2.1	2.6
Totals	8.4	10.7	12.2	7.9	10.3

Sources: U.S. Department of State, USAID, U.S. Department of Agriculture.

Table 9. U.S. Assistance to Malaysia, 2001-2005 (millions of U.S. dollars)

Account	FY2001	FY2002	FY2003	FY2004 estimate	FY2005 estimate
IMET	0.8	0.8	0.8	1.2	1.1
NADR	0.1	0.2	1.3	0.1	1.0
Totals	0.9	1.0	2.1	1.3	2.1

Sources: U.S. Department of State, USAID, U.S. Department of Agriculture.

Table 10. U.S. Assistance to Somalia (millions of U.S. dollars)

	Account FY2003	Actual FY2004 Est.	FY2005 Req.*
CSH	0.3	0.1	-
DA	3.1	0.9	1.0
NADR-HD	0.5	-	-
Totals	3.8	1.0	1.0
P.L._480_Title_II Food Aid	136.4	89.0	–

Source: "Somalia," Request by Region: Africa, FY2005 Congressional Budget Justification for Foreign Operations, Feb. 10, 2004.

***Note:** No Somalia-specific appropriations were enacted for FY2005. Data on levels of any U.S. assistance for Somalia will become available after the Administration has notified the appropriate Congressional committees of its functional account allocations, in accordance with the Foreign Assistance Act of 1961, as amended. Overall assistance to sub-Saharan Africa rose slightly over FY2004 levels.

List of Aid-Related Abbreviations

CSH	Child Survival and Health Programs
DA	Development Assistance Programs
ESF	Economic Support Fund Programs
IMET	International Military Education and Training Programs
NADR-EXBS	Nonproliferation, Anti-Terrorism, Demining, and Related – Export

Control and Related Border Security Assistance Programs

P.L.480 Title II	Emergency and Private Assistance food aid (grants)
Section 416(b)	The Agricultural Act of 1949, as amended (surplus agricultural commodity donations)

Appendix 2. Child Protection Issues in Tsunami- Affected Countries

Indonesia

- One confirmed case of four-year-old boy taken out of Banda Aceh by a couple claiming to be his parents. (We cannot confirm the child was trafficked.)
- There may be other possible cases of child-trafficking: media reports sighting by an "NGO worker" of about 100 infants carried in a speed boat in the middle of the night.
- The government has imposed a moratorium on adoptions of children from Aceh.
- Children from Aceh under 16 cannot leave the country at this time.
- Surveillance will be increased at airports and seaports in North Sumatra and Aceh.

- The government has placed many Acehnese children in orphanages in Medan and other towns across Sumatra Island.
- Children being placed with Acehnese families under a temporary foster care scheme.
- Twenty child-friendly centers for unaccompanied children will soon be opened in major displacement camps in Aceh.
- Registration of children has begun.
- When adoptions become possible, Achenese residents will be given priority.

Thailand

- The government reports no cases of trafficking or abduction.
- The government has ruled out adoptions for unaccompanied children at this time.
- Specific measures being taken to prevent trafficking include registration of children, provision of temporary accommodation for unaccompanied children in government reception homes and family tracing.
- Child rights volunteers deployed in Ranong and Phuket to conduct community surveillance. UNICEF will work with provincial and district authorities to mobilize NGO partners, communities, and the media to be more vigilant on child protection issues.

Sri Lanka

- No reports of trafficking or abuse of children (in camps) received by UNICEF.
- UNICEF and partners providing additional support to grandparentheaded families and unaccompanied children.
- Reports of Sri Lankan citizens wanting to adopt children. Process for adoption takes up to five years. UNICEF is advocating for foster system. (Foster care is not a tradition in Sri Lanka.)
- UNICEF and partners have mobilized teams to identify and register all unaccompanied and separated children.
- Police and authorities are not yet present in camps, raising concerns that children will be more vulnerable to sexual and other abuse.
- UNICEF and the NCPA are conducting an emergency assessment to identify children in displaced camps who are without parents or otherwise vulnerable.
- UNICEF will support authorities in the investigation of all incidences of abuse of children.
- Data collection on unaccompanied and separated children is ongoing in all districts.

India

- No reports of trafficking or abuse of children received by UNICEF.

- UNICEF is seeking the views of the government of India on the adoption policy announced by the government of Tamil Nadu.
- UNICEF is providing psychosocial support to traumatized children in 13 districts.
- Unaccompanied children have been identified in camps in two districts in Tamil Nadu.
- Special orphanages for unaccompanied children have been opened in Tamil Nadu.

Malaysia

- The text message offering 300 Acehnese "orphans" for adoption is under investigation.
- UNICEF working with the government and UNICEF Indonesia as necessary to strengthen the monitoring capacity of immigration controls to prevent trafficking into Malaysia.

Source: Reported by UNICEF on January 12, 2005.

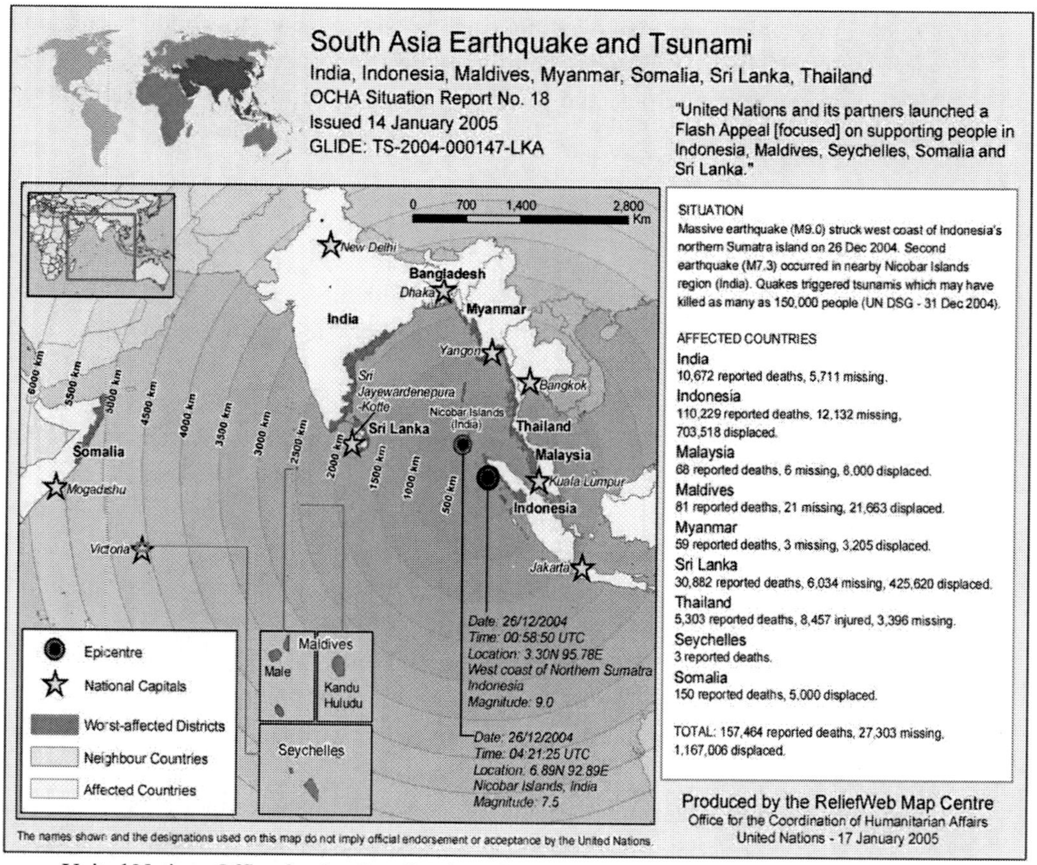

Source: United Nations Office for the Coordination of Humanitarian Affairs – 1/17/05

Figure 2. Countries Affected by the Tsunami.

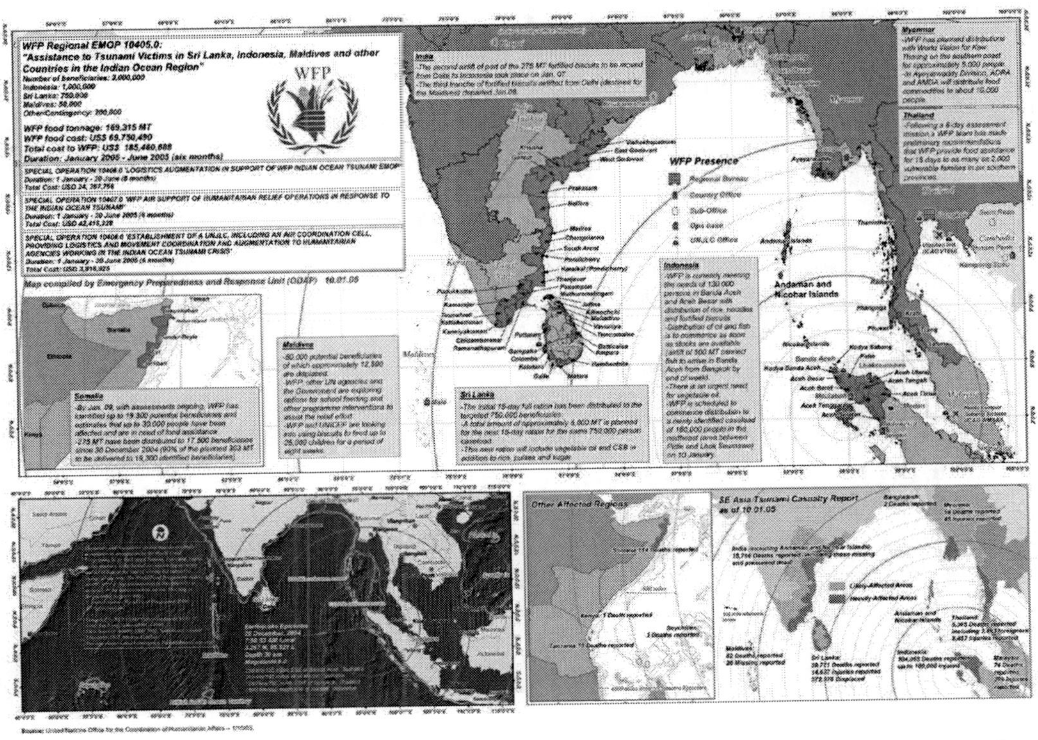

Figure 3. Regional Assistance and Food Aid Requirements.

References

[1] See USAID, "Indian Ocean — Earthquake and Tsunamis," Fact Sheet #32, February 8, 2005.

[2] Information for this section was drawn from interviews, the USAID fact sheets, reports by various U.N. agencies, international organizations, and non-governmental organizations available at [http://www.reliefweb.org].

[3] Prepared by Nicolas Cook, African Affairs Specialist.

[4] To assist the reader, this section repeats some legislation mentioned in the January 21, 2005 version of this report.

[5] See "Bush Seeks Additional $600 Million for Tsunami Relief: Aid to fund infrastructure projects, early warning systems," February 10, 2005 at [http://www.us info.state.gov].

[6] "Indian Ocean-Earthquake and Tsunamis," *Agency for International Development*, February 8, 2005.

[7] "Indian Ocean-Earthquake and Tsunamis," *Agency for International Development*, February 8, 2005.

[8] "US Estimates Tsunami Killed 33 Americans," *Washington Post*, February 9, 2005.

[9] "In Asia Trying to Look Ahead," *The New York Times*, January 14, 2005.

[10] "Indian Ocean-Earthquake and Tsunamis," *Agency for International Development*, February 8, 2005.

[11] Colin Gonsalves, "The Deadly Bureaucracy in the Andamans," *Indian Express* (Bombay), January 26, 2005.

[12] Prepared by Rhoda Margesson, Foreign Affairs Analyst.

[13] Early estimates of deaths from natural disasters are difficult to calculate and usually quite different from the final count. In this disaster the final number likely will never be known with any accuracy given the number of countries involved, the long, populous coastlines that were struck by the tsunamis, and the number of villages completely destroyed. Numbers fluctuate. See Donald G. McNeil, Jr., "Experts Say Accurate Toll is Hard to Calculate," *New York Times*, December 29, 2004.

[14] Prepared by Mark Manyin, Specialist in Asian Affairs.

[15] "Indonesia: Preliminary Damage and Loss Assessment: The December 26, 2004 Natural Disaster," The Consultative Group on Indonesia, Government of Indonesia (State Minister for National Planning Development Agency/BAPPENAS) and World Bank (for the international donor community), January 19-20, 2005.

[16] Prepared by Rhoda Margesson, Foreign Affairs Analyst.

[17] "Response to Enormity," *The Washington Post*, December 29, 2004.

[18] See "Questions and Answers: South Asia Earthquake and Tsunami," World Health Organization, January 14, 2005.

[19] "Relief: Massive Effort, Massive Need," *Christian Science Monitor*, January 3, 2005.

[20] See maps in Figures 2 and 3 at the end of this report for a regional overview of affected countries and assistance requirements.

[21] For example, according to USAID, a road north of Meulaboh, Indonesia is open to trucks and passable for 67 kilometres, but the journey now takes 12 hours instead of the 2 it used to take prior to the tsunami.

[22] Prepared by Rhoda Margesson, Foreign Affairs Analyst.

[23] "Anywhere between 20,000 and 50,000 people are trafficked into the United States each year, depending on the source. In addition, there are around 200,000 young people in America who may be victims of trafficking within the United States." Remarks of Under Secretary of State for Global Affairs, Paula Dobriansky, in Helsinki, Finland on June 3, 2003 at [http://www.usembassy.fi/servlet/PageServer?Page=trafficking /dobriansky.html] For background see also CRS Report RL30545 *Trafficking in Persons: The U.S. and International Response* by Francis T. Miko.

[24] *Trafficking in Children for Sexual Purposes: an Analytical Review*, p. 17 at [http://www.csecworldcongress.org/PDF/en/Yokohama/Background_reading/Theme_p a pers/ThemepaperTrafficking] in Children.pdf

[25] *A Child-Rights Approach on International Migration and Child Trafficking: a UNICEF Perspective* U.N. document: UN/POP/MIG/2004/9, October 18, 2004.

[26] Prepared by Rhoda Margesson, Foreign Affairs Analyst.

[27] Secretary Colin Powell, Briefing with Assistant Administrator for United States Agency for International Development Ed Fox," U.S. Department of State, December 27, 2004.

[28] John Harris and Robin Wright, "Aid Grows Amid Remarks About President's Absence," *The Washington Post*, December 29, 2004.

[29] Also see CRS Report RS22027 *Indian Ocean Earthquake and Tsunamis: Food Aid Needs and the U.S. Response* by Charles Hanrahan.

[30] Additional information is available on a U.S. Pacific Command Fact Sheet at [http://www.pacom.mil].

[31] "U.S. International Leaders Work to Coordinate Tsunami Relief," January 11, 2005 at [http://usinfo.state.gov]

[32] See also CRS Report RL32738 *Charitable Contributions for Tsunami Relief: P.L. 109-1* by Pamela L. Jackson.

[33] The Denton program, named after former Member of Congress Jeremiah Denton, authorizes shipment of privately donated humanitarian goods on U.S. military aircraft on a space-available basis. The donated goods must be certified as appropriate for the disaster by USAID's OFDA and can be bumped from the transport if other U.S. government aid must be transported.

[34] For background information see CRS Report RL32714, *International Disasters and Humanitarian Assistance: U.S. Governmental Response*, by Rhoda Margesson.

[35] There are a number of variables that make reading the United States government numbers and drawing accurate conclusions problematic. Questions about authority, definitions and categories of services make up part of the reason it is a challenge to grasp the concept and function of humanitarian assistance. Another factor has to do with how the numbers are generated in budgets within the U.S. government. Each agency has its own budget, with its own criteria, accounting detail and regional specificity. The fact that an urgent response to humanitarian crises is often required only compounds the problem. Budgets may reflect regional support, a certain area, specific countries, or a combination thereof over time and with changing events. Particularly in comparing assistance levels with other countries, financial sources may be compared against other forms of assistance (blankets, etc.) or they may reflect commitments of support rather than overall obligations.

[36] Private donations may be made to the private agencies working the area which are listed on the internet at [http://www.interaction.org.]

[37] Authorized in Sec. 491-493 of P.L. 87-195, the Foreign Assistance Act of 1961.

[38] Governed by P.L. 103-326, the maximum amount is $100 million. Authorized in sections 2 and 3 or P.L. 87-510 of the Migration and Refugee Assistance Act of 1962.

[39] When there is functional or programmatic overlap between USAID and PRM, they coordinate with each other and define partners. Traditionally PRM is a funder of UNHCR and other multilateral actors; USAID creates bilateral arrangements with NGOs. There is now a shift in partnering due to funding and resources required.

[40] Prepared by Nicolas Cook, African Affairs Specialist.

[41] The information is only as complete as the various governments' willingness to report the information. It does not include non-cash contributions in services or in kind (such as trucks and aircraft, crews, and emergency and medical personnel).

[42] "World Scrambles to Help Asia Tidal Victims," *Agence France Presse*, December 27, 2004.

[43] "Officials in Asia Concede That They Failed to Issue Warnings," *Associated Press*, December 27, 2004.

[44] Colum Lynch, "Billions in Aid Needed for Devastated Areas, U.N. Official Says," *The Washington Post*, December 28, 2004.

[45] ASEAN is comprised of Brunei, Cambodia, Indonesia, Laos, Malaysia, Myanmar, the Philippines, Singapore, Thailand and Vietnam.

[46] Special ASEAN Meeting Thursday to Coordinate Tsunami Response," Agence France-Presse, January 3, 2005.

[47] Prepared by Larry Niksch, Specialist in Asian Affairs, and Bruce Vaughn, Analyst in Asian Affairs.
[48] Interview with Zachary Abuza, January 13, 2005. Abuza, Jachary. *Muilitant Islam in Southeast Asia.* Boulder and London, Lynne Rienner Publishers, 2003. p. 140-158.
[49] Prepared by Bruce Vaughn, Analyst in Asian Affairs.
[50] "State of Play in Tsunami-hit Countries," *Reuters*, January 14, 2005.
[51] Brian Knowlton, "Officials Gather in Jakarta," *International Herald Tribune*, January 6, 2005.
[52] Deb Riechman, "Bush Sends Condolences to Asia, Offers Aid," *Associated Press*, December 27, 2004.
[53] "US Official Wolfowitz Visits Tsunami Hit Area in Sri Lanka," *BBC News,* January 17, 2005.
[54] Marc Grossman, "News Briefing on Indian Ocean Disaster Relief," *Federal Document Clearing House*, December 29, 2004.
[55] Marc Grossman, "News Briefing on Indian Ocean Disaster Relief," *Federal Document Clearing House*, December 29, 2004.
[56] "After the Tsunami the Rising Cost," *The Age*, December 30, 2004.
[57] Paddy Murphy, "Call for Choppers," *The Australian,* December 30, 2004.
[58] Joss White, "Wolfowitz Cites Sri Lanka's Progress on Reconstruction," *The Washington Post*, 1/18/05.
[59] "State of Play in Tsunami-hit Countries," *Reuters*, January 14, 2005.
[60] Amy Waldman and James Brooke, Disaster's Damage to Economies may be Minor," *The New York Times*, January 3, 2005.
[61] For additional information see CRS Report RL31707, *Sri Lanka: Background and U.S. Relations,* by Bruce Vaughn.
[62] Amy Waldman and David Rohde, "A Once-cherished Sea Gave Life, Then Took It," The New York Times, January 6, 2004.
[63] Prepared by Alan Kronstadt, Analyst in Asian Affairs.
[64] "India Tsunami Costs 'Hit $1.6 Billion,'" BBC News, January 7, 2005. On January 13, the Indian Ministry of Home Affairs was reporting 10,672 Indians confirmed dead and another 5,711 missing.
[65] T.S. Subramanian, "Killer Waves," *Frontline* (Madras), January 14, 2005.
[66] S. Anand, "The Big Churn," *Outlook India* (Delhi), December 30, 2004; USAID Fact Sheet #7, FY2005, January 2, 2005.
[67] Pankaj Sekhsaria, "Andaman's Agony," *Frontline* (Madras), January 14, 2005; Janaki Kremmer, "No easy Access For Remote Islands," *Christian Science Monitor*, January 4, 2005.
[68] "India Turns Down Foreign Relief Aid," ANSA English Media Service, December 29, 2004; "Tsunami-Hit India Struggles to Channel Flood of Aid to Needy," Agence France Presse, January 2, 2005.
[69] S. Anand, "The Big Churn," *Outlook India* (Delhi), December 30, 2004; "Tsunami Washes Away Tourism," *Times of India* (Delhi), December 27, 2004; Chris Tomlinson, "World's Second-Longest Beach, Center of Madras Life, Abandoned After Tsunami," Associated Press Newswires, January 5, 2004.
[70] Marc Grossman, "News Briefing on Indian Ocean Disaster Relief," Federal Document Clearing House, December 29, 2004.

[71] Rajesh Moudgil, "'A Wake-Up Call for India," *Hindustan Times* (Delhi), January 2, 2005; "Tsunami-Hit India Struggles to Channel Flood of Aid to Needy," Agence France Presse, January 2, 2005; "Aid to Indian Islands 'Hijacked,'" BBC News, January 13, 2005; "Rights Body Says India's Tsunami Relief Efforts 'Pathetic,'" Agence France Presse, January 10, 2005; "India: End Caste Bias in Tsunami Relief," Human Rights Watch Press Release, January 14, 2005. See also K.P.S. Gill, "Combined Muddled Group," *Outlook India* (Delhi), January 14, 2005.

[72] World Bank Press Release, January 11, 2005; USAID Fact Sheet #20, FY2005, January 15, 2005. Some observers believe that New Delhi's reliance on indigenous capabilities and sizeable aid contributions to neighboring states grow at least partly from a desire to have India seen as a major and self-sufficient power. India's rejection of external aid brought criticism from some quarters and reportedly has caused skepticism about motives among some diplomats. At least one report suggested that a U.S. military presence in Sri Lanka was being viewed by New Delhi as a symbolic intrusion into India's sphere of influence ("Post-Tsunami India's Image Rises Globally," *Hindustan Times* (Delhi), January 5, 2005; Edward Luce, "India Aims to Be Part of the Solution," *Financial Times* (London), January 6, 2005; "US-India Struggle For Control in Disaster Zone," *Telegraph* (Calcutta), January 4, 2005).

[73] "Earthquake and Tsunamis Wreak Devastation in Indian Ocean Region," Embassy of India Press Release; "The Indian Relief Effort," Embassy of India Press Release.

[74] "Statement by External Affairs Minister Shri K. Natwar Singh at the Special Meeting of Leaders Convened by ASEAN in the Aftermath of the Earthquake and Tsunami," Embassy of India Press Release, January 6, 2005.

[75] Government of Tamil Nadu, "Rescue and Relief Operations" at [http://www.tn.gov.in/tsunami/rescue.htm].

[76] "Foreign NGOs Seek Andamans Access," BBC News, January 3, 2005.

[77] C.I.A. World Factbook 2004; UNDP Human Development Report 2004.

[78] "India to Install Tsunami System," *Asia Pulse* (Sydney), January 4, 2005; T.V.R. Shenoy, "How Not to Respond to a Tsunami," *Indian Express* (Delhi), January 13, 2005.

[79] Prepared by Emma Chanlett-Avery, Analyst in Asian Affairs.

[80] *CNN News*. January 13, 2005.

[81] "Warning Rejected to Protect Tourism," *The Nation*. December 28, 2004.

[82] "Hopes Fade on Identifying Missing Foreigners," *Washington Post*. January 2, 2005.

[83] "MASS EXHUMATIONS: ID Operation Starts Again From Scratch," *The Nation*. January 14, 2005.

[84] "US Begins Shuttle of Aid to Victims Along Thai Coast," *New York Times*. January 1, 2005.

[85] "Thailand Death Toll Could Reach 2,000," *CNN.com*. December 28, 2004.

[86] Prepared by Larry Niksch, Specialist in Asian Affairs.

[87] Prepared by Bruce Vaughn, Analyst in Asian Affairs.

[88] "Americans Told to Avoid Travel to Sri Lanka and Thailand," *Agence France Presse*, December 27, 2004.

[89] James Hookway, "Tourism Thrives in the Maldives," *The Wall Street Journal,* January 3, 2004.

[90] "Quake Prompts Enormous Aid Effort," *BBC News*, December 28, 2004.

[91] "Ghost Island of the Maldives," *The Australian*, January 4, 2005.
[92] Maldives Leader Names Ministers," *BBC News*, September 1, 2004.
[93] "Country Profile: The Maldives," *BBC News*, August 14, 2004.
[94] "The Maldives: Introductory Survey," in *The Europa World Yearbook 2004*, (London: Europa Publications, Taylor and Francis Group, 2004). See also "Maldives: Quarterly Forecast Analysis," *Global Insight*, [http://www.globalinsight.com]
[95] Prepared by Bruce Vaughn, Analyst in Asian Affairs.
[96] M. Kayal and M. Wald, "Tracking Tsunamis: Why was There No Warning?" *The New York Times*, December 29, 2004.
[97] Prepared by Bruce Vaughn, Analyst in Asian Affairs.
[98] "After the Tsunami the Rising Cost," *The Age*, December 30, 2004 and "Malaysia Economic and Corporate News Summary," *AFX*, January 3, 2005.
[99] Asian Tsunami Causes Patchy Damage," *WMRC Daily*, December29, 2004.
[100] "Malaysians Do Care," *New Straits Times*, December 29, 2004.
[101] "PM Urges Cooperation in Providing Information," *New Straits Times*, December 29, 2004.
[102] Prepared by Bruce Vaughn, Analyst in Asian Affairs.
[103] "After the Tsunami the Rising Cost," *The Age*, December 30, 2004.
[104] "Major Natural Disasters," *US News and World Report*, January 10, 2005.
[105] "SAARC Urged to Organize Help for Tsunami-battered Countries," *Xinhua News Agency*, January 5, 2005.
[106] The remainder of the individual country entries were prepared by Nicolas Cook, African Affairs Specialist.
[107] "Somali Tsunami Victim Toll Rises," *BBC News*, January 5, 2005.
[108] Rodrique Ngowi, "Somalia still waiting for food, shelter, medical help for victims of the tsunamis," Associated Press, Jan. 3, 2005.
[109] Reuters, "Schroeder urges debt relief for Indonesia, Somalia," December 29, 2004; *Agence France Presse*, "US 'open' to debt relief for tsunami victims," December 29, 2004; White House, Office of the Press Secretary, "President Discusses Support for Earthquake and Tsunami Victims," December 29, 2004.
[110] BBC, "Many missing..."; Voice of America, "Tidal Wave Hits Somalia, Kenya," Dec. 27, 2004; Kenyan KBC radio, "Kenya sets up "crisis desk" to monitor tidal waves," BBC Monitoring Newsfile, Dec. 27 2004; Adrian Blomfield, "Evacuation from beaches cut deaths by hundreds in Kenya East Africa," The Daily Telegraph, Dec. 29 2004; Voice of America, "Government Officials to Travel Around Somalia to Assess Damage," Dec. 28, 2004.
[111] BBC, "Many missing..."; Pflanz, "Waves kill..."; Tom Maliti, "U.N. Struggles to Get Aid to Somali Town," Associated Press, Dec. 29 2004.
[112] BBC, "Many missing..."; Pflanz, "Waves kill..."; The Irish Examiner," Seychelles Caught in Tsunami's Path," Dec. 27, 2004; State Department communications.
[113] Mohamed Ali Bile, "Waves kill 38 Somalis, UN fears toll may rise," Reuters, Dec. 27 2004.
[114] Agence France Presse, "Over 100 feared dead in Somalia from killer Asian tidal waves," Dec. 27, 2004; State Department personal communication.
[115] BBC, "Many missing..."
[116] Cape Argus, "Somalia asks for UN help," Dec. 29, 2004.

[117] Prepared by Nicolas Cook, Mark Manyin, Rhoda Margesson, Larry Niksch, Larry Nowels, Bruce Vaughn, and Wayne Morrissey, Senior Research Assistant.

[118] John Harris and Robin Wright, "Aid Grows Amid Remarks About President's Absence," *The Washington Post*, December 29, 2004.

[119] December 29, 2004 Interview on the PBS TV Program, *The News Hour*. For more information on donor contribution comparisons, see CRS Report RS22032, *Foreign Aid: Understanding Data Used to Compare Donors*, by Larry Nowels.

[120] David Sanger, "It's About Aid, and an Image," *New York Times*, December 30, 2004.

[121] James Darcy, "The Indian Ocean Tsunami Crisis: Humanitarian Dimensions," *Overseas Development Institute*, January 11, 2005.

[122] Prepared by Larry Nowels, Foreign Affairs Specialist.

[123] Elizabeth Becker, "No New Funds Needed For Relief, Bush Aides Say," *New York Times*, January 4, 2005.

[124] Edward Clay, "Lessons for Life," *The Guardian Review*, January 12, 2005.

[125] "Debt Freeze for Tsunami Nations Gets Boost at Summit," Reuters News Service, January 6, 2005; BBC News, "Brown pushes tsunami debt relief," Jan. 4 2005; Reuters, "Schroeder urges debt relief for Indonesia, Somalia," December 29, 2004.

[126] *Agence France Presse*, "US 'open' to debt relief for tsunami victims," December 29, 2004; White House, Office of the Press Secretary, "President Discusses Support for Earthquake and Tsunami Victims," December 29, 2004.

[127] At their January 12, 2005 meeting, Paris Club members "shared the view" that "with immediate effect and consistent with the national laws of the creditor countries, they will not expect debt payments from affected countries that request such forbearance until the World Bank and the IMF [International Monetary Fund] have made a full assessment of their reconstruction and financing needs." Following such assessments, the Paris Club "will consider what further steps are necessary." See Paris Club, "Paris Club communique on Tsunami affected countries," January 12, 2005. For background, see CRS Report RS21482, *The Paris Club and International Debt Relief,* by Martin A. Weiss.

[128] Dan Gardner, "Bush is Losing the War for Hearts and Ninds," *The Ottawa Citizen*, March 13, 2004 and Ellen Nakashima, "U.S. Policy Censured in Indonesia," *The Washington Post*, October 21, 2003.

[129] Prepared by Wayne Morrissey, Senior Research Assistant, Resources, Science, and Industry Division. See also CRS Report RL32739 *Tsunamis: Monitoring, Detection, and Early Warning Systems* by the same author.

[130] *The Washington Times*, Jan. 7, 2005: A10.

[131] See "Off W Coast of Northern Sumatra, Can It Happen in the United States?" See USGS Earthquake Hazards Program: FAQ, Jan. 4, 2005 at [http://earthquake.usgs.gov/eqinthenews/2004/usslav/canit.html], visited Jan. 5, 2005.

[132] "Officials in Asia Concede That They Failed to Issue Warnings," *Associated Press*, December 27, 2004.

[133] Jonathan Weisman, "Funds Ready for Tsunami Aid, but Hill Seeks to Do More," *Washington Post*, January 6, 2005.

[134] State Department, "Secretary Colin L. Powell Remarks to the Traveling Press in Indonesia," Press Filing Center, Jakarta, Indonesia, January 6, 2005.

[135] Alan Sipress and Noor Huda Ismail, "Relief Transcends U.S.-Indonesia Divide," *Washington Post*, January 4, 2005.

In: Tsunamis: Causes, Characteristics, Warnings and Protection ISBN: 978-1-60876-360-3
Editors: N. Veitch and G. Jaffray, pp. 215-224

Chapter 9

TSUNAMIS: MONITORING, DETECTION, AND EARLY WARNING SYSTEMS*

Wayne A. Morrissey

Abstract

Recently, some in Congress have become concerned about the possible vulnerability of U.S. coastal areas to tsunamis, and about the adequacy of early warning for coastal areas of the western Atlantic Ocean. Those concerns stem from the December 26, 2004, tsunami that devastated many coastal areas around the northern Indian Ocean, where few tsunami early warning systems currently operate. Caused by a strong underwater earthquake off the coast of Sumatra, Indonesia, the tsunami disaster is estimated to have claimed at least 150,000 lives. Affected nations, assisted by others, are pursuing a multilateral effort to develop a detection and warning network for the Indian Ocean. Also, some Members of Congress and the Bush Administration have proposed a tsunami warning network for the U.S. Atlantic seaboard. Although instrumentation costs could run into the millions of dollars, existing weather buoys and state and local coastal and ocean observation networks might serve as possible platforms for the instrumentation. The European Union, Canada, and the United States may consider multilateral efforts to establish coverage for the North Atlantic. This report will be updated as warranted.

Introduction

Recently, numerous congressional inquiries have asked about the possibility of tsunamis occurring in U.S. coastal areas; the extent to which these areas are currently monitored; how tsunamis can be detected; and whether there is a national capacity to issue evacuation warnings for tsunamis. These concerns stem from the December 26, 2004, tsunami triggered by an underwater earthquake off the west coast of northern Sumatra in Indonesia. That earthquake was measured at M_w 9.0. [1] The ensuing tsunami devastated many coastal areas around the northern Indian Ocean, and caused loss of life and damages in other areas.

* Excerpted from CRS Report RL32739

International disaster agencies currently estimate that at least 150,000 people lost their lives to the tsunami.

The National Oceanic and Atmospheric Administration (NOAA) of the Department of Commerce and various international science agencies have indicated that there were few, if any, tsunami early warning systems monitoring the Indian Ocean on December 26, 2004. However, nations bounded also by the Pacific Ocean, including Australia and Indonesia, had tsunami early warning systems monitoring the Pacific shores where they perceived a threat. [2] Because of the lack of infrastructure to receive tsunami warnings rapidly, some have pointed out that for people on Indonesia's Indian Ocean shores, emergency communications were useless in many cases.

Although most deadly tsunamis have occurred historically in the western Pacific Ocean, there are examples of recoded events in the Atlantic. In 1692, a tsunami generated by massive landslides in the Atlantic Puerto Rican Trench reached Jamaica's coast, causing an estimated 2,000 deaths. In 1775, a tsunami struck in the eastern Atlantic Ocean on the coast of Portugal, killing an estimated 60,000 people. More recently, in 1929, a tsunami generated in the Grand Banks region of Canada hit Nova Scotia, killing 51. It was the third lethal tsunami for Canada's Atlantic Coast within 150 years. [3]

On January 5, 2005, the House Science Committee, House Coastal Caucus, and House Oceans Caucus co-sponsored a briefing organized by the U.S. Geological Survey (USGS) of the Department of the Interior. One purpose of the briefing was to consider the possible implications of the Indian Ocean tsunami for the United States. Experts from USGS and NOAA delivered presentations on the circumstances surrounding that tsunami disaster, and discussed current capabilities for monitoring, detection, and early warning around the globe. [4]

Proposals for International Tsunami Early Warning Systems

Currently, most experts agree that considerable challenges must be overcome to establish an extensive tsunami early warning network in the Indian Ocean.

Challenges

Few nations would question that development of such a system, including localized warnings, will require involving many nations with widely varying technological capabilities. Reports indicate that political leaders expect that most of the responsibility for paying for such a system will likely fall on the wealthiest nations. The costs of procuring, operating, and maintaining those instruments and platforms, and the challenge of obtaining international cost sharing, are likely to be the most critical factors for sustaining a long-term international effort for global tsunami detection and warning. International science agencies are calling for an inventory of existing global capacity for tsunami monitoring, detection, and warning systems to use as a baseline from which to determine what may still be needed.

Some U.S. policy experts also have suggested that technological challenges and possible national security issues may arise, including multinational sharing of international telecommunication networks and international standardization for tsunami warning instrumentation on data platforms. Intelligence experts suggest that some data could be considered sensitive and perhaps compromising to U.S. or other nations' intelligence-gathering operations. Gregg Withee, Assistant Director of NOAA Satellite and Information Services, has noted that some nations, including India, maintain proprietary rights to all of their real-time satellite data. Some of these data, he asserted, could be important for tsunami detection in the Indian Ocean.

Proposals

On January 6, 2005, the United Nations Environment Program (UNEP) announced an international effort to develop a tsunami early warning capacity for nations bounding the Indian Ocean. The Australian government announced it would develop and fund its own effort to guard its Indian Ocean coastlines. [5] With respect to Atlantic coast vulnerabilities, possible approaches could include multilateral agreements among the United States, Canada, and European Union so as to establish more comprehensive tsunami coverage for the North Atlantic.

Some Members of Congress have proposed a "global" tsunami detection/ warning system in the aftermath of the Indian Ocean disaster. Senator Lieberman, for example, has called for expanding the U.S. tsunami early warning program in the Pacific to include sites in the Indian and Atlantic Oceans. [6] Based on the costs of current NOAA operations, the Senator estimates that the cost for expanding from the existing six dedicated tsunami warning platforms in the Pacific Ocean to 50 globally would be approximately $30 million for implementation, with operations and maintenance costs an additional $8 million annually. [7] These figures do not take into account costs of an emergency management infrastructure to deliver regional tsunami warnings directly to the public in the wider region of the Indian Ocean, however. [8]

Representative Pallone called for establishing a tsunami detection and warning network for the U.S. Atlantic coast, the Gulf of Mexico, and the Caribbean Sea. [9] Others question whether the risks for tsunamis on the U.S. Atlantic coast would justify such expenditures. NOAA reported that the Puerto Rican Trench, which is the deepest point in the western Atlantic Ocean, is a great concern. [10] As noted above, massive landslides and sloughing have occurred on the North American continental shelf, generating deadly tsunamis. One U.S. Atlantic coast state, New Hampshire, already has an emergency contingency plan for tsunamis, and a clearinghouse for information about historical tsunami disasters. [11]

On January 14, 2005, the White House Office of Science and Technology Policy (OSTP) announced the Bush Administration's plan for an improved tsunami warning and detection system for the United States. [12] The plan initially includes deploying 32 new dedicated tsunami warning and detection buoys by mid-2007 in the Pacific and Atlantic Oceans, Gulf of Mexico, and Caribbean Sea to protect U.S. coastal areas. (See **Figure 1**.) The President would commit $37.5 million over the next two years to implement the plan. The Director of OSTP noted that the system will "ultimately include the Indian Ocean." Some question whether the risks for tsunamis on the U.S. Atlantic coast would justify such expenditures.

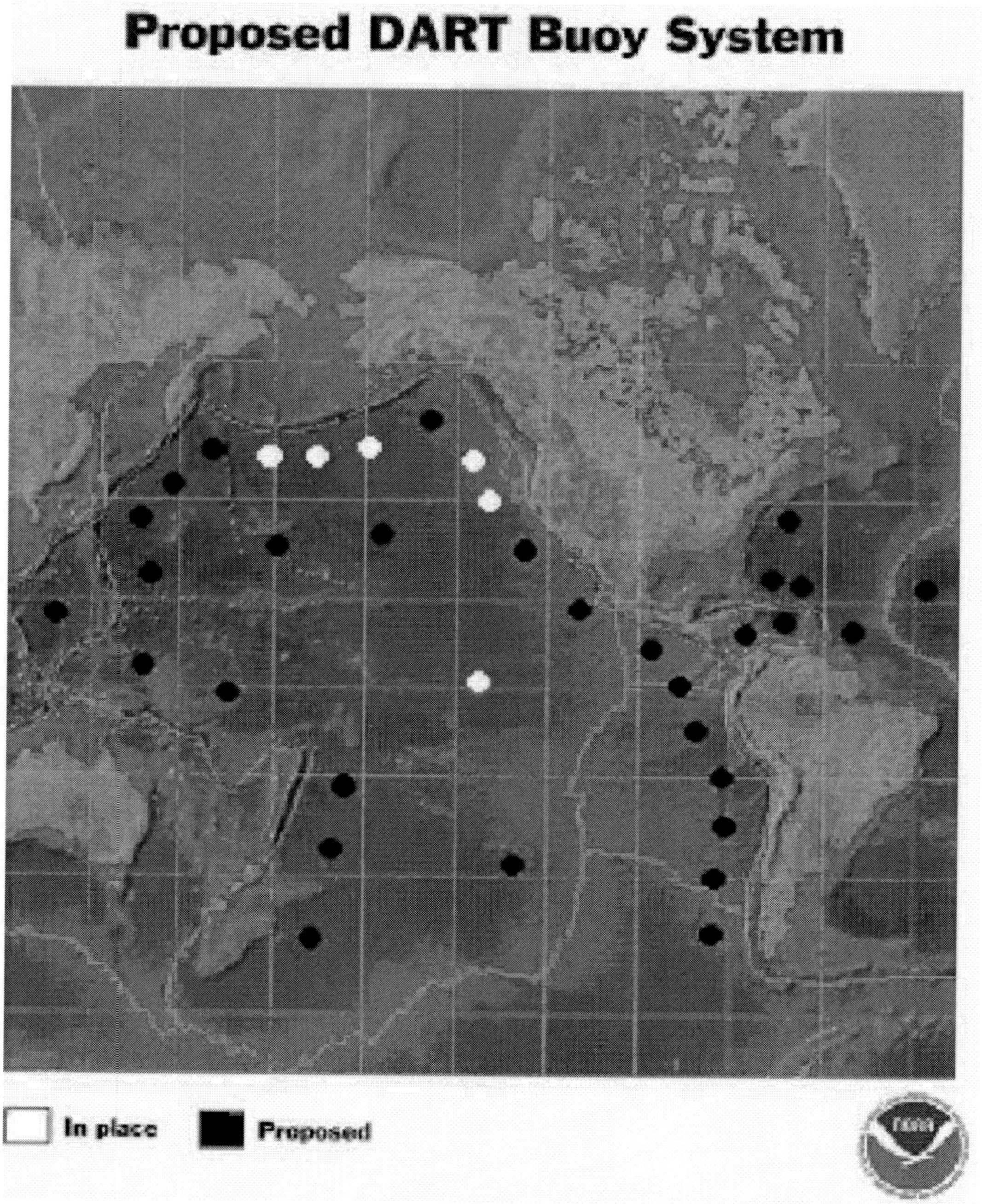

Source: National Oceanic and Atmospheric Administration, from "U.S. Announces Plans for an Improved Tsunami Warning and Detection System" (modified by CRS for contrast), at [http://www.noaanews.noaa.gov/stories2005/s2369.htm], visited Jan. 18, 2005.

Figure 1. U.S. Proposal for Tsunami Detection/Warning System.

The Director of NOAA's National Weather Service (NWS) has emphasized that in addition to needing the capacity to monitor and detect possible tsunamis, a telecommunications infrastructure for issuing tsunami warnings, such as that presently in place in the Pacific Ocean, would be critical for the Indian and Atlantic Oceans operations.

He noted that NOAA's Administrator, Admiral Lautenbacher, has promoted the development of an international Global Earth Observing System of Systems (GEOSS). One component of the system would rely on existing platforms and telecommunications capabilities of other observation systems currently operating, including the International Global Ocean Observing System (IGOOS) and Argo (climate monitoring) floats, helping to achieve a global tsunami detection and warning capacity. [13]

U.S. Tsunami Programs

Presently, NOAA has a national program to warn Pacific coastal areas of tsunamis, consisting of two regional U.S. tsunami warning centers in the Pacific Ocean, a cooperative program to reduce false tsunami alarm rates in the Pacific Ocean, monitoring and detection operations, and tsunami research activities.

Tsunami Warnings

The NWS operates the West Coast/Alaska Tsunami Warning Center (WC/AKTWC) at Palmer, AK, and the Pacific Tsunami Warning Center (PTWC), at Ewa Beach, HI.

The PTWC monitors for tsunamis and issues warnings for the Hawaiian Islands, the U.S. Pacific territories, and other U.S. and international interests in the Pacific Basin. It was established in 1949, after a strong earthquake and massive landslides off the coast of southwest Alaska caused a tsunami disaster in the Hawaiian Islands hours later. The WC/AKTWC was established in 1967, after a devastating earthquake in Anchorage, AK, in 1964 caused localized tsunami damages. This center is responsible for issuing warnings to emergency managers in Alaska, British Columbia, Washington, Oregon, and California. In addition, in 1992, NOAA launched a National Tsunami Hazard Mitigation Program to address Pacific tsunami warnings, which, at that time, were being issued with a 75% false alarm rate, causing significant social upheaval and economic disruption. This program also focused on the potential that a sizable earthquake in the Pacific Northwest Cascadia Region could generate devastating tsunamis that would damage U.S. Pacific coastal regions. [14]

U.S. Operations and Research

NOAA currently has a network of six dedicated tsunami detection and relay stations, operating as part of its Deep-Ocean Assessment and Reporting of Tsunamis (DART) program. [15] (See **Figure 1**, above, for their location, and **Figure 2**, below, for the components.) These are equipped for an early warning capability, but their emergency communications are only effective if there are emergency managers to receive them and, in turn, alert the public.

NOAA officials estimate that the cost of adding tsunami detection instruments on Atlantic Ocean platforms, such as weather buoys, or building dedicated DART platforms, could vary depending upon the scale of the project — for example, the number of instruments to be included and the costs of operation and maintenance. [16] At a minimum, NOAA anticipates that the cost for three new DART platforms it has proposed for the western

Atlantic Ocean and Gulf of Mexico/Caribbean Sea, including costs of operation and maintenance, and construction of a new regional center, would be comparable to annual funding for the two Pacific regional tsunami early warning operations centers — approximately $8 million for FY2005. [17] (For a schematic of the DART buoy platform, see **Figure 2**.)

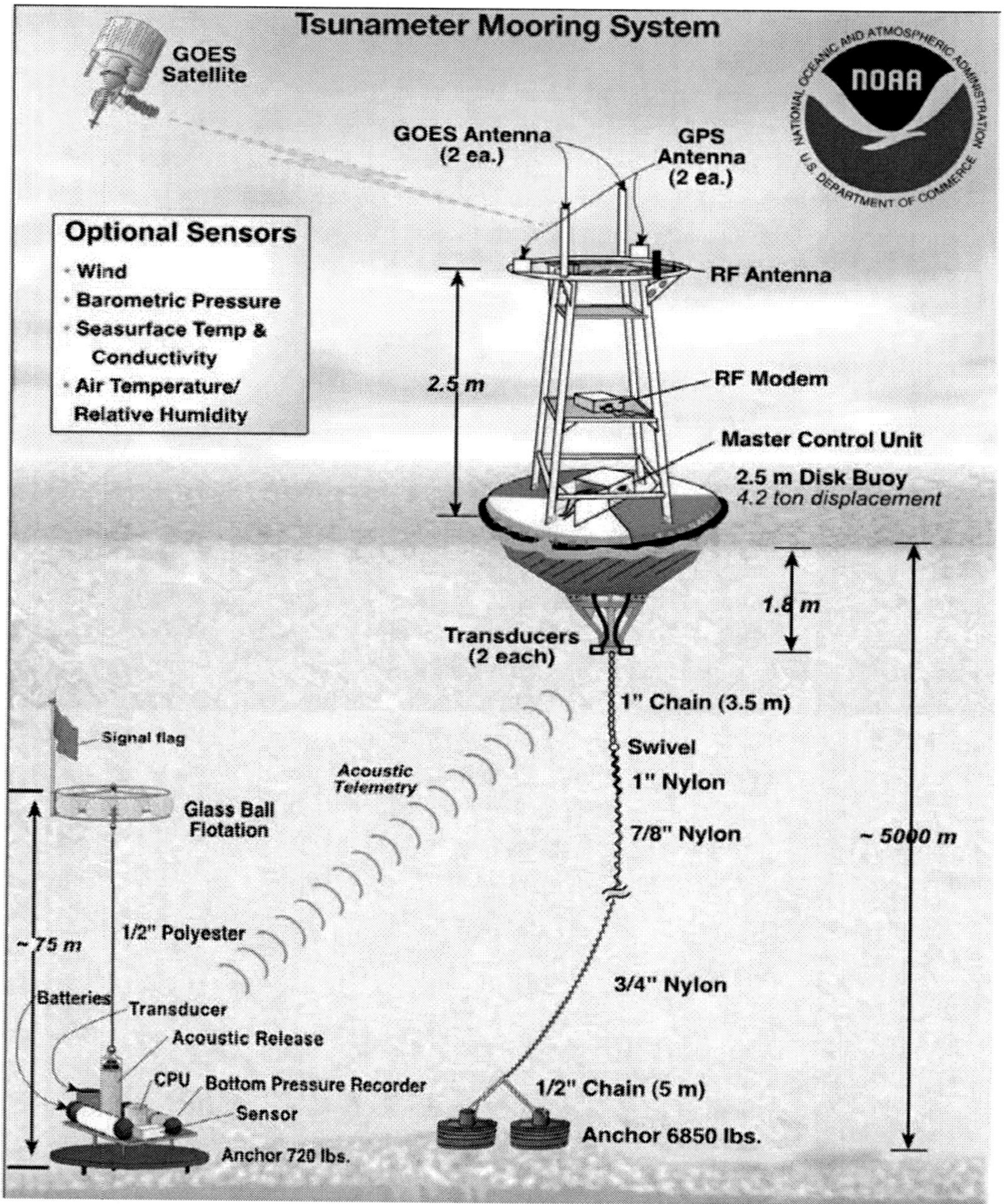

Source: National Oceanic and Atmospheric Administration, from "U.S. Announces Plans for an Improved Tsunami Warning and Detection System." See [http://www.noaanews.noaa.gov/stories2005/s2369.htm], visited Jan. 18, 2005.

Figure 2. NOAA DART Platform.

Related U.S. Programs

To reduce costs for a U.S. Atlantic coast tsunami early warning system, engineers at NOAA say that it is technologically possible to modify weather and marine data buoys, such as those currently situated off the United States, to serve as platforms for mounting tsunami monitoring and detection instrumentation. Others suggest taking advantage of existing international communications networks for issuing tsunami warnings to local emergency managers. [18]

Hundreds of NWS weather buoys operating off the coasts of the United States already record various meteorological data; while marine data buoys measure speed of ocean currents, temperature, salinity, and pressure change. Sea surface height (or sea level) also is measured by satellite-GPS (global positioning system) by NOAA's National Ocean Service tidal monitoring network, which is responsible for issuing warnings.

All of these buoys are equipped to relay data and emergency communications for navigational purposes. [19] In addition, an array of 3,000 data buoys, known as Argo floats, currently deployed in the equatorial waters of the Pacific Ocean, are being used to detect conditions for El Niños and La Niñas, which are three- to seven-year climate variations that affect global weather. Argo floats might also be considered as possible platforms for situating tsunami detection instrumentation. [20] These floats have been advocated by NOAA as "the next step in global [Earth] observations." [21] In the Atlantic Ocean, other possible platforms for tsunami monitoring and detection include a growing number of regional and local coastal and ocean monitoring networks in development along the coasts of Canada and the United States. A proposal to use such systems for tsunami warning was introduced in the 108th Congress. [22]

Additionally, NOAA and other international weather agencies issue warnings of meteorological conditions that primarily affect commercial air traffic, but which also might put human lives in danger and cause significant economic disruption for global nations. The U.N. World Weather Watch (WWW) is a cooperative program organized and administered by the World Meteorological Organization (WMO). [23] NOAA plays a leadership role in the WWW, representing the United States in scientific research, weather data collection and management, meteorological forecast and warning. The Department of State also plays an important role for achieving and maintaining international agreements to sustain WWW operations globally. The WWW has an established international telecommunications network for receiving and distributing weather data and warnings, including those for the United States and its trust territories. NOAA Satellite Services now manages two WWW data centers for weather data analysis and forecasting. [24]

Currently, the U.S. Department of Homeland Security and the NWS are developing a National All Hazards Warning Network using NWS's NOAA Weather Radio network as the initial infrastructure for communicating public warnings. In the United States, Congress has expanded NOAA Weather Radio so that this emergency telecommunications infrastructure is able to provide adequate coverage of weather services and support local forecasting and warning of extreme weather. NOAA has improved technology of weather instrumentation to increase lead time of emergency warnings; constructed transmission towers; added repeaters to expand ranges of emergency notification; and distributed individual NOAA Weather Radio receivers to the public, particularly in rural areas, so as many U.S. citizens as possible can receive disaster warnings and emergency communications. [25]

Conclusion

Decisions about whether and how to proceed with establishing an international tsunami early warning system for the Indian Ocean (and elsewhere) will likely be complicated for a number of reasons. One reason is because of the number of different potential international parties that would be involved with the need to coordinate data collection and warning dissemination, and a second is the funding needed to establish a tsunami warning system in that region. A third is that nations, including some in the Indian Ocean, might charge for access to critical satellite data that may help in warning potential victims. Senator Lieberman and others contend that the costs of acquiring those data may be well worth it, in terms of lives saved. Others assert that the costs of accessing and using those proprietary data could be prohibitive. They are of the opinion that nations should provide free access to global environmental data, especially when the United States and other nations are providing disaster relief and plan on funding tsunami detection and warning activities for the region. [26]

Still others foresee challenges to standardize instrumentation and other related technology. In addition, there are concerns about national security and compromising U.S. intelligence-gathering operations if international telecommunications networks are used. Finally, some U.S. lawmakers question the risks of a tsunami hitting the U.S. Atlantic coast. [27] They believe the probability is low, and assert that risk should be an important factor for guiding development of and investment in a cooperative early tsunami warning system for the U.S. eastern seaboard.

References

[1] Mw, the moment of magnitude, is a way to measure the force of an earthquake's total seismic energy released as a function of rock rigidity in the fault, the total area of contact where friction occurs, and the amount of slippage (or displacement). It is used for earthquakes greater than M8.2 on the Richter scale.

[2] General David L. Johnson, "NOAA Tsunami and Natural Disaster Information," presentation at House Science Committee briefing, Jan. 5, 2005.

[3] Statistics on deaths resulting from tsunamis were compiled by CRS from online sources, and include data from the Tsunami Laboratory of Novosibirsk, NOAA's National Geophysical Data Center, the University of Southern California, Tsunami Research Group, and others. See [http://geology.about.com/library/bl/bltsunamideath table.htm], visited Jan. 11, 2005.

[4] Presenters at that briefing included, David Applegate, Science Advisor for Earthquake and Geological Hazards at the USGS; General David Johnson, Assistant Director of NOAA's National Weather Service; Gregg Withee, Assistant Director for NOAA Satellite and Information Services; and, Eddie Bernard, Associate Director of NOAA's Pacific Marine Environmental Laboratory (teleconferencing from Seattle, WA).

[5] Briefing for the House Science Committee by USGS and NOAA, Jan. 5, 2005.

[6] Sen. Joseph Lieberman, "Lieberman Unveils Tsunami Early Warning Legislation, Senator's Bill Would Deploy 40-50 New Sensors and Fund System at $30 million," news release, Jan. 6, 2005.

[7] For current NOAA tsunami-related funding, see U.S. Congress, House, "Conference Report on H.R. 4818, the Consolidated Appropriations Act, 2005 (P.L.108-447)," H.Rept. 108-792, Title II, Commerce, Justice, State, and the Judiciary and Other Related Agencies Appropriations Act, 2005, *Congressional Record*, Nov. 19, 2004: H10457.

[8] Associated Press, "Hill Eyes Tsunami Warning System — Lieberman calls for Global Net," *Washington Times*, Jan. 7, 2005: A10.

[9] *Congressional Record*, Jan. 4, 2005: H40.

[10] House Science Committee briefing, Jan. 5, 2005.

[11] State of New Hampshire, "Disaster Plan 409," Sect. II, Geological Hazards, Seismic Hazards, at [http://www.nhoem.state.nh.us/mitigation/state_of_new_hampshire.asp], visited Jan. 11, 2005. See also "Is your Community Ready for the Next Tsunami," National Weather Service Tsunami Ready program, at [http://tsunami.gov], visited Jan. 11, 2005.

[12] U.S. Office of Science and Technology Policy, "U.S. Announces Plan for Improved Tsunami Detection and Warning System," press release, *OSTP News*, Jan. 14, 2005.

[13] Gen. David Johnson, House Science Committee briefing, Jan. 5, 2005.

[14] Eddie Bernard of NOAA, House Science Committee briefing, Jan. 5, 2005.

[15] Hugh B. Milburn et al., "Real-Time Tsunami Reporting from the Deep Ocean," NOAA Pacific Marine Environmental Laboratory (1996), at [http://www.ndbc.noaa.gov/Dart/milburn_1996.shtml], visited Jan. 4, 2005.

[16] NOAA officials estimated the cost to produce the existing six experimental DART platforms, instrument them, provide a telecommunications capability, and maintain them at approximately $125,000 each, but suggested there would be an economy of scale if their proposed total of 23 platforms for the United States in the Pacific and Atlantic Oceans were produced.

[17] This amount also includes funding for NOAA's U.S. tsunami-related research activities. Annual funding requested for U.S. tsunami monitoring, early warning, and research is found in the NWS budget under Operations and Research. Appropriations for these activities are provided in Title II of the annual Commerce, Justice, State, Judiciary, and Related Agencies appropriations acts.

[18] Kenneth B. Allen, Director of the Partnership for Public Warning, "Letter to President Bush,"Jan. 3, 2005, at [http://www.partnershipforpublicwarning.org/ppw/], visited Jan. 21, 2005.

[19] Eddie Bernard, House Science Committee briefing, Jan. 5, 2005.

[20] NOAA/Woods Hole Oceanographic Institute, *Observing the Ocean in Real-Time: Argo, a Global Array of Profiling Floats to Understand and Forecast Climate*, ed. Stan Wilson (1996). Funded in part by private academic institutions.

[21] Ibid.

[22] On January 5, 2005, Representative Curt Weldon circulated a "Dear Colleague" letter advocating the reintroduction of H.R. 5001 (108th Congress), the Ocean and Coastal Observation System Act, in the 109th Congress. This legislation promoted development of an "Integrated Ocean Observation System," to protect U.S. citizens in coastal communities from tsunamis.

[23] U.S. Department of Commerce, NOAA, Office of the Federal Coordinator for Meteorology, "World Weather Program," *The Federal Plan for Meteorological*

Services and Supporting Research: Fiscal Year 2004, Report FCM P1-2003, Appendix B, p. 219 (Washington, DC: Oct. 2003).

[24] NOAA's Satellite and Information Services, which operates the two U.S. WWW data Centers, reviews weather satellite data, which has since provided valuable information about the Indian Ocean tsunami. See "NOAA Scientists Able to Measure Tsunami Height from Space," at [http://www.noaanews.noaa.gov/stories2005/s2365.htm], visited Jan. 11, 2005.

[25] See NOAA Weather Radio (NWR) at [http://www.nws.noaa.gov/nwr/], visited Jan. 10, 2005.

[26] *The Washington Times*, Jan. 7, 2005: A10.

[27] See "Off W Coast of Northern Sumatra, Can It Happen in the United States?" See USGSEarthquake Hazards Program: FAQ, Jan. 4, 2005 at [http://earthquake.usgs.gov/eqinthenews/2004/usslav/canit.html], visited Jan. 5, 2005.

In: Tsunamis: Causes, Characteristics, Warnings and Protection ISBN 978-1-60876-360-3
Editors: N. Veitch and G. Jaffray, pp. 225-246

Chapter 10

TSUNAMIS, GENERATION AND MATHEMATICAL MODELING

M.A. Helal
Department of Mathematics, Faculty of Science,
Cairo University, Giza, Egypt

Abstract

The word "tsunami" comes from the Japanese language. It means harbor (" tsu") and wave ("namis"). A tsunami is a very long wave generated by different methods.

In this work, we are looking for the causes and the generation of tsunamis, as well as the mathematical modeling that simulates the tsunamis [*and the water waves impacts*]. The most famous origin of generation of tsunamis comes from a sudden displacement of the ocean bottom. This happens due to some kind of earthquake in the ocean. Two different models exist for the generation of tsunamis by the underwater earthquake. Another origin for the generation of tsunamis is falling cosmic bodies.

The second part of this work deals with the mathematical modeling. We start with a very simple model, then we continue with other different more complicated models, like the Boussinesq model and other nonlinear gravitational wave equations. Another model represented by the well-known KdV equation is also studied. This represents the famous nonlinear equation that describes the nonlinear behavior of this killer physical phenomenon, i.e., tsunamis.

1. Introduction

Until December 2004, the phenomena of tsunami was not on the minds of most of us. That changed on the morning of December 24, 2004 when an earthquake of magnitude 9.1 occurred along the oceanic trench off the coast of Sumatra in Indonesia. This large earthquake resulted in vertical displacement of the sea floor and generated a tsunami that eventually killed 280,000 people and affected the lives of another several million. Although people living on the coastline near the epicenter of the earthquake had little time or warning of the approaching tsunami, those living farther away along the coasts of Thailand, Sri Lanka, India, and East Africa had plenty of time to move to higher ground and escape. But, there was no tsunami warning system in place in the Indian Ocean, and although other tsunami warning centers attempted to provide a warning, there was no effective communication system. Unfortunately, it has taken a disaster of great magnitude to point out the failings of the world's scientific community and to educate almost everyone about the tsunami.

What is a Tsunami?This word, which comes from the Japanese language, is a very long-wavelength wave of water that is generated by sudden displacement of the seafloor or disruption of any body of standing water. Tsunamis are sometimes called "seismic sea waves", although, as we will see, these waves can be generated by mechanisms other than earthquakes.

Tsunamis have also been called "tidal waves", but this term should not be used because they are not in any way related to the tides of the Earth. Since tsunamis occur suddenly, often without warning, they are extremely dangerous to coastal communities. The most important thing to be learned about tsunamis is that they are waves. Specifically, they are water waves that are formed in the ocean by a huge displacement of water in the ocean, where the depths of the water average 4 km. Displacement of water following a huge release of energy from, say, an earthquakes, landslides, and volcanic activity or a cosmic body impact creates a wave or a series of waves that have wavelengths on the order of hundreds of kilometers long. Although tsunamis usually have small amplitudes (on the order of 1 M.), the volume of the water that gets displaced and the speeds reached by these waves allow them to carry enough energy to wipe out towns and cities.

A water wave is a combination of both transverse and longitudinal waves. Some of the winds' energy can be transferred to the water. This causes waves on the surface of the ocean (storm waves/sea surges). These are wind-driven waves created on top of normal tides, and are often caused by hurricanes and cyclones. They reach 30–40 ft. in height, producing a constant pounding motion as opposed to the tsunami's characteristic single huge wave.

Another important kind of water waves is the seiche. This is the rhythmic vibration of water in an enclosed water body. Water moves slowly back and forth from shore to shore in waves not higher than 5 ft. They are created by either seismic action or storms.

Tsunamis are characterized as shallow-water waves. These are different from the waves most of us have observed on the beach, which are caused by the wind blowing across the ocean's surface. Wind-generated waves usually have a period (time between two successive waves) of five to twenty seconds and a wavelength of 100 to 200 meters (cf. Fig. 1).

A tsunami can have a period in the range of 10 min. to 2 h and wavelengths greater than 500 km. A wave is characterized as a shallow-water wave when the ratio of the water depth and wavelength is very small (cf. Fig. 2). The velocity V of a shallow-water wave is also equal to the square root of the product of the acceleration of gravity, g, (10m/s^2) and the depth of the water, d. This means that

$$V = \sqrt{g * d} \tag{1}$$

The rate at which a wave loses its energy is inversely related to its wavelength. Since a tsunami has a very large wavelength, it will lose little energy as it propagates. Thus, in very deep water, a tsunami will travel at high speeds with little loss of energy. For example, when the ocean is 6100 m deep, a tsunami will travel about 890 km/h, and thus can travel across the Pacific Ocean in less than one day. As a tsunami leaves the deep water of the open sea and arrives at the shallow waters near the coast, it undergoes a transformation.

Since the velocity of the tsunami is also related to the water depth, as the depth of the water decreases, the velocity of the tsunami decreases. The change of total energy of the tsunami, however, remains constant.

There is an average of two destructive tsunamis per year in the Pacific basin. A Pacific-wide tsunami is a rare phenomenon, occurring every 10–12 years on average. Most of these tsunamis are generated by earthquakes that cause displacement of the seafloor, but, as we shall see, a tsunami can be generated by volcanic eruptions, landslides, underwater explosions, and meteorite impacts [1].

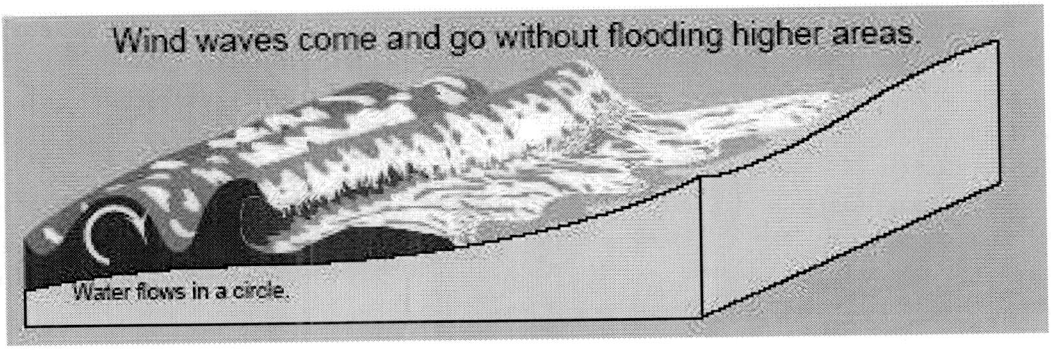

Figure 1.

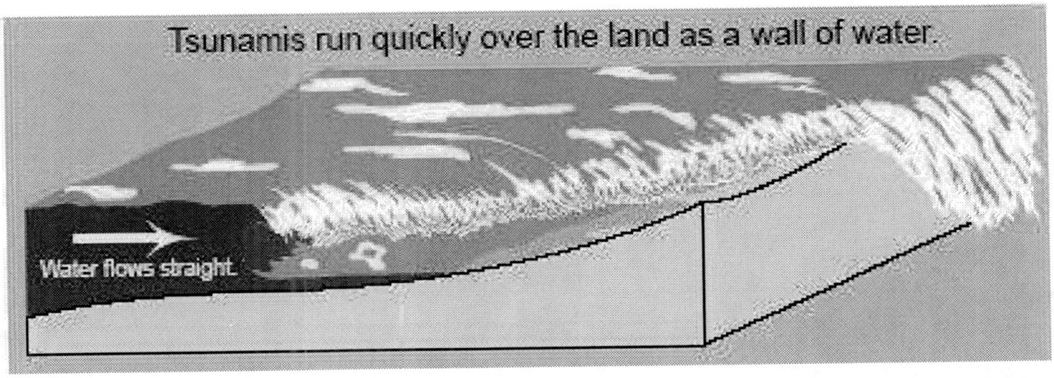

Figure 2.

Generation of Tsunamis

2. Preliminary

If we are going to assess the risk of a tsunami at some particular place on the planet, we must first understand how to make a tsunami. Submarine earthquakes and volcanoes generate the majority of tsunamis, and the theory of plate tectonics explains the cause of earthquakes and volcanoes.

The basic concept is that the outermost part the Earth consists of several large and fairly stable slabs of solid and relatively rigid rock, called plates (see Figure 3). These plates are constantly moving (very slowly), and rub against one another along the plate boundaries, which are also called faults. Consequently, stress and strain build up along these faults, and eventually they become too great to bear and the plates move abruptly so as to release the stress and strain, creating an earthquake. Most of tsunamigentic earthquakes occur in subduction zones around the Pacific Ocean rim, where the dense crust of the ocean floor dives beneath the edge of the lighter continental crust and sinks down into the Earth's mantle. These subduction zones include the west coasts of North and South America, the coasts of East Asia (especially Japan), and many Pacific island chains . There are different types of faults along subduction margins. The interplate fault usually accommodates a large relative motion between two tectonic plates and the overlying plate is typically pushed upward. This upward push is impulsive; it occurs very quickly, in a few seconds. The ocean water surface responds immediately to the upward movement of the seafloor and the ocean surface profile usually mimics the seafloor displacement. The interplate fault in a subduction zone has been responsible for Ridge axis most of the largest tsunamis in the twentieth century. For example, the 1952 Kamchatka, 1957 Aleutian, 1960 Chile, 1964 Alaska, and 2004 Sumatra earthquakes all generated damaging tsunamis not only in the region near the earthquake epicenter, but also on faraway shores. For most of the interplate fault ruptures, the resulting seafloor displacement can be estimated based on the dislocation theory. Using the linear elastic theory, analytical solutions can be derived from the mean dislocation field on the fault. Several parameters defining the geometry and strength of the fault rupture are needed to be specified.

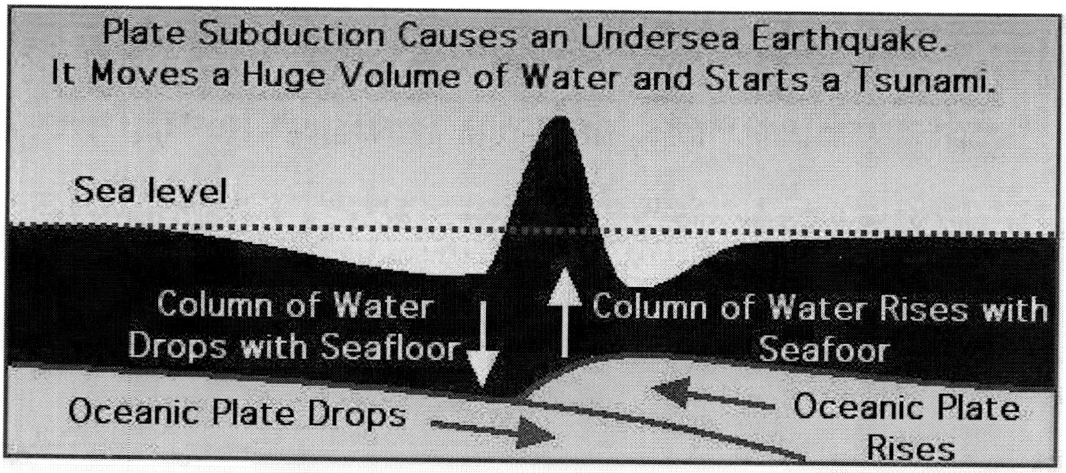

Figure 3.

Waves at the surface of a liquid can be generated by various mechanisms: wind blowing on the free surface, wavemaker, moving disturbance on the bottom or the surface, or even inside the liquid, fall of an object into the liquid, inside a moving container, etc. In this chapter, we concentrate on the case where the waves are created by a given motion of the bottom. One example is the generation of a tsunami by a sudden seafloor deformation. There are different natural phenomena that can lead to a tsunami; for example, one can mention submarine slumps, slides, volcanic explosions, etc. In this work, we use a submarine faulting generation mechanism as a tsunami source. The resulting waves have some well-known features, such as: characteristic wavelengths are large and wave amplitudes are small compared with water depth. Two factors are usually necessary for an accurate modelling of tsunamis: information about the magnitude and distribution of the displacements caused by the earthquake, and a model of surface gravity waves generated by the seafloor motion. Most studies of a tsunami's generation assume that the initial free-surface deformation is equal to the vertical displacement of the ocean bottom. The details of the wave's motion are neglected during the time that the source operates. While this is often justified because the earthquake rupture occurs very rapidly, there are some specific cases where the time scale of the bottom deformation may become an important factor. During the 26 December 2004 Sumatra-Andaman event, there was—in the northern extent of the source—a relatively slow faulting motion that led to significant vertical bottom motion but left little record in the seismic data.

Our study is restricted to the water region where the incompressible Euler equations for potential flow can be linearized. The wave propagated away from the source can be investigated by shallow water models which may or may not take into account nonlinear effects and frequency dispersion. Such models include the Korteweg-de Vries equation for unidirectional propagation, nonlinear shallow-water equations and Boussinesq type models. Several authors have modeled the incompressible fluid layer as a special case of an elastic medium. In our opinion it may be convenient to model the liquid by an elastic material from a mathematical point of view, but it is questionable from a physical point of view. "The crust was modeled as an elastic isotropic half-space", this assumption will also be adopted in the present study.

Here we essentially follow the framework proposed by Hammack and others. The tsunami generation problem is reduced to a Cauchy-Poisson boundary value problem in a region of constant depth. The main extensions given in the present work consist of three-dimensional modelling and more realistic source models.

Most analytical studies of linearized wave motion use integral transform methods. The complexity of the integral solutions forced many authors to use asymptotic methods, such as the method of stationary phase, to estimate the far-field behavior of the solutions.

3. Moving Bottom

Tsunamis are often generated by a moving sea bottom. There exists different works to deal with the case where the tsunami source is an earthquake [2](cf. Dutykh et al.).

The linearized water-wave equation has been presented and solved analytically for various sea bottom motions.

In their study, they have also obtained asymptotic formulas for integral solutions.

All the numerical results presented in this section were obtained in this manner. One should use asymptotic solutions with caution since they approximate exact solutions of the linearized problem.

In their works, illustrated numerical results based on the analytical solutions are shown. All the numerical results presented in this work were obtained in this manner.

The free-surface profiles, the horizontal and vertical velocities as well as the bottom pressure are presented.

4. Tsunamis from Physical Point of View

Tsunamis have intrigued us, and frightened us, ever since the existence of the phenomena became known. Furthermore, the most recent disaster (December 2004) has taught us that, even in the age of almost immediate world-wide awareness, and modern scientific skills, we are unable to predict the occurrence of a tsunami and suitably get prepared for its arrival. However, we are able to use appropriate fundamental equations of fluid mechanics, together with some carefully chosen mathematical models, to construct a theory, and asymptotic solutions, for this phenomenon. This will enable us to highlight the essential mechanisms that exist and to identify the factors that govern these factors, for example, the number of wave fronts that appear at a shoreline.

A tsunami generally consists of a series of waves, often referred to as the tsunami wave train. The amount of time between successive waves, known as the wave period, is only a few minutes; but in some instances, waves are over an hour apart. Many people have lost their lives after returning home in between the waves of a tsunami, thinking that the waves had stopped coming.

Tsunamis now used internationally to refer to a series of waves traveling across the ocean with extremely long wavelengths (up to hundreds of miles between wave crests in the deep ocean). As the waves strike the shore they may inundate low-lying coastal areas resulting in mass destruction and—in many instances—loss of life. Often a tsunami was incorrectly referred to as a tidal wave. Tidal waves are simply the periodic movement of water associated with the rise and fall of the tides produced by the gravitational attraction of the sun and moon. Tsunamis have no connection with the weather nor the tides.

Oceanographers often refer to tsunamis as seismic sea waves as they are usually the result of a sudden rise or fall of a section of the earth's crust under or near the ocean. A seismic disturbance can displace the water column, creating a rise or fall in the level of the ocean above.

This rise or fall in sea level is the initial formation of a tsunami wave. Tsunami waves can also be created by volcanic activity and landslides occurring above or below the sea surface. These types of activities produce tsunamis with much less energy than those produced by submarine faulting. The size and energy of these tsunamis dissipates rapidly with increasing distance from the source, thus resulting in more local devastation.

Tsunamis have been recorded in all the major oceans of the world. However, this phenomenon is mainly restricted to the Pacific basin, an area surrounded by volcanic island arcs, mountain chains and subduction zones earning the nickname the "ring of fire", as it is the most geologically active area on the planet. The amount of activity in this region makes it much more susceptible to submarine faulting and subsequent tsunami events, whereas the Indian and Atlantic oceans are far less geologically active, with some exceptions, and therefore the occurrence of tsunamis is rare.

An important question arise: How does a tsunami wave differ from a normal wave? The waves you see at the beach are generated by wind blowing over the sea surface. The size of these waves depends on the strength of the wind creating them and the distance over which it blows. Generally the distance between these waves, known as the wavelength, ranges from a couple of feet to perhaps a thousand feet. The speed of these waves as they travel across the ocean ranges from a few miles per hour up to 60 miles per hour in some instances. Tsunami waves resulting from physical mechanisms behave much differently than wind-generated waves. The magnitude of the disturbance causing the tsunami is the primary factor influencing the size and strength of the waves. The height of the wave when it is generated is very small, usually less than a few feet. The distance between successive wave crests or the wavelength, is much larger than that of a normal wave and may be hundreds of miles apart. Depending on the depth of the water in which the tsunami is traveling, it may attain speeds of up to 500 miles per hour (see Fig. 1 and Fig. 2).

Another rare event that may result from a tsunami is a standing wave or seiche. A seiche occurs in bodies of water that are partially or completely enclosed, such as Hilo Bay, creating a standing wave that continually sloshes back and forth. The appearance of a seiche would be very similar to what happens when you place a glass of water on the table; the water rocks back and forth before settling. When a seiche is generated by a tsunami, subsequent tsunami waves may arrive in unison with a seiche resulting in an increase in the drawdown in sea level and a dramatic increase in wave height. Seiche waves may continue several days after a tsunami.

When a tsunami approaches a coastline, the wave begins to slow down and increase in height, depending on the topography of the sea floor. Often the first sign of a tsunami is a receding water level caused by the trough of the wave. However, in some instances, a small rise in the water level just before the recession has been observed. Regardless, the incoming wave approaches much like the incoming tide though on a much faster scale. The maximum vertical height to which the water is observed with reference to sea level is referred to as run-up. The maximum horizontal distance that is reached by a tsunami is referred to as inundation[3,4].

The wave height of a tsunami can be highly variable in a local area depending on the underwater topography, orientation of the oncoming wave, the tidal level, and the magnitude of the tsunami. Because direct physical measurement of a tsunami wave would be a life threatening event, the most common method for determining tsunami wave height is by measuring the runup, the highest vertical point reached by the wave. Runup heights are measured by looking at the distance and extent of salt-killed vegetation, and the debris left once the wave has receded. This distance is referred to as a datum level, usually being the mean sea level or mean lower low water level. The reference to mean lower low water is more significant in areas with greater tidal ranges such as in Alaska where a smaller tsunami wave can be more devastating during a high tide than a larger wave at low tide [26].

When a tsunami is generated offshore, the wave will behave as a shallow water wave. A shallow water wave is one that travels through water having a depth less than 1/20 of its wavelength. Knowing that the average ocean depth is roughly three miles, oceanographers can determine the speed of the tsunami, and calculate the time it will take to travel between any two points.

This information has led to the development of travel-time charts that make it possible to predict the arrival time of a tsunami wherever it is generated (Fig. 4). Areas near the epicenter of earthquakes, landslides or volcanic activity are most vulnerable to the effects of a tsunami as they cannot be properly warned by the Tsunami Warning Center of the coming danger [3].

5. Some Important Concepts and Illustrative Questions

5.1. Some Concepts about Tsunamis

(1) When breaking occurs in very shallow water, the amplitude of tsunami waves grows to such an extent that typically an undulation appears on the long wave, which develops into a progressive bore.

(2) This turbulent front can be quite high and travel onto the beach at great speed. Then the front and the turbulent current behind it move onto the shore, past buildings and vegetation until they are finally stopped by rising ground. The water level can rise rapidly, typically from 0 to 3 m in 1.5 min.

(3) Trajectory of currents and their velocity are quite unpredictable, especially in the final stages because they are sensitive to small changes in the topography, and to the stochastic patterns of the collapse of buildings, and to the accumulation of debris such as trees, cars, logs, furniture.

(4) Dynamics of tsunami waves are somewhat similar to dynamics of flood waves caused by a dam breaking, dyke breaking or overtopping of dykes (cf. the recent tragedy of hurricane Katrina in August 2005).

(5) Research on flooding events and measures to deal with them may be able to contribute to improved warning and damage reduction systems for tsunami waves in the areas of the world where these waves are likely to occur as shallow surge waves.

(6) Longer waves travel faster than shorter waves (for gravity-induced surface water waves), but all very long waves all travel at speed $\sqrt{gh}$ g: gravity , h: fluid depth

(7) Korteweg and Vries derived their equation to describe the motion of long waves of moderate amplitude.

(8) The tsunami changes its shape primarily in shallower coastal waters, where $h(x, y)$ decreases.

(9) A tsunami that is barely noticeable in the open ocean becomes deadly in coastal waters because:

$c = \sqrt{gh}$- (locally); and wave volume is conserved

5.2. Illustrative Answers for the Most Frequently Asked Questions

Here, we introduce some famous questions with their answers (cf. Helal & Mehanna [2008]).

The first question

As a tsunami begins to evolve into a large-amplitude wave near shore, what controls the wave evolution?

This gives the following answer

(i) $c = \sqrt{g(h(x, y) + \zeta(x, y, t))}$for a long nonlinear wave ($z = \zeta$ represents the equation of bottom);

(ii) conservation of wave volume;

(iii) wave reflection by the changing bathymetry.

The second question

Not all underwater earthquakes create tsunamis. What information about an earthquake determines whether it generates a tsunami?

This gives the following answer

The time and place of the earthquake.

Claim: The volume of water displaced by the earthquake is the next most important piece of information about the quake.

The third question

Are there "immediate" seismic measurements of the earthquake that determine the volume of water displaced?

This give the following answer

(i) A "strike-slip fault" displaces very little water.

(ii) A "thrust fault" or a "normal fault" can displace much more water.

(iii) From historical records, geologists can classify known fault regions into one of these types.

The fourth question

If a tsunami occurs, where and when will it reach shore?

This give the following answer

(1) Simplest approximate answer

Where? If there is a straight line from the epicenter of the quake to your beach, then you will experience some part of the tsunami. (Sufficient, not necessary)

When? Until the tsunami reaches shallow coastal waters,$c = \sqrt{gh(x,y)}$ locally.

Total time: $T(x,y) = \int_0^x \int_0^y \frac{ds\,dw}{\sqrt{gh(s,w)}}$along each path.

(2) More accurate answer:

Either solve the wave equation in 2-D, or use geometric optics with a spatially varying "index of refraction", $\sqrt{gh(x,y)}$

Along each curve from the epicenter to your beach, the total propagation time along that path is:

$$T(x,y) = \int_0^x \int_0^y \frac{ds\,dw}{\sqrt{gh(s,w)}}$$

Minimize this for the warning system.

Note that the tsunami can diffract around the objects.(See Fig. 4)

6. The Greatest Historical Catastrophes

Hereafter we cite the most catastrophic tsunamis that had caused the maximum damaged.

The following table illustrate the date, region, magnitude and casualties of each specific destructive tsunamis.

[http://www.tsunami-alarm-system.com/en/phenomenon-tsunami/
phenomenon-tsunami_occurrences.html]

Date	Sea region	Affected region	Mag.	Max. wave height	Casualties
17.06.2006	Pacific	Indonesia, Java	7,7	2,00 m	700
26.12.2004	Indian Ocean	Indonesia, Off w. Coast of Sumatra	9,0	34,90 m	283.100
23.06.2001	Pacific	Peru	8,4	7,00 m	26
17.07.1998	Pacific	Papua New Guinea	7,0	15,00 m	2.182
17.02.1996	Indian Ocean	Indonesia, Irian Jaya	8,2	7,70 m	127
14.11.1994	Pacific	Philippines, Philippines Islands	7,1	7,30 m	62
02.06.1994	Indian Ocean	Indonesia, Java	7,8	13,00 m	222
12.07.1993	Pacific	Sea of Japan	7,7	31,70 m	330
12.12.1992	Pacific	South Pacific, Indonesia	7,5	26,20 m	1.000
12.12.1979	Pacific	Colombia, Colombia-Ecuador	7,7	5,00 m	500
19.08.1977	Indian Ocean	Indonesia, Sunda Islands	8,0	15,00 m	190
16.08.1976	Pacific	Celebes Sea, Philippines, Moro Gulf	8,1	5,00 m	5.000
26.07.1971	Pacific	Papua New Guinea	7,9	10,00 m	n.a.
22.11.1969	Pacific	Bering Sea, Russia, Bering Strait	7,7	15,00 m	n.a.
04.02.1965	Pacific	USA, Rat Islands, Alaska	8,7	10,70 m	n.a.
28.03.1964	Pacific	Usa, Prince William Sound, Alaska	9,2	70,00 m	123
22.05.1960	Pacific	Chile, Central Chile	9,5	25,00 m	1.260
09.03.1957	Pacific	USA, Fox Islands, Andreanof Islands	9,1	15,00 m	n.a.
09.07.1956	Mediterranean	Greece, Amorgos Island, Aedean	7,5	20,00 m	50
23.06.1946	Pacific	Northeast Pacific, USA, Unimak Island,	7,3	30,00 m	n.a.
27.11.1945	Indian Ocean	Pakistan, Makran Coast	8,3	15,30 m	n.a.
07.12.1944	Pacific	Japan, Off Southeat Coast Kii	8,1	10,00 m	40
02.03.1933	Pacific	Japan, Sanriku	8,4	30,00 m	3.000
22.06.1932	Pacific	Eastern Pacific, Mexico	7,0	10,00 m	75
03.10.1931	Pacific	Solomon Islands	7,9	10,00 m	50
02.02.1931	Pacific	South Pacific, New Zealand	7,7	15,30 m	n.a.
16.11.1925	Pacific	Eastern Pacific, Mexico	7,0	11,00 m	n.a.
01.09.1923	Pacific	Japan, Tokaido	7,9	12,00 m	2.144
13.04.1923	Pacific	Western Pacific, Russia, Kamchatka	7,2	30,00 m	20
03.02.1923	Pacific	Western Pacific, Russia, Kamchatka	8,3	8,00 m	3
11.11.1922	Pacific	Chile, North Chile	8,5	9,00 m	100
11.10.1918	Atlatic	USA, Puerto Rico, Mona Passage	7,5	6,00 m	40
01.05.1917	Pacific	New Zealand, Kermadec Islands	8,0	12,00 m	n.a.
28.12.1908	Mediterranean	Italy, Messina	7,2	n.a.	n.a.
31.01.1906	Pacific	South Pacific, Ecuador, Colombia	8,8	5,00 m	1.000
30.09.1899	Banda Sea	Indonesia	7,8	12,00 m	3.620
10.09.1899	Pacific	Usa, Yakutat Bay, Alaska	8,2	60,00 m	n.a.
15.06.1896	Pacific	Japan, Sanriku	7,6	38,00 m	26.360
27.08.1883	Indian Ocean	Indonesia, India	n.a.	35,00 m	36.500
10.05.1877	Pacific	Peru	8,3	24,00 m	500
13.08.1868	Pacific	Chile, North Chile	8,5	21,00 m	25.000
28.06.1859	Pacific	Indonesia, N.Moluccas Islands	7,0	9,00 m	n.a.
24.12.1854	Pacific	Japan, Nankaido	8,4	28,00 m	3000
26.11.1852	Banda Sea	Indonesia	8,2	8,00 m	60
07.05.1842	Caribbain Sea	Haiti, Cap-Haitian	7,7	5,00 m	300
24.04.1771	Pacific	Japan, Ryukyu Islands	7,4	85,00 m	13.500
01.11.1755	Atlantic	Portugal, Lisbon	9,0	12,00 m	60.000
29.10.1746	Pacific	Peru	8,0	24,00 m	3.800
28.10.1707	Pacific	Japan	8,4	11,00 m	30.000
31.12.1703	Pacific	Japan, Tokaido-Kashima	8,2	10,50 m	5.200
20.10.1687	Pacific	Peru	8,5	8,00 m	500
04.11.1677	Pacific	Japan, Kashima	7,4	8,00 m	500
26.09.1650	Mediterranean	Greece, Thera Island,	n.a.	16,00 m	n.a.
02.12.1611	Pacific	Japan, Sanriku	8,0	25,00 m	5.000
24.11.1604	Pacific	Peru	8,5	16,00 m	80
09.07.1586	Pacific	Peru	8,5	24,00 m	n.a.
20.09.1498	Pacific	Japan, Nankaido	8,6	17,00 m	31.000
12.09.1495	Pacific	Sagami Bay, Japan, Tokaido	7,1	5,00 m	200

Source: National Geographic Data Center- Tsunami Event Database

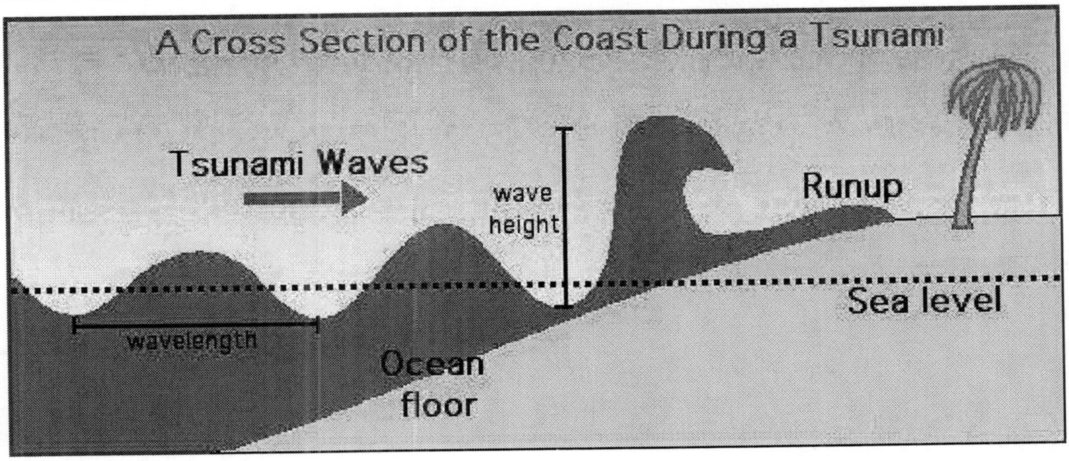

Figure 4.

7. Visualization Collections of Tsunamis

Some illustrative photos of the effective disaster happened due to different tsunamis are shown in the reference [4, 24].

Mathematical Modelling

8. Linear Theory of Tsunamis Generated by Moving Bottom

The three-dimensional fluid flow in open seas, under an impulsive pressure on the bottom, leads to a nonlinear complicated model. Solving such a problem is so complicated and haas two difficulties, i.e., the nonlinearity and the unknown domain of solution.

Dutykh, Dias and Kervella; linearize this complicated problem in dimensional variable as follows:

$$\nabla^2 \Phi = 0, \qquad (x, y, z) \in \mathbb{R}^2 \mathrm{X}\, (-h, 0) \tag{2}$$

$$\frac{\partial \Phi}{\partial z} = \frac{\partial \eta}{\partial t}, \qquad \frac{\partial \Phi}{\partial t} + g\,\eta = 0, \qquad z = 0 \tag{3}$$

$$\frac{\partial \Phi}{\partial z} = \frac{\partial \zeta}{\partial t}, \qquad z = -h \tag{4}$$

(where $z = \zeta$ represent the equation of the bottom)

This leads to one equation that describe the potential function Φ as follows:

$$\frac{\partial^2 \Phi}{\partial t^2} + g\,\frac{\partial \Phi}{\partial z} = 0\,, \quad z = 0 \tag{5}$$

The authors used the famous mathematical technique of integral transforms to get the analytical solution of this linearized problem. An illustrative numerical computations with some comparisons with the physical data were also presented.

9. Non-linear Mathematical Modeling of Tsunamis

Tsunamis form "Solitary waves", or waves with crests but no troughs-more like sand dunes than sine waves. Tsunamis waves are also called "N-waves".,as they resemble the English letter "N". (See Fig. 5)

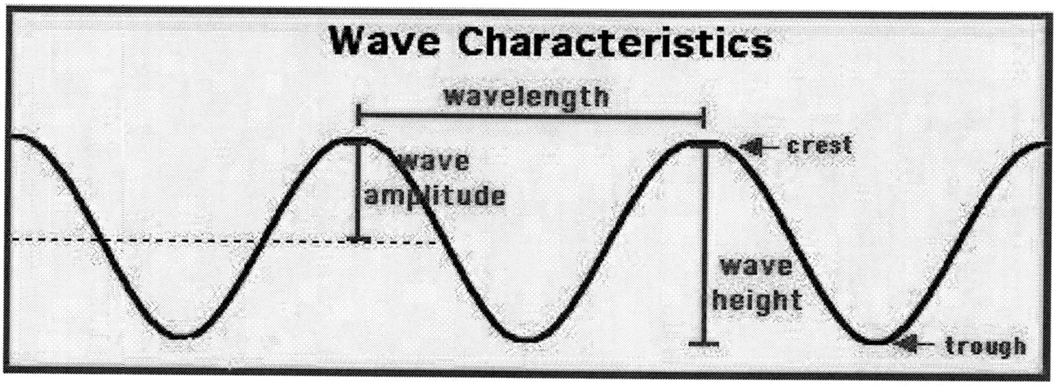

Figure 5.

Tsunamis have three stages: formation, mid-ocean propagation, and breaking and run-up on the beach. We discuss mid-ocean propagation. We formulate mathematically the fully nonlinear water wave problem, weakly nonlinear dispersive long waves are known to satisfy the Korteweg-de Vries(KdV) nonlinear partial differential equation. The KdV equation has solitary pulse traveling wave solutions which have coherent structure, and the energy does not spread out.

Hereafter, we introduce a very simple mathematical model to illustrate the physics and mathematics of the tsunamis Nonlinear Theories. The Boussinesq's theory of long waves makes the hypothesis of an horizontal ground (origin for the z axis) and a bi-dimensional movement in the x and z directions. It is based on the solution of the Laplace's equation satisfying the potential $\Phi\ (x, z, t)$.

We start with Bernoulli's equation

$$\frac{1}{2g}[(\frac{\partial\Phi}{\partial x})^2 + (\frac{\partial\Phi}{\partial z})^2] + h + \frac{1}{g}\frac{\partial\Phi}{\partial t} = d \tag{6}$$

Where: the variable d is the depth at the initial moment.

The conditions on the free surface at the coordinate h(x,t) above the ground sea level are:

$$\frac{dh}{dt} = w \tag{7}$$

Which can be expressed as :

$$\frac{\partial h}{\partial t} + u\frac{\partial h}{\partial x} = w \tag{8}$$

The potential $\Phi\ (x, z, t)$ is developed through the following series:

$$\Phi(x,z.t) = \varphi(x,t) - \frac{z^2}{2!}(\frac{\partial^2\varphi}{\partial x^2}) + \frac{z^4}{4!}(\frac{\partial^4\varphi}{\partial x^4}) + ... \tag{9}$$

$$\varphi(x,t) = \Phi(x,0,t) \tag{10}$$

$$\left.\frac{\partial\varphi}{\partial x}\right|_{t=0} = 0 \tag{11}$$

We can introduce the potential $\Phi(x,z,t)$:

$$u = \frac{\partial\Phi}{\partial x} \quad , w = \frac{\partial\Phi}{\partial z} \tag{12}$$

$$\frac{\partial h}{\partial t} + \frac{\partial h}{\partial x}\frac{\partial\Phi}{\partial x} = \frac{\partial\Phi}{\partial z} \tag{13}$$

Let's introduce some non-dimensional variables (x^1, z^1, t^1, Φ^1)

$$x = Lx^1 \quad , z = dz^1 \quad , t = \frac{Lt^1}{\sqrt{gd}} \quad , \Phi = \Phi^1 \varepsilon L\sqrt{gd} \tag{14}$$

$$\varepsilon = H/d \tag{15}$$

$$\mu = d^2/L^2 \tag{16}$$

$$U = HL^2/d^3 \tag{17}$$

Where ε is the wave amplitude corresponding to the depth d; L is the wave length; $\eta(x,t)$ is the relative movement of the free surface relating to the maximum height; U is the Ursell number very characteristic of the waves propagation.

For simplification, the superscript (1) over the symbols will bo omitted in the sequel.

By introducing the following definition

$$h = d\,(1 - \varepsilon\eta) \qquad \textit{this implies} \tag{18}$$

$$\frac{\partial\eta}{\partial t} + \frac{\partial^2\varphi}{\partial x^2} = -\varepsilon[\frac{\partial^2\varphi}{\partial x^2} + \frac{\partial\eta}{\partial x}\frac{\partial\varphi}{\partial x}] + \frac{\mu}{6}\frac{\partial^4\varphi}{\partial x^4} \tag{19}$$

$$\eta + \frac{\partial\varphi}{\partial t} = -\frac{\varepsilon}{2}(\frac{\partial\varphi}{\partial x})^2 + \frac{\mu}{2}\frac{\partial^3\varphi}{\partial x^2\partial t} \tag{20}$$

Let's now introduce the non-dimensional horizontal velocity u defined as follows:

$$u = \frac{\partial\varphi}{\partial x} \tag{21}$$

Thus, the mass equation and the pressure condition lead to the following system:

$$\frac{\partial \eta}{\partial t} + \frac{\partial u}{\partial x} = -\varepsilon \frac{\partial}{\partial x}(\eta u) + \frac{\mu}{6}\frac{\partial^3 u}{\partial x^3} \tag{22}$$

$$\frac{\partial \eta}{\partial x} + \frac{\partial u}{\partial t} = -\varepsilon u \frac{\partial u}{\partial x} + \frac{\mu}{2}\frac{\partial^3 u}{\partial x^2 \partial t} \tag{23}$$

Hence, it also leads to the following first order equation:

$$\frac{\partial \eta}{\partial x} - \frac{\partial \eta}{\partial t} = \frac{\partial u}{\partial x} - \frac{\partial u}{\partial t} = 0 \qquad Hence, \tag{24}$$

$$\eta = f(x - ct), \qquad u = g(x - ct) \tag{25}$$

$$\eta = u \tag{26}$$

Finally, we obtain the Kortweg-de-Vries equation introducing the variation of the free surface according to the variables x and t.[5]

$$\frac{\partial \eta}{\partial t} + \frac{\partial \eta}{\partial x} + \frac{3}{2}\varepsilon \eta \frac{\partial \eta}{\partial x} + \frac{\mu}{6}\frac{\partial^3 \eta}{\partial x^3} = 0 \tag{27}$$

10. Tsunami and the Boussinesq Equation

In this section we will study the (2+1)-dimensional nonlinear Boussinesq equation

$$u_{tt} - (u_{xx} + u_{yy}) - \lambda(u^2)_{xx} - u_{xxxx} = 0, \lambda \neq 0, \tag{28}$$

with $u(x, y, t)$ is a sufficiently often differentiable function. The last presented equation (28) was introduced by Boussinesq to describe the propagation of long waves in shallow water under gravity propagating in both directions. It also arises in other physical applications such as nonlinear lattice waves, ion sound waves in a plasma, and in vibrations in a nonlinear string. It is used in many physical applications such as the percolation of water in porous subsurface of a horizontal layer of material. This particular form is of special interest because it is a completely integrable equation that admits an infinite number of conservation laws of energy. For $\lambda = -1$ we obtain the good or well-posed Boussinesq equation. For $\lambda = 1$ we obtain the bad or ill-posed classical Boussinesq equation. However, if the dispersive term u_{xxxx} is replaced by the term u_{xxtt} we obtain the so-called *improved* Boussinesq equation.

It is important to notice that for $\lambda = 0$ we obtain the linear Boussinesq equation given by

$$u_{tt} - (u_{xx} + u_{yy}) - u_{xxxx} = 0, \tag{29}$$

where $u = u(x, y, t)$ is a sufficiently often differentiable function. Boussinesq equation (29) contains the linear dispersion term u_{xxxx}. The linear Boussinesq equation (29) should be used to simulate the tsunami waves because the nonlinear Boussinesq equation (28) models long waves in a shallow water under gravity. When a tsunami propagates over a long distance the dispersion and Coriolis force play a role in forming tsunamis. Coriolis force is defined as an apparent force that as a result of the earth's rotation deflects moving objects to the right in the northern hemisphere and to the left in the southern hemisphere.

This means that the linear Boussinesq equation (29) with the Coriolis force should be the most proper model to describe the propagation of distant typical tsunami waveforms.

In the following section, we will follow some famous method to solve the linear Boussinesq equation, namely the tanh method, for theoretical analysis and the Adomian decomposition method for numerical purposes.

11. Solution of Nonlinear Boussinesq Equation

11.1. Using the tanh Method

Here, we follow the tanh method well-known technique, and all symbols appears here (M, μ, c) are defined in the standard text (cf. Wazwaz).

The Boussinesq equation

$$u_{tt} - (u_{xx} + u_{yy}) - \lambda(u^2)_{xx} - u_{xxxx} = 0, \tag{30}$$

will be converted to the ODE

$$(c^2 - 2)u - \lambda u^2 - u'' = 0, \tag{31}$$

obtained upon using $\xi = x + y - ct$ and integrating twice. Balancing the nonlinear term u^2 with the highest order derivative u'' gives $M = 2$. The tanh method uses the substitution

$$u(x, y, t) = S(Y) = \sum_{i=0}^{2} a_i Y^i. \tag{32}$$

Substituting (32) into (31), collecting the coefficients of each power of $Y^i, 0 \leq i \leq 2$, setting each coefficient to zero, and solving the resulting system of algebraic equations, we found that $a_1 = 0$ and we also obtain the following sets of solutions:

(i)

$$a_0 = \frac{3(c^2 - 2)}{2\lambda}, \; a_2 = -\frac{3(c^2 - 2)}{2\lambda}, \; \mu = \frac{1}{2}\sqrt{c^2 - 2}, c^2 > 2. \tag{33}$$

(ii)

$$a_0 = -\frac{(c^2 - 2)}{2\lambda}, \; a_2 = \frac{3(c^2 - 2)}{2\lambda}, \; \mu = \frac{1}{2}\sqrt{2 - c^2}, c^2 < 2 \tag{34}$$

As a result we obtain following the solitons solutions

$$u_1(x, y, t) = \frac{3(c^2 - 2)}{2\lambda} \operatorname{sech}^2 \left[\frac{1}{2}\sqrt{c^2 - 2}(x + y - ct)\right], c^2 > 2, \tag{35}$$

and

$$u_2(x,y,t) = -\frac{c^2-2}{2\lambda}(1-3\tanh^2\left[\frac{1}{2}\sqrt{2-c^2}(x+y-ct)\right], c^2 < 2. \quad (36)$$

11.2. Using Adomian Decomposition Method

The well-known and famous Adomian decomposition method, (cf. Wazwaz), will be used here to determine the solitary wave solutions of the (2+1)-dimensional Boussinesq equation

$$u_{tt} - (u_{xx} + u_{yy}) - \lambda(u^2)_{xx} - u_{xxxx} = 0, \lambda \neq 0, \quad (37)$$

where the initial conditions are given by

$$u(x,y,0) = \frac{6k^2 e^{k(x+y)}}{\lambda(1+e^{k(x+y)})^2}, \qquad u_t(x,y,0) = \frac{6ck^3 e^{k(x+y)}(e^{k(x+y)}-1)}{\lambda(1+e^{k(x+y)})^3}, \quad (38)$$

where $k = \sqrt{c^2-2}$.

Applying the inverse operator L_t^{-1} to Eq. (37) gives

$$\begin{aligned} \sum_{n=0}^{\infty} u_n(x,y,t) &= \frac{6k^2 e^{k(x+y)}}{\lambda(1+e^{k(x+y)})^2} + \frac{6ck^3 e^{k(x+y)}(e^{k(x+y)}-1)}{\lambda(1+e^{k(x+y)})^3} t \\ &+ L_t^{-1}\left((\lambda \sum_{n=0}^{\infty} A_n) + (\sum_{n=0}^{\infty} u_n)_{xx} + (\sum_{n=0}^{\infty} u_n)_{yy} + (\sum_{n=0}^{\infty} u_n)_{xxxx}\right) \end{aligned} \quad (39)$$

where A_n refers to the Adomian polynomials that represent the nonlinear term $(u^2)_{xx}$. The Adomian polynomials A_n can be computed as presented before in other sections.

$$\begin{aligned} u_0(x,y,t) &= \frac{6k^2 e^{k(x+y)}}{\lambda(1+e^{k(x+y)})^2} + \frac{6ck^3 e^{k(x+y)}(e^{k(x+y)}-1)}{\lambda(1+e^{k(x+y)})^3} t, \\ u_{j+1}(x,t) &= L_t^{-1}\left(\lambda A_j + (u_j)_{xx} + (u_j)_{yy} + (u_j)_{xxxx}\right), \quad j \geq 0. \end{aligned} \quad (40)$$

Consequently, we obtain

$$\begin{aligned} u_0(x,y,t) &= \frac{6k^2 e^{k(x+y)}}{\lambda(1+e^{k(x+y)})^2} + \frac{6ck^3 e^{k(x+y)}(e^{k(x+y)}-1)}{\lambda(1+e^{k(x+y)})^3} t, \\ u_1(x,y,t) &= \frac{3c^2k^4 e^{k(x+y)}(1-4e^{k(x+y)}+3e^{2k(x+y)})}{\lambda(1+e^{k(x+y)})^4} t^2, \\ u_2(x,y,t) &= \frac{c^3k^5 e^{k(x+y)}(-1+11e^{k(x+y)}-11e^{2k(x+y)}+e^{3k(x+y)})}{\lambda(1+e^{k(x+y)})^5} t^3, \end{aligned} \quad (41)$$

the solution in a series form is given by

$$\begin{aligned} u(x,y,t) &= \frac{6k^2 e^{k(x+y)}}{\lambda(1+e^{k(x+y)})^2} \frac{6ck^3 e^{k(x+y)}(e^{k(x+y)}-1)}{\lambda(1+e^{k(x+y)})^3} t \\ &+ \frac{3c^2k^4 e^{k(x+y)}(1-4e^{k(x+y)}+3e^{2k(x+y)})}{\lambda(1+e^{k(x+y)})^4} t^2 \\ &+ \frac{c^3k^5 e^{k(x+y)}(-1+11e^{k(x+y)}-11e^{2k(x+y)}+e^{3k(x+y)})}{\lambda(1+e^{k(x+y)})^5} t^3 + \cdots \end{aligned} \quad (42)$$

12. Numerical Applications

The problem of tsunamis reduced to a governing equation (in our work we obtained the well-known KdV equation) and initial condition. We can solve this problem numerically by different techniques, e.g., Cranck Nicolson method, Hopscotch method. The reader may consult the work of Helal & Mehanna [4] as well as Helal [23].

Another famous numerical method is based on the Finite element method. A lot of papers have illustrated this technique. The numerical results are similar to many works and recent works of Helal [22] and Helal & Mehanna [4].

Finally, we must note that this disaster has much numerical data that comes from the centers of predictions. These data can be used in comparison between numerical and experimental results. This part is out of our interest and needs other types of research work.

During the last five decades, an exciting and active area of research has been devoted to the construction of exact solution for a wide class of nonlinear equations. This includes the most famous nonlinear one of Korteweg and de Vries.

Certain important and novel subareas of research, such as: the applications of the Jacobian elliptic function [6], the powerful inverse scattering transform (IST), the Bācklund transformations technique (BTs) [7,8], the Painleve analysis [9], the Lie group theoretical methods [10], the direct algebraic method [11] and tangent hyperbolic method [12] are leading to the construction of exact solutions to the KdV equation.

Regarding formal mathematical approaches: Sawada and Kotera [13], Rosales [14], Whitham [15], Wadati and Sawada [16,17] and Hickernel [18]; they all have employed perturbation techniques. Their methods lead, after some tedious algebraic manipulations, to the Known N solitary wave solutions of some nonlinear partial differential equations , e.g., KdV, Burgers, Boussinesq equations, etc.

In more recent work, Helal and El Eissa [19], Khater et al. [20,21] and Helal [22,23] have studied analytically and numerically some physical problems that lead to the KdV equations.

In the next higher order of the perturbation theory, a KdV equation can be obtained, which includes in general cubic nonlinearity, fifth-order linear dispersion, and nonlinear dispersion. The contribution of the various high-order terms depends on the parameters of the model (e.g., density stratification and shear flow), and sometimes several of them may be more important [23].

In 2007, Helal & Mehanna [5] presented a paper containing a comparative study of two different methods for solving the general KdV equation, namely the numerical Crank Nicolson method, and the semi analytic Adomian decomposition method. The paper conducted a comparative study between these two methods. Moreover, this work presented the enhancement provided by Adomian decomposition method.

Conclusion

For all bodies in motion, there is an associated fluid flow, against the surrounding medium (airplanes, birds, frisbees, ships, submarines, dolphins, fish). The body may be at rest in a moving fluid (mountains, islands, skyscrapers, ocean platforms, flagpoles, bridge towers, pylons, trees, mussels). The fluid motion may be internal that involves transport processes (such as: internal combustion engines, cooling and ventilation systems, flows in the blood vessels and lungs). The scales can be very large (planetary scale motions, atmospheric and ocean circulations on earth, and turbulence on galactic scales), or very small (circulation and transport in micromachines, bubble and particle motions).

In many practical applications, a very large range of scales spanning many orders of magnitude are simultaneously present. Flow at interfaces between different fluids can generate a rich set of phenomena (ocean surface waves, oil/water mixtures in petroleum extraction, bubbly flows, combustion and chemical mixing).

Fluid mechanics is the subdiscipline of continuum mechanics that studies fluids, i.e., liquids and gases. It is the physical science dealing with the action of fluids at rest (fluid statics) or in motion (fluid dynamics), and their interaction with flow devices and applications in engineering. It has several subdisciplines itself, including aerodynamics (the study of gases in motion) and hydrodynamics (the study of liquids in motion). Fluid dynamics has a wide range of applications, including calculating forces and moments on aircrafts, determining the mass flow rate of petroleum through pipelines, predicting weather patterns and reportedly modelling fission weapon detonation. Some of its principles are even used in traffic engineering, where traffic is treated as a continuous fluid.

Geophysical fluid dynamics—fluid phenomena associated with the dynamics of the atmosphere and the oceans such as hurricane and weather systems, Bio-fluid mechanics—fluid mechanics involved in biophysical processes such as blood flow in arteries and many others are subareas of fluid studies.

Modern applications use the computational approach to develop solutions to fluid mechanics problems. The discipline concerned with this is called the CFD, Computational Fluid Dynamics.

Computational fluid dynamics (CFD) is one of the fluid mechanics branches that uses numerical analysis methods and algorithms to solve and analyze problems that involve fluid flows. Computers are used to perform the millions of calculations required to simulate the interaction of fluids and gases with the complex surfaces used in engineering. However, even with simplified equations and high-speed supercomputers, only approximate solutions can be achieved in many cases. More accurate codes that can accurately and quickly simulate even complex scenarios such as transonic or turbulent flows are an ongoing area of research. Validation of such codes is often performed using a wind tunnel.

In this work, "Tsunamis, generation and mathematical modeling", our main goal is to study almost everything concerning tsunamis. Tsunamis are characterized as shallow-water waves. These are different from the waves most of us have observed on a beach, which are caused by the wind blowing across the ocean's surface.

Tsunamis have been recorded in all the major oceans of the world. However, this phenomenon is mainly restricted to the Pacific basin, an area surrounded by volcanic island arcs, mountain chains and subduction zones earning the nickname the "ring of fire", as it is the most geologically active area on the planet.

There is an average of two destructive tsunamis per year in the Pacific basin. A Pacific-wide tsunami is a rare phenomenon, occurring every 10–12 years on average. Most of these tsunamis are generated by earthquakes that cause displacement of the seafloor, but, as we have seen, a tsunami can be generated by volcanic eruptions, landslides, underwater explosions, and meteorite impacts.

we conclude that :

1- All large earthquakes do not cause a tsunami, but many do. If the quake is located near or directly under the ocean, the probability of a tsunami increases.

2- A tsunami can occur at any time, day or night. They can travel up rivers and streams that lead to the ocean.

3- A small tsunami at one beach can be a giant a few miles away.

4- Never go down to the beach to watch a tsunami! When you can see the wave, you are too close and it is too late to escape. A tsunami can move faster than a person can run!

Finally, the study clearly showed that each of the four methods has its own characteristics. However, the Adomian decomposition method introduced several enhancements over existing techniques such as the finite differences method or the Crank Nicolson method. Sometimes the solution can be obtained easily with few iterations. For concrete problems, we usually use few components to get an approximation of a high degree of accuracy. On the other hand, the $tanh$ method is a powerful one to handle the nonlinear evolution equation. Its power lies in how it makes it easy to convert the PDE to a system of algebraic equations. An interesting point should be made here. If solutions are in the form of, $sech$, $sech^2$, or $\tanh^2$, which is related to $sech^2$, this will give the bell-shaped solitons. However, if the solution in the form of a tanh, then the solution is a kink that approaches a constant at $\pm\infty$.

References

[1] Nelson SA, Natural disasters, *EENS* **204**, Tulane University, 2006.

[2] Dutykh D, Dias F, Kervella Y, Linear theory of wave generation by a moving bottom, *C. R. Acad. Sci. Paris, Ser. I* **343** (2006) 499–504

[3] Segur H. Waves in shallow water. In: International workshop on tsunami and nonlinear waves, Calcutta, India, March 2006.

[4] Helal MA, Mehanna MS, Tsunamis from nature to physics, *Chaos, Solitons & Fractals*, 2008, 36, 4, 787-796

[5] Helal MA, Mehanna MS, A comparative study between two different methods for solving the general Korteweg–de Vries equation (GKdV), *Chaos, Solitons & Fractals,* 2007, 33, 3, 725-739

[6] Lamb GL. *Elements of soliton theory*. New York: Wiley; 1980.

[7] Khater AH, Helal MA, El Kalaawy OH. Bäcklund transformations: exact solutions for the KdV and Calogero-Degasperis-Fokas mKdV equations. *Math Meth Appl Sci* 1998;21:719-31.

[8] Wahlquist HD, Estabrook FB. Ba¨cklund transformation for solutions of the KdV equation. *Phys Rev Lett* 1973;31:1386–90.

[9] Khater AH, El- Sabbagh MF. The Painleve´ property and coordinates transformations. *Nuovo Cimento B* 1989;104(2):123–9.

[10] Olver PJ. Applications of Lie groups to differential equations. *Graduate texts in mathematics*. Berlin: Springer; 1986. p. 107.

[11] Hereman W, Korpel A, Banerjee PP. A general physical approach to solitary wave construction from linear solutions. *Wave Motion* 1985;7(3):283–9.

[12] Malfliet W. Solitary wave solutions of nonlinear wave equations. *Am J Phys* 1992;60(7):650–4.

[13] Sawada K, Kotera T. A method for finding N-solitons of the KdV equation and KdV like equation. *Prog Theor Phys* 1974;51:1355–67.

[14] Rosales RR. Exact solutions of some nonlinear evolution equations. *Stud Appl Math* 1978;59:117–57.

[15] Whitham GB. *Linear and nonlinear waves.* New York: Wiley/Interscience; 1974.

[16] Wadati M, Sawada K. New representations of the soliton solution for the KdV equation. *J Phys Soc Jpn* 1980;48:312–8.

[17] Wadati M, Sawada K. Application of the trace method to the modified KdV equation. *J Phys Soc Jpn* 1980;48:319–26.

[18] Hickernel FJ. The evolution of large horizontal scale disturbance in marginally stable, inviscid, shear flows: II. Solution of the Boussinesq equation. *Stud Appl Math* 1983;69:23–49.

[19] Helal MA, El Eissa HN. Shallow water waves and KdV equation (oceanographic application). *PUMA* 1996;7(3–4):263–82.

[20] Khater AH, El- Kalaawy OH, Helal MA. Two new classes of exact solutions for the KdV equation via Bäcklund transformations. *Chaos, Solitons & Fractals* 1997;8(12):1901–9.

[21] Khater AH, Helal MA, Seadawy AR. General soliton solutions of an n-dimensional nonlinear Schro¨dinger equation. *Nuovo Cimento B* 2000;115(11):1303–11.

[22] Helal MA. Chebyshev spectral method for solving KdV equation with hydrodynamical application. *Chaos, Solitons & Fractals* 2001;12:943–50.

[23] Helal MA. Soliton solution of some nonlinear partial differential equations and its applications in fluid mechanics. *Chaos, Solitons & Fractals* 2002;13:1917–29.

[24] <http://en.wikipedia.org/wiki/Tsunami>.

[25] <http://www.tsunami.org/references.html>.

[26] <http://www.tsunami.org/links.html>.

General Bibliography

Ablowitz MJ, Segur H. *Solitons and the inverse scattering transforms.* Philadelphia: SIAM; 1981.

Abramowitz M, Stegun IA. *Hanbook of mathematical functions.* New York: Dover; 1965.

Adomian G., *Solving Frontier Problems of Physics: The Decomposition method,* Kluwer, 1994.

Boussinesq MJ. *C R Acad Sci Paris* 1871;72:755-9.

Boussinesq MJ. *J Math Pures Appl Ser 2* 1872;17:55-108.

Canuto C, Hussaini MY, Quarteroni A, Zang TA. *Spectral methods in fluid dynamics.* Berlin: Springer; 1988.

Constantine A, Johnson RS. Modelling tsunamis. *J Phys A: Math Gen* 2006;39:L215–7.

Dodd RK, Eilbeck JC, Gibbon JD, Morris HC. *Solitons and nonlinear wave equations.* London: Academic Press;1984.

Drazin PG. *Solitons.* Cambridge: Cambridge University Press;1983.

Drazin PG, Johnson RS. *Solitons: an introduction.* Cambridge: Cambridge University Press;1989.

Fermi E, Pasta J, Ulam S. *Studies of nonlinear problems,* I. Los Alamos Report, Los vAlamos, NM, 1940.

Gardner CS, Greene JM, Kruskal MD, Miura RM. Methods for solving the KdV equation. *Phys Rev Lett* 1967;19:1095-7.

Greig IS, Morris JL. A hopscotch method for the KdV equation. *J Comput Phys* 1976;20:64-80.

Korteweg DJ, De Vries G. On the change of form of long waves advancing in a rectangular canal, and on a new type of long stationary wave. *Philos Mag* 1895;5(39):422-43.

[15] Dutykh D, Dias F, Water waves generated by a moving bottom, in: Kundu A. (Ed.), Tsunami and Nonlinear Waves, *Geosciences*, Springer-Verlag, 2006.

Lamb GL. Elements of soliton theory. New York: Wiley 1980.

Miles JW. The KdV equation: a historical essay. *J Fluid Mech* 1981;106:131-47.

Russel SJ. Report on waves, *Report to The 14th Meeting (1844) of the British Ass. for Adv. of Science,* London, 1845;311.

Stoker JJ. *Water waves,* Interscience Pub., New York, 1957.

Su CH, Gardner CS. Korteweg-de Vries equation and generalizations. III. Derivation of the Korteweg-de Vries equation and Burgers equation. *J Math Phys* 1969;10:536-9.

Taha TR, Ablowitz MJ. Analytical and numerical aspects of certain nonlinear evolution equations, (I) analytical. *J Comput Phys* 1984;55:192-202.

Taha TR, Ablowitz MJ. Analytical and numerical aspects of certain nonlinear evolution equations, (III) numerical, KdV equation. *J Comput Phys* 1984;55:231-53.

Wazwaz AM. *Partial Differential Equations: Methods and Applications*, Balkema, The Netherlands, 2002.

Whitham GB. *Linear and Nonlinear Waves,* Interscience Pub., New York, 1974.

INDEX

B

C

D

E

G

H

I

J

K

L

M

N

O

P

Q

R

S

T

U

V

W